AF400298

Progress in Mathematical Physics

Volume 36

Dariusz Chruściński

Andrzej Jamiołkowski

Geometric Phases in Classical and Quantum Mechanics

Springer Science+Business Media, LLC

Dariusz Chruściński
Nicholas Copernicus University
Torun 87-100
Poland

Andrzej Jamiołkowski
Nicholas Copernicus University
Torun 87-100
Poland

Library of Congress Cataloging-in-Publication Data
Chruściński, Dariusz.
  Geometric phases in classical and quantum mechanics / Dariusz Chruściński, Andrzej Jamiołkowski.
    p. cm. – (Progress in mathematical physics ; v. 36)
  Includes bibliographical references and index.
  ISBN 978-1-4612-6475-0   ISBN 978-0-8176-8176-0 (eBook)
  DOI 10.1007/978-0-8176-8176-0
    1. Geometric quantum phases. 2. Quantum theory. 3. Mathematical physics. I.
Jamiołkowski, Andrzej, 1946- II. Title. III. Series.

QC20.7.G44C47 2004
530.12-dc22   2004046278
                                           CIP

AMS Subject Classifications: 22E05, 22E70, 37J05, 37J10, 37J15, 37J35, 53-01, 53A45, 53B05, 53B20, 53B21, 53B35, 53B50, 53C05, 53C07, 53C10, 53C15, 53C22, 53C29, 53C30, 53C55, 53C80, 53D05, 53D20, 53D22, 53Z05, 55R05, 55R10, 55R25, 55R91, 58A05, 58A10, 58A12, 58B05, 70E15, 70H05, 70H06, 70H11, 70H15, 78A25, 78A35, 81P68, 81Q05, 81Q70

ISBN 978-1-4612-6475-0   Printed on acid-free paper.

©2004 Springer Science+Business Media New York
Originally published by Birkhäuser Boston in 2004
Softcover reprint of the hardcover 1st edition 2004

9 8 7 6 5 4 3 2 1   SPIN 10863573

www.birkhasuer-science. com

Contents

Preface

Twenty years ago Michael Berry (Berry 1984) demonstrated that the standard description of adiabatic processes in quantum mechanics is incomplete. Berry noticed that if the Hamiltonian of the system depends on a number of parameters which vary adiabatically during the evolution, then a cyclic variation of the parameters is accompanied by a change of wave function by an additional phase factor, which hitherto had been completely ignored. This additional factor, known today as the *Berry phase*, possesses a remarkable geometrical property — it depends only upon the geometric structure of the space of parameters and does not depend on the duration of the evolution.

Almost simultaneously, a similar phenomenon was observed by J. Hannay (Hannay 1985) within the framework of classical mechanics.[1] The classical counterpart of the phase of the wave function is the phase of quasi-periodic motion — the so-called angle variable in the action-angle representation of integrable systems. These classical geometric phases are called *Hannay's angles*.

It was soon realized that Berry's remarkable observation finds a surprisingly broad spectrum of applications. An analog of Berry's geometric phase is manifested in many apparently unrelated phenomena like, for example, the Foucault pendulum, the passage of photons through optical fibres, the spectra of molecules, the quantum Hall effect, and anomalies in quantum field theory. Why is this so? It turns out that they have a universal mathematical description. B. Simon (Simon 1983)[2] was the first who observed that Berry's phase may be interpreted as a purely geometric object, namely as a

---

[1] Interestingly, many important discoveries connected with the notion of geometric phase were made in Bristol. For the *Bristol Anholonomy Calendar* see Berry 1991, and the review Berry 1990b.

[2] Actually, due to the referee's delay Simon's paper appeared in 1983 and Berry's in 1984.

holonomy in a certain fibre bundle. This way an elegant and fairly sophisticated mathematical theory of bundles and connections enters elementary quantum mechanics. The bundle constructed by Simon, called by physicists the *spectral bundle*, is uniquely determined by the spectral properties of the system's Hamiltonian. Simon showed that the adiabatic evolution considered by Berry defines the so-called parallel transport of a vector from the system's Hilbert space along a curve in the parameter space. Now, a vector transported along a closed curve does not in general return to its original form but acquires a phase factor predicted by Berry. This geometrical phenomenon is well known in classical geometry: a vector parallel transported along a closed curve on a two-dimensional sphere does in general change its orientation by an angle of rotation equal to the solid angle subtended by the curve. As we shall see, this 19th century observation finds new, interesting applications in various branches of physics.

Actually, the theory of fibre bundles was successfully applied in the seventies (of the last century) in the mathematical formulation of gauge theories — electrodynamics and its generalization known as Yang–Mills theory. Therefore, it is not surprising that there are many analogies between gauge theories and geometric phases. For example, such classical topics as magnetic poles and instantons find new and fresh illustration when dealing with geometric phases.

There are several reviews that focus on various aspects of geometric phases. Berry's articles *Quantum Adiabatic Holonomy* (Berry 1989a) and *The quantum phase, five years after* (Berry 1989b) constitute a beautiful overview and summarize the first, most exciting period in the development of the subject. The articles *Topological phases in quantum mechanics and polarization optics* (Vinitskii et al. 1990) and *Polarization of light and topological phases* (Bhandari 1997) presents the application of geometric phases in optics. Phases in molecular physics are reviewed in *The geometric phase in molecular systems* (Mead 1992) and *The Geometric Phase in Quantum Mechanics* (Bohm 1993b). Actually, the third edition of Bohm's *Quantum Mechanics* (Bohm 1993a) contains an introduction to the subject together with examples from molecular physics. Moreover, there are two collections of papers: *Topological Phases in Quantum Theory*, edited by Markowski and Vinitskii, and *Geometric Phases in Physics* with excellent commentaries by Shapere and Wilczek. Finally, there are the reviews *Berry phase* (Zwanzinger et al. 1990) and *Geometric phases in physics* (Anandan et al. 1997), which provide very useful guides to the literature on geometric phases. Although interest in the geometric phase dates to the mid 1980s, the beginnings of this problem came much earlier. Exciting historical reviews may be found in *Anticipations of the geometric phase* (Berry 1990b) and the appendix of *Topological phases in quantum mechanics and polarization optics* (Vinitskii et al. 1990). Actually, when this book was completed there appeared a monograph, *The Geometric Phase in Quantum Systems* by Bohm et al. (Springer 2003), with a thorough introduction and applications from molecular and condensed matter physics.

Why write yet another book on geometric phases? What distinguishes this book from other texts is that it covers both quantal and classical geometric phases from a unified, geometric point of view and at a rather sophisticated level. Moreover, it provides insights into the relationships between quantal and classical phases which have not been emphasized previously at the textbook level. This book is addressed to graduate students in mathematical and theoretical physics, as well as theoretical physicists and applied mathematicians. We hope that it helps the reader to enter the exciting world of geometric phases in classical and quantum mechanics and to feel their elegant, coherent mathematical description. It can certainly be used as a supplementary textbook in a course on differential geometry for physicists, as well.

To achieve our goal we start with the mathematical background in Chapter 1 and plunge the reader into the arena of differential geometry. This introductory chapter also includes basic facts from Lie groups and algebras and concentrates on fibre bundles and connections, which serve as the main tool to study geometric phases. The reader will find a detailed exposition of the celebrated Hopf fibrations and several physical illustrations as magnetic poles and instantons in Yang–Mills theory. Having at hand the mathematical tools, we start in Chapter 2 to investigate the physics of geometric phases. The reader learns about adiabatic theorem in quantum mechanics, the quantal *adiabatic phases* of Berry, and their non-abelian generalization due to Wilczek and Zee. The presentation of the physical side of the problem is simple and requires only basic notions from quantum mechanics; it should be accessible for mathematicians interested in theoretical physics. The mathematical side uses the previously introduced notions of fibre bundles and connections, and stresses the geometric aspects of adiabatic evolution.

Chapter 3 deals with adiabatic geometric phases in classical mechanics. Here we introduce basic facts from symplectic geometry, Hamiltonian mechanics and integrable systems. It is shown how the classical adiabatic theorem leads to classical geometric phases — Hannay's angles. This chapter also includes many examples of classical systems displaying geometric phases. In Chapter 4 we present the geometric approach to classical phases using the mathematical language of bundles and connections. It introduces elegant geometric constructions: momentum maps, the celebrated Marsden–Weinstein reduction procedure and finally the Hannay–Berry connection. This chapter is illustrated by the dynamics of the rigid body — a system where the analog of the geometric phase was already observed in the 19th century.

Chapter 5 describes the natural geometric structure of quantum evolution. The standard approach to nonrelativistic quantum mechanics is based on a complex Hilbert space. However, as is well known, the Hilbert space is not an appropriate phase space for the quantum system. Any two unit vectors differing by a phase factor define the same physical state and hence they are physically equivalent. Therefore, the *true* quantum space of states is a *projective Hilbert space* — the space of equivalence classes (or the space of rays). A projective Hilbert space is endowed with two geometric structures — a Riemannian metric and a symplectic form. Hence, the geometric structure of quantum evolution is much richer than its classical counterpart. We show that symplectic structure is responsible for the so-called *Aharonov–Anandan geometric phase* whereas metric structure is closely related to the beautiful notion of the *Pancharatnam phase*. Finally, we present the geometric framework for the quantum evolution of mixed states and the corresponding nonabelian *Uhlmann geometric phase*. The recent proposal of interferometric measurement of the geometric phase for mixed states is also included. The geometric framework applied to quantum mechanics enables one to get more insight into the structure of the quantum space of states and deserves to be more widely known. We stress that this topic has not been so extensively discussed in the literature on geometric phases. Recently, it received considerable attention due to the rapid development in the field of quantum information theory.

Finally, Chapter 6 shows the geometric phases "in action." It includes several standard examples such as the appearance of geometric phases in optics and molecular physics. We present coherent derivations of the Aharonov–Bohm and Aharonov–Casher effects using the underlying symmetries of nonrelativistic quantum mechanics. We show how the geometry of fibre bundles enters the highly nontrivial physics of quantum Hall effects and show how topology explains the quantization of Hall conductance. Moreover, we review Berry and Robbins' (Berry and Robbins 1997) approach to the spin-statistics theorem. We close this chapter with a discussion of the recent surprising application of geometric phases to quantum computation — geometric phases are used to model quantum gates in a quantum computer. This example shows that the subject of geometric phases is still alive and perhaps one can see just another manifestation of geometric phases in the near future.

Acknowledgements. The authors thank Professor Józef Szudy, the dean of the Faculty of Physics, Astronomy and Informatics at Nicolaus Copernicus University in Toruń for his interest in this project and his kind support. We thank Ann Kostant and the staff at Birkhäuser Boston for their advice and help. Special thanks are due to David Winters for his first rate copyediting and to Elizabeth Loew for her excellent production of this book.

D.C. would like to thank Jerzy Kijowski and Gerry Kaiser for their encouragement during this project. I thank professor Iwo Białynicki-Birula for his remarks on the history of geometric phase in optics. I am grateful to Hartmann Römer for many interesting discussions and his kind hospitality in Freiburg, where part of the work was done. Special thanks are due to Jacek Jurkowski who carefully read the manuscript and pointed out many errors. For financial support I thank the DFG-Graduiertenkolleg *Nichtlineare Differentialgleichungen* in Freiburg and the Nicolaus Copernicus University in Toruń (Grant nr M/7/2001). Most of all, I thank my wife Alicja for her love, patience and understanding.

Dariusz Chruściński
Andrzej Jamiołkowski
Toruń, October 2003

Geometric Phases
in Classical and
Quantum Mechanics

1
Mathematical Background

The mathematical background required for the study of geometric phases in classical and quantum mechanics is rather extensive. The aim of this introductory chapter is to provide a background of some basic notions of classical differential geometry and topology. Classical differential geometry is now a well established tool in modern theoretical physics. Many classical theories like mechanics, electrodynamics, Einstein's General Relativity or Yang-Mills gauge theories are well known examples where the geometrical methods enter in the natural and very effective way. As we shall see throughout this book, also quantum physics shows its intricate beauty when one applies an appropriate geometric framework. All this proves Wigner's celebrated statement about the "unreasonable effectiveness" of mathematics in natural sciences.

1.1 Manifolds, forms and all that

1.1.1 Basic notions

The concept of a manifold generalizes the concept of a smooth surface or a curve in $\mathbb{R}^3$. Manifolds occur in all areas of physics. Whenever one speaks about spaces, like a space of states, configuration space, parameter space or a physical space-time, one usually deals with manifolds.

A topological space M is called an n-dimensional *topological manifold* if it looks locally like a Euclidean space $\mathbb{R}^n$, or, more precisely if there exists a family of open subsets $(U_i, i \in I)$ of M such that

1. it covers M, i.e., $\bigcup_{i \in I} U_i = M$,

2. for each $i \in I$ there is a homeomorphism (continuous, invertible map)

$$\varphi_i \; : \; U_i \; \longrightarrow \; \varphi(U_i) \subset \mathbb{R}^n \, .$$

A pair (U_i, φ_i) is called a *chart* or more often a local coordinate system. Since $\varphi_i : U_i \longrightarrow \mathbb{R}^n$, we may represent φ_i as follows:

$$\varphi_i = (x_{(i)}^1, x_{(i)}^2, \dots, x_{(i)}^n) \, ,$$

where $x_{(i)}^k : U_i \longrightarrow \mathbb{R}$, for $k = 1, 2, \dots, n$. The maps $(x_{(i)}^1, x_{(i)}^2, \dots, x_{(i)}^n)$ are called local coordinates on the patch U_i. Moreover, any two charts on M have to be compatible, that is, the overlap map

$$\varphi_{ji} := \varphi_j \circ \varphi_i^{-1} \; : \; \varphi_i(U_i \cap U_j) \; \longrightarrow \; \varphi_j(U_i \cap U_j) \tag{1.1}$$

defines a homomorphism for any $i, j \in I$, cf. Fig. 1.1. An overlap map φ_{ji} describes how the different charts are glued together. One calls a collection of compatible charts covering M an *atlas*. If the maps φ_{ji} are C^N diffeomorphisms, i.e., they are N times differentiable, then M is called a C^N manifold. In particular, C^1 and C^∞ manifolds are called *differentiable* and *smooth* manifolds, respectively. Throughout this book all manifolds are assumed to be smooth.

The most trivial example of a differentiable manifold is an open subset U of $\mathbb{R}^n$. An atlas consists just of one chart (U, id_U), where id_U stands for the identity map on U. However, usually one needs more than one chart to cover M. An important example is provided by

Example 1.1.1 (*n*-dimensional sphere S^n) S^n is defined as the following subset in $\mathbb{R}^{n+1}$:

$$S^n := \left\{ (x^1, \dots, x^{n+1}) \in \mathbb{R}^{n+1} \; \bigg| \; \sum_{i=1}^{n+1} (x^i)^2 = 1 \right\} \, .$$

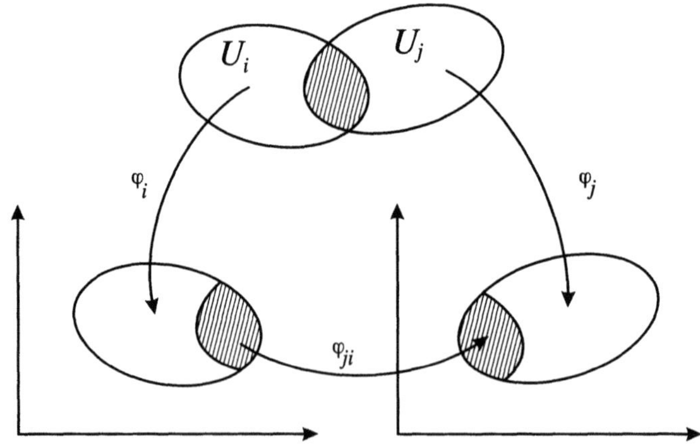

Figure 1.1: Overlapping charts on a manifold.

Define two patches U_N and U_S covering S^n:

$$U_N := \left\{ (x^1, \ldots, x^{n+1}) \in S^n \,\middle|\, x^{n+1} > -1 \right\} = S^n - \{(0, 0, \ldots, 0, -1)\},$$

and

$$U_S := \left\{ (x^1, \ldots, x^{n+1}) \in S^n \,\middle|\, x^{n+1} < 1 \right\} = S^n - \{(0, 0, \ldots, 0, 1)\}.$$

In analogy with $\mathbb{R}^3$ we may call the points $\{(0, 0, \ldots, 0, -1)\}$ and $\{(0, 0, \ldots, 0, 1)\}$ the south and north poles of S^n, respectively. Now, let us define two maps: $\varphi_N : U_N \longrightarrow \mathbb{R}^n$ and $\varphi_S : U_S \longrightarrow \mathbb{R}^n$ by stereographic projections on the equatorial plane $x^{n+1} = 0$ from the south and north poles, respectively. That is,

$$\varphi_N(x^1, \ldots, x^{n+1}) := \frac{1}{1 + x^{n+1}} (x^1, \ldots, x^n),$$

and

$$\varphi_S(x^1, \ldots, x^{n+1}) := \frac{1}{1 - x^{n+1}} (x^1, \ldots, x^n).$$

One easily finds that the overlap map φ_{NS}, given by

$$\varphi_{NS}(x^1, \ldots, x^n) = \left(\sum_{i=1}^{n} (x^i)^2 \right)^{-1} (x^1, \ldots, x^n)$$

defines a smooth map on $\mathbb{R}^n - \{0\}$. This proves that the two charts (U_N, φ_N) and (U_S, φ_S) provide an atlas on S^n. $\diamondsuit$

An atlas on a manifold M enables us to develop a differential calculus. A function f on M, i.e., a map $f : M \longrightarrow \mathbb{R}$, is differentiable if

$$f \circ \varphi_i^{-1} : \varphi_i(U_i) \longrightarrow \mathbb{R}^n \tag{1.2}$$

is differentiable for each chart (U_i, φ_i). We shall denote the space of smooth functions on M by $C^\infty(M)$. Consider now a chart $(U, \varphi = (x^1, \ldots, x^n))$ on M and let $x \in U$. Define a differential operator V_x at x by

$$V_x(f) := \sum_{i=1}^{n} V_x^i \frac{\partial f}{\partial x^i}(x), \tag{1.3}$$

for any differentiable function f on U. This construction shows that V_x is uniquely determined by n numbers $(V_x^1, \ldots, V_x^n)$ and hence that differential operators at x span an n-dimensional linear space called the *tangent space* $T_x M$ at the point x. Elements from $T_x M$ are called tangent vectors (attached) at the point x. Thus, there is a one-to-one correspondence between differential operators and tangent vectors. Clearly, the set of differentials

$$\left\{ \frac{\partial}{\partial x^1} \bigg|_x, \ldots, \frac{\partial}{\partial x^n} \bigg|_x \right\} \tag{1.4}$$

defines a basis in $T_x M$ — the so-called *coordinate basis*. The n numbers $(V_x^1, \ldots, V_x^n)$ are components of a tangent vector V_x with respect to a coordinate basis. From its definition it is clear that $V_x(f)$ is the directional derivative of a function f at a point x in the direction of a vector V_x. An assignment

$$M \ni x \longrightarrow V_x \in T_x M , \tag{1.5}$$

is called a *vector field* on M. We shall denote the space of vector fields on M by $\mathfrak{X}(M)$.

Denote by $T_x^* M$ the algebraic dual of $T_x M$, called a *cotangent space* at x. Physicists often call elements from $T_x M$ covectors (dual vectors). Having a coordinate basis (1.4) in $T_x M$ let us introduce the corresponding dual basis in $T_x^* M$:

$$\left\{ dx^1, \ldots, dx^n \right\} , \tag{1.6}$$

such that,

$$dx^i \left(\frac{\partial}{\partial x^j} \right) = \delta^i{}_j . \tag{1.7}$$

The above constructions of tangent and cotangent spaces enable one to introduce an arbitrary tensor field on M. We shall call a smooth map

$$M \ni x \longrightarrow T(x) \in T_x^{(k,l)} M := \overbrace{T_x M \otimes \ldots \otimes T_x M}^{k} \otimes \overbrace{T_x^* M \otimes \ldots \otimes T_x^* M}^{l} \tag{1.8}$$

a tensor field of type (k, l). Clearly, a vector field is a tensor field of type $(1, 0)$. Any tensor field is uniquely defined by its components. In particular, using a coordinate basis in $T_x M$ and $T_x^* M$ we have

$$T = T_{j_1 \ldots j_l}^{i_1 \ldots i_k} \frac{\partial}{\partial x^{i_1}} \otimes \ldots \otimes \frac{\partial}{\partial x^{i_k}} \otimes dx^{j_1} \otimes \ldots \otimes dx^{j_l} , \tag{1.9}$$

where we use the Einstein summation convention.

Let us consider two manifolds M and N together with a smooth map

$$\varphi : M \longrightarrow N .$$

The derivative $T_x \varphi$ (one calls it also a tangent map) of φ at a point $x \in M$ is a linear map

$$T_x \varphi : T_x M \longrightarrow T_{\varphi(x)} N ,$$

such that

$$[T_x \varphi(v_x)](f) := v_x(\varphi \circ f) , \tag{1.10}$$

for any $f \in C^\infty(N)$ and $v_x \in T_x M$. Representing φ in local coordinates $(x^1, \ldots, x^m)$:

$$\varphi^i = \varphi^i(x^1, \ldots, x^m) , \quad i = 1, \ldots, n = \dim N ,$$

the tangent map $T_x\varphi$ is represented by the $n \times m$ matrix

$$[T_x\varphi]^i{}_j := \frac{\partial \varphi^i}{\partial x^j}(x) , \quad i = 1, \ldots, n , \quad j = 1, \ldots, m .$$

Using a map $\varphi : M \longrightarrow N$ one may transport tensor fields between M and N. Let ω be a tensor field of type $(0, k)$ on N. A *pull-back* $\varphi^*\omega$ of ω is a $(0, k)$-tensor field on M defined by

$$(\varphi^*\omega)_x(v_1, \ldots, v_k) := \omega_{\varphi(x)}(T_x\varphi(v_1), \ldots, T_x\varphi(v_k)) , \tag{1.11}$$

for any $x \in M$ and $v_1, \ldots, v_k \in T_x M$. Conversely, a tensor field U of type $(l, 0)$ may be *pushed forward* from M to N, giving rise to an $(l, 0)$-tensor field $\varphi_* U$:

$$(\varphi_* U)_{\varphi(x)}(\alpha_1, \ldots, \alpha_l) := U_x(\varphi_x^*\alpha_1, \ldots, \varphi_x^*\alpha_l) , \tag{1.12}$$

where $\alpha_1, \ldots, \alpha_l \in T^*_{\varphi(x)} N$. It is easy to show that if $\varphi : M \longrightarrow N$ is a diffeomorphism, then

$$\varphi_* = (\varphi^{-1})^* . \tag{1.13}$$

Example 1.1.2 Let V be a $(1, 0)$-tensor on M, that is, a vector field $V = V^j \partial/\partial x^j$. Then the pushed-forward vector field $\varphi_* V$ is given by

$$\varphi_* V = (\varphi_* V)^i \frac{\partial}{\partial y^i} ,$$

where $\{\partial/\partial y^j\}$ denotes the coordinate basis on N and

$$(\varphi_* V)^i(\varphi(x)) = \frac{\partial \varphi^i}{\partial x^j}(x) \, V^j(x) .$$

Similarly, if α is a $(0, 1)$-tensor on N, i.e., a covector field $\alpha = \alpha_i \, dy^i$, then the pulled-back covector field $\varphi^*\alpha$ on M is given by

$$\varphi^*\alpha = (\varphi^*\alpha)_j \, dx^j ,$$

with

$$(\varphi^*\alpha)_j(x) = \frac{\partial \varphi^i}{\partial x^j}(x) \, \alpha_i(\varphi(x)) .$$

This way, one recovers the well-known transformation rules for vectors and covectors.

1.1.2   Differential forms

Now we are going to introduce an important class of tensor fields on a differential manifold that play a prominent role in physical applications.

Definition 1.1.1 *A skew-symmetric tensor of type $(0, k)$ is called a differential form of order k (or simply a k-form).*

Denote by $\Lambda^k(M)$ the space of k-forms on M. Evidently, $\Lambda^k(M) = \{\emptyset\}$ for $k > n$. Therefore, the space of differential forms on M, denoted by $\Lambda(M)$, splits into the following direct sum:

$$\Lambda(M) = \bigoplus_{k=0}^{n} \Lambda^k(M) \, ,$$

with $\Lambda^0(M) = C^\infty(M)$. The space $\Lambda(M)$ is equipped with two basic operations: *wedge product* and *exterior derivative*.

Definition 1.1.2 *A wedge product $\wedge$ (called also the exterior or Grassmann product):*

$$\wedge \, : \, \Lambda^k(M) \, \times \, \Lambda^l(M) \, \longrightarrow \, \Lambda^{k+l}(M) \, , \tag{1.14}$$

is defined by

$$\alpha \wedge \beta \, := \, \frac{(k+l)!}{k!\,l!} \, \mathbf{A}(\alpha \otimes \beta) \, , \tag{1.15}$$

where $\mathbf{A}$ is an alternation operator which selects the skew-symmetric part of the $(0, k+l)$-tensor $\alpha \otimes \beta$.

For example, if α and β are one-forms, then

$$(\alpha \wedge \beta)(v_1, v_2) = \alpha(v_1)\beta(v_2) - \alpha(v_2)\beta(v_1) \, , \tag{1.16}$$

for any vectors v_1 and v_2. In terms of local coordinates $(x^1, \dots, x^n)$ any k-form α has the following component representation:

$$\alpha = \frac{1}{k!} \, \alpha_{i_1 \dots i_k} \, dx^{i_1} \wedge \dots \wedge dx^{i_k} \, . \tag{1.17}$$

One easily shows that

$$(\alpha \, \wedge \, \beta)_{i_1 \dots i_{k+l}} = \alpha_{[i_1 \dots i_k} \beta_{i_{k+1} \dots i_{k+l}]} \, , \tag{1.18}$$

where the square bracket stands for anti-symmetrization.

Example 1.1.3 If $\alpha = \alpha_i \, dx^i$ and $\beta = \beta_j \, dx^j$, then

$$\alpha \wedge \beta = \alpha_i \beta_j \, dx^i \wedge dx^j = \frac{1}{2}(\alpha_i \beta_j - \alpha_j \beta_i) \, dx^i \wedge dx^j \, ,$$

and hence

$$(\alpha \wedge \beta)_{ij} = \alpha_i \beta_j - \alpha_j \beta_i \, ,$$

in agreement with formula (1.18). $\diamond$

The properties of the wedge product are summarized in the following

Proposition 1.1.1 *The wedge product satisfies*

1. $(\alpha \wedge \beta) \wedge \gamma = \alpha \wedge (\beta \wedge \gamma)$,

2. $\alpha \wedge \beta = (-1)^{kl} \beta \wedge \alpha$, *where* $\alpha \in \Lambda^k(M)$ *and* $\beta \in \Lambda^l(M)$.

Since there is a natural pairing between vectors and forms it is convenient to introduce the so-called *interior product*, which is a contraction of a vector field $v \in \mathfrak{X}(M)$ and a k-form α:

$$i_v \, : \, \Lambda^k(M) \longrightarrow \Lambda^{k-1}(M) \,,$$

that is,

$$(i_v \, \alpha)_{i_1 \ldots i_{k-1}} \; := \; v^j \alpha_{j i_1 \ldots i_{k-1}} \,. \tag{1.19}$$

Moreover, we declare that $i_v \, f = 0$ for $f \in \Lambda^0(M)$.

The next operation we are going to introduce enables one to differentiate k-forms.

Definition 1.1.3 *The exterior derivative*

$$d \, : \, \Lambda^k(M) \longrightarrow \Lambda^{k+1}(M)$$

is defined as follows:

$$d\alpha \; = \; \frac{1}{k!} \frac{\partial \alpha_{i_1 \ldots i_k}}{\partial x^j} \, dx^j \wedge dx^{i_1} \wedge \ldots \wedge dx^{i_k} \,, \tag{1.20}$$

for any k-form α represented by the formula (1.17).

Example 1.1.4 Consider a function f on a differential manifold M. Since a function is a zero-form one defines its exterior derivative df to be the following one-form:

$$df \; = \; \frac{\partial f}{\partial x^i} \, dx^i \,.$$

Note that

$$d(df) \; = \; \frac{\partial^2 f}{\partial x^j \partial x^i} \, dx^j \wedge dx^i = 0 \,,$$

because the partial derivatives $\partial_i \partial_j f$ are symmetric in (ij) whereas $dx^j \wedge dx^i$ is anti-symmetric. In particular, if $M = \mathbb{R}^n$ and $(x^1, \ldots, x^n)$ are cartesian coordinates, then df reproduces the components of grad f. $\diamond$

Example 1.1.5 (Differential forms in $\mathbb{R}^3$) Let us choose cartesian coordinates (x^1, x^2, x^3) in $\mathbb{R}^3$ and consider a one-form α and a two-form β. Clearly,

$$\alpha = \sum_{i=1}^{3} \alpha_i \, dx^i \,, \qquad \beta = \frac{1}{2} \sum_{i,j=1}^{3} \beta_{ij} \, dx^i \wedge dx^j \,. \tag{1.21}$$

Now, since $\beta_{ij} = -\beta_{ji}$, one has $\beta_{ij} = \sum_{k=1}^{3} \epsilon_{ijk}\beta_k$, where ϵ_{ijk} is the Levi–Civita tensor in $\mathbb{R}^3$, and hence

$$\beta = \frac{1}{2} \sum_{i,j,k=1}^{3} \epsilon_{ijk}\beta_i \, dx^j \wedge dx^k = \beta_1 \, dx^2 \wedge dx^3 + \beta_2 \, dx^3 \wedge dx^1 + \beta_3 \, dx^1 \wedge dx^2 \, .$$

$$(1.22)$$

We therefore find that

$$d\alpha = \psi_1 \, dx^2 \wedge dx^3 + \psi_2 \, dx^3 \wedge dx^1 + \psi_3 \, dx^1 \wedge dx^2 \, , \qquad (1.23)$$

with

$$\psi_1 = \frac{\partial \alpha_3}{\partial x^2} - \frac{\partial \alpha_2}{\partial x^3} \, , \qquad \psi_2 = \frac{\partial \alpha_1}{\partial x^3} - \frac{\partial \alpha_3}{\partial x^1} \, , \qquad \psi_3 = \frac{\partial \alpha_2}{\partial x^1} - \frac{\partial \alpha_1}{\partial x^2} \, , \qquad (1.24)$$

and hence the cartesian components of $d\alpha$ represent the curl of a vector field $\boldsymbol{\alpha} = (\alpha_1, \alpha_2, \alpha_3)$:

$$d\alpha \quad \longleftrightarrow \quad \boldsymbol{\psi} = \mathbf{curl}\,\boldsymbol{\alpha} \, .$$

The exterior derivative of β gives

$$d\beta = \sum_{i=1}^{3} \frac{\partial \beta_i}{\partial x^i} \, dx^1 \wedge dx^2 \wedge dx^3 \, , \qquad (1.25)$$

and hence, it represents the divergence of the vector field $\boldsymbol{\beta} = (\beta_1, \beta_2, \beta_3)$:

$$d\beta \quad \longleftrightarrow \quad \mathbf{div}\,\boldsymbol{\beta} \, .$$

This way one recovers standard vector analysis in $\mathbb{R}^3$. Note that $d(d\alpha) = 0$, which reproduces the well-known identity

$$\mathbf{div}\,\mathbf{curl}\,\boldsymbol{\alpha} \equiv 0 \, .$$

Clearly, $d(d\beta) = 0$ since $d(d\beta)$ as a four-form vanishes identically in $\mathbb{R}^3$. Moreover, if f is a smooth function, then the identity

$$\mathbf{curl}\,\mathbf{grad}\,f = 0$$

follows from $d(df) = 0$ (see the previous Example). $\diamond$

These simple observations in $\mathbb{R}^3$ may be immediately deduced from the following

Proposition 1.1.2 *The exterior derivative satisfies*

   1. $d^2\alpha = d(d\alpha) = 0$, for any $\alpha \in \Lambda(M)$,

   2. $d(\alpha \wedge \beta) = d\alpha \wedge \beta + (-1)^k \alpha \wedge d\beta$, for $\alpha \in \Lambda^k(M)$.

Consider two manifolds M and N and let $\varphi : M \longrightarrow N$ be a smooth map. A pull-back operation induces a map

$$\varphi^* : \Lambda(N) \longrightarrow \Lambda(M) .$$

Proposition 1.1.3 *The pull-back operation commutes with wedge product and exterior derivative. That is,*

$$\varphi^*(\alpha \wedge \beta) = \varphi^*\alpha \wedge \varphi^*\beta , \tag{1.26}$$

and

$$\varphi^*(d\alpha) = d(\varphi^*\alpha) , \tag{1.27}$$

for any differential forms α and β on N. In the last equation we use the same letter 'd' to denote the exterior derivative on N and M.

1.1.3   Integration of forms

Differential forms occur implicitly in all branches of physics because they are natural objects appearing as integrands of line, surface, and volume integrals as well as their n-dimensional generalizations. Consider $\mathbb{R}^n$ with cartesian coordinates $(x^1, \ldots , x^n)$. Having a function $f : \mathbb{R}^n \longrightarrow \mathbb{R}$, one defines an n-dimensional "volume" integral

$$f \longrightarrow \int f\, dV = \int \ldots \int f(x^1, \ldots , x^n)\, dx^1 \ldots dx^n .$$

Clearly, the value of the integral can not depend upon the particular coordinates chosen to parametrize $\mathbb{R}^n$. In particular, changing coordinates from $(x^1, \ldots , x^n)$ to $(\widetilde{x}^1, \ldots , \widetilde{x}^n)$ one finds $\int f\, d\widetilde{V}$, with

$$d\widetilde{V} = J\, dV ,$$

where J stands for the Jacobian of the transformation, i.e.,

$$J = \det \left(\frac{\partial \widetilde{x}^i}{\partial x^j} \right) .$$

What is the origin of J? Note that on $\mathbb{R}^n$ we have a natural n-form $dx^1 \wedge \ldots \wedge dx^n$. One immediately sees that

$$d\widetilde{x}^1 \wedge \ldots \wedge d\widetilde{x}^n = J\, dx^1 \wedge \ldots \wedge dx^n , \tag{1.28}$$

which shows that the convenient notation $dV = dx^1 \ldots dx^n$ actually denotes an n-form, and hence, it should rather be written as follows

$$dV = dx^1 \wedge \ldots \wedge dx^n .$$

Hence, it is clear that to perform n-dimensional integration one needs n-forms.

Definition 1.1.4 *An n-dimensional manifold M is orientable iff there exists a nowhere-vanishing n-form τ on it.*

One shows that a manifold M is orientable if we can cover it by coordinate patches (U_i, φ_i) having positive Jacobians in each overlap, i.e., $\det(\varphi_{ji}(x)) > 0$ for any $x \in U_i \cap U_j$. It should be clear that if M is orientable, then there are exactly two different ways to orient it. Of course, if M can be covered by a single chart then it is orientable. For example, any open subset of $\mathbb{R}^n$ is an n-dimensional orientable manifold.

Example 1.1.6 (Riemannian manifold) A Riemannian manifold (M, g) is a smooth manifold M together with a smooth tensor g of type $(0, 2)$, called a metric tensor, such that

1. g is symmetric,

2. for each $x \in M$, the bilinear form $g_x : T_x M \times T_x M \longrightarrow \mathbb{R}$ is nondegenerate.

A Riemannian manifold is called proper if

$$g_x(v, v) > 0 \quad \text{for all } v \in T_x M , \ v \neq 0 .$$

Otherwise a manifold is called pseudo-Riemannian. For example, Euclidean space $\mathbb{R}^n$ is proper Riemannian, whereas the Minkowski space $\mathbb{R}^{1,3}$ is pseudo-Riemannian. Note that on a Riemannian orientable manifold one may define a canonical volume form. In the space of differential forms on M one introduces so called Hodge operation (or Hodge star),

$$\star \ : \ \Lambda^k(M) \ \longrightarrow \ \Lambda^{n-k}(M) , \tag{1.29}$$

by the following formula:

$$(\star\alpha)_{i_1 \dots i_{n-k}} \ := \ \frac{1}{k!} \sqrt{|g|} \, \epsilon_{i_1 \dots i_{n-k} j_1 \dots j_k} \, \alpha^{j_1 \dots j_k} , \tag{1.30}$$

with $g := \det(g_{ij})$ and

$$\alpha^{j_1 \dots j_k} \ := \ g^{j_1 m_1} \dots g^{j_k m_k} \, \alpha_{m_1 \dots m_k} . \tag{1.31}$$

Here $\epsilon_{i_1 \dots i_n}$ stands for the Levi-Civita tensor in $\mathbb{R}^n$, and g^{ij} denotes the inverse of g_{ij}. The form $\star\alpha$ is usually called the *Hodge dual* of α. A Hodge star induces a natural volume form

$$\tau := \star 1 , \tag{1.32}$$

where '1' is a constant function on M, i.e., $1(x) = 1$ for any $x \in M$. Evidently

$$\tau = \sqrt{|g|} \, dx^1 \wedge \dots \wedge dx^n . \tag{1.33}$$

If M is compact then

$$\text{Vol}(M) \ := \ \int_M \tau \tag{1.34}$$

is called the volume of M (with respect to τ). $\diamond$

Consider now an n-dimensional manifold M and let K be a k-dimensional orientable submanifold of M.[1] Denote by $j : K \hookrightarrow M$ a canonical embedding. If α is a k-form on M, then $j^*\alpha$ is a k-form on K, and hence one may define an integral of $j^*\alpha$ over K:

$$(K, \alpha) \longrightarrow \int_K j^*\alpha \,.$$

Let $(y^1, \dots, y^k)$ denote local coordinates on K, and let the embedding j be described by

$$j \longrightarrow \begin{cases} x^1 &= x^1(y^1, \dots, y^k) \\ x^2 &= x^2(y^1, \dots, y^k) \\ \quad\vdots \\ x^n &= x^n(y^1, \dots, y^k). \end{cases}$$

Then if $\alpha = \frac{1}{k!}\alpha_{i_1 \dots i_k} dx^{i_1} \wedge \dots \wedge dx^{i_k}$, the above integral may be rewritten in a more familiar form:

$$\int_K j^*\alpha = \frac{1}{k!} \int \dots \int \alpha_{i_1 \dots i_k} \frac{\partial x^{i_1}}{\partial y^1} \dots \frac{\partial x^{i_k}}{\partial y^k} dy^1 \wedge \dots \wedge dy^k \,, \tag{1.35}$$

which is a generalization of the line and surface integrals in $\mathbb{R}^3$.

To formulate one of the most important results in the theory of integration of differential forms we need a notion of a manifold with boundary. Let $\mathbb{R}^n_+ = \{(x^1, \dots, x^n) \in \mathbb{R}^n \mid x^1 \geq 0\}$. Then the boundary, $\partial\mathbb{R}^n_+$, of $\mathbb{R}^n_+$ is defined by $\partial\mathbb{R}^n_+ = \{(x^1, \dots, x^n) \mid x^1 = 0\}$. An n-dimensional manifold, M, has the structure of a *manifold with a boundary* when there exists an open covering (U_i, φ_i) such that $\varphi_i(U_i)$ defines an open subset of $\mathbb{R}^n_+$. The boundary, ∂M, of M is defined by

$$\partial M := \bigcup_i \varphi_i^{-1}(\partial\mathbb{R}^n_+) \,.$$

One sees that the boundary, ∂M, of M is an $(n-1)$-dimensional differential manifold.

Example 1.1.7 Let B^n be a unit ball in $\mathbb{R}^n$:

$$B^n = \left\{ (x^1, \dots, x^n) \in \mathbb{R}^n \;\middle|\; \sum_{i=1}^n (x^i)^2 \leq 1 \right\} \,.$$

---

[1] $K \subset M$ is a k-dimensional submanifold iff for any point $x \in K$ there exists a chart (U, φ) on M, such that

$$\varphi|_{U \cap K} : U \cap K \longrightarrow (x^1, \dots, x^k, 0, \dots, 0) \,.$$

Then the boundary of B^n is an $(n-1)$-dimensional sphere,

$$\partial B^n = S^{n-1} \, ,$$

which agrees with our intuition of a boundary.   ◇

Note that the notion of a boundary satisfies

$$\partial^2 M = \partial(\partial M) = \{\emptyset\} \, , \tag{1.36}$$

for any manifold M, that is, a boundary does not have a boundary. In particular, the boundary of an n-dimensional sphere $\partial S^n = \partial^2 B^{n+1} = \{\emptyset\}$. A manifold without a boundary is called *closed*.

Theorem 1.1.4 (Stokes theorem) *Let M be an n-dimensional manifold with boundary, and let $\omega \in \Lambda^{n-1}(M)$. Then*

$$\int_M d\omega = \int_{\partial M} \omega \, , \tag{1.37}$$

where ∂M denotes the boundary of M.

Example 1.1.8 The Stokes theorem generalizes well-known theorems from vector analysis in $\mathbb{R}^3$:

1. If Σ is a two-dimensional surface in $\mathbb{R}^3$ and $\mathbf{A}$ a vector field, then

$$\int_\Sigma \mathbf{curl\,A} \cdot d\mathbf{S} = \oint_{C=\partial\Sigma} \mathbf{A} \cdot d\mathbf{l} \, ,$$

    where $d\mathbf{S}$ denotes a surface element on Σ, and $d\mathbf{l}$ stands for a line element along the closed curve $C = \partial\Sigma$.

2. If V is a three-dimensional region in $\mathbb{R}^3$ and $\mathbf{A}$ a vector field, then

$$\int_V \mathbf{div\,A}\, dV = \oint_{\partial V} \mathbf{A} \cdot d\mathbf{S} \, ,$$

    where dV denotes the volume element in $\mathbb{R}^3$. This formula is usually called a *Gauss theorem*.   ◇

1.1.4   De Rham cohomology

Recall from the vector analysis in $\mathbb{R}^3$ that $\mathbf{div\,(curl\,A)} = 0$ for any (smooth enough) vector field $\mathbf{A}$. However, the converse statement is in general not true, that is, the vanishing of $\mathbf{div\,B}$ does not imply the existence of $\mathbf{A}$ such that $\mathbf{B} = \mathbf{curl\,A}$. In the case of differential forms, the nilpotency of the exterior derivative, that is, $d^2 = 0$, leads to the important notion of *cohomology*. We shall call a form α

- a closed form if $d\alpha = 0$,

- an exact form if $\alpha = d\beta$ for some form β.

Clearly, any exact form is also closed but the converse statement is in general not true.

Example 1.1.9 Consider the following one-form on $\mathbb{R}^2$:

$$\beta = \frac{x\,dy - y\,dx}{x^2 + y^2} \, . \tag{1.38}$$

A simple calculation shows that $d\beta = 0$. Is β also exact? Note first, that β is not defined in all $\mathbb{R}^2$ — certainly we must omit the origin. Thus the manifold in question is $M = \mathbb{R}^2 - \{0\}$. Introducing polar coordinates (r, φ) in $\mathbb{R}^2$ we easily find that

$$\beta = d\varphi \, , \tag{1.39}$$

which seems to prove that β is exact. But this is not so. Integrating β over a closed curve $C := \{x^2 + y^2 = 1\}$, we obtain

$$\oint_C \beta = \int_0^{2\pi} d\varphi = 2\pi \, , \tag{1.40}$$

which shows that β is not exact, since, due to the Stokes theorem, the integral of an exact form over a closed manifold vanishes. Is there any contradiction? Certainly not. Note that the zero-form φ does not define a function on M since it is not single-valued, i.e., $\varphi + 2\pi \cong \varphi$. $\diamond$

Let M be a differentiable manifold and denote by $Z^k(M)$ and $B^k(M)$ the sets of closed and exact k-forms on M, respectively:

$$Z^k(M) = \left\{ \alpha \in \Lambda^k(M) \,\middle|\, d\alpha = 0 \right\} ,$$

and

$$B^k(M) = \left\{ \alpha \in \Lambda^k(M) \,\middle|\, \exists \, \beta \in \Lambda^{k-1}(M) \, , \, \alpha = d\beta \right\} .$$

Define the following relation in $\Lambda^k(M)$:

$$\alpha_1 \sim \alpha_2 \iff \exists \, \beta \in \Lambda^{k-1}(M) \, , \, \alpha_1 - \alpha_2 = d\beta \, .$$

Clearly,

$$\alpha_1 \sim \alpha_2 \implies d\alpha_1 = d\alpha_2 \, .$$

Evidently it is an equivalence relation, and hence we may define the space of equivalence classes

$$H^k(M) = Z^k(M)/B^k(M) \, , \tag{1.41}$$

i.e., $H^k(M)$ is the set of closed k-forms which differ only by an exact k-form. It is called the kth *de Rham cohomology group* of M ($H^k(M)$ is an abelian group, where the group operation is the addition of k-forms). The equivalence class containing ω will be denoted by $[\omega]$ and called a cohomology class of ω. For $[\omega_1] \in H^k(M)$ and $[\omega_2] \in H^l(M)$ we have

$$[\omega_1 \wedge \omega_2] = [\omega_1] \wedge [\omega_2] \in H^{k+l}(M) \,. \tag{1.42}$$

It turns out that on $\mathbb{R}^n$ all closed forms are exact, that is, all $H^k(\mathbb{R}^n)$ are trivial for $k > 0$. Moreover,

Proposition 1.1.5 (Poincaré Lemma) *Any closed form on a differentiable manifold M is locally exact, that is, if $d\alpha = 0$, then for any $x \in M$ there is a neighborhood U containing x such that $\alpha = d\beta$ on U.*

Hence, only global properties of M decide whether or not de Rham cohomology groups are trivial. Now, if φ is a smooth map

$$\varphi \,:\, M \longrightarrow N \,,$$

then it induces a linear transformation

$$\varphi^\sharp \,:\, H^k(N) \longrightarrow H^k(M) \,,$$

defined by

$$\varphi^\sharp([\omega]) := [\varphi^* \omega] \,. \tag{1.43}$$

It turns out that when ϕ is a homeomorphism, i.e., M and N are topologically equivalent, the induced map $\varphi^\sharp$ is an isomorphism. That is, topologically equivalent manifolds have isomorphic cohomology groups. In particular

$$b_k(M) = b_k(N) \,,$$

where

$$b_k(M) := \dim H^k(M) \,, \qquad k = 0, 1, \ldots, n \,, \tag{1.44}$$

are the so-called Betti numbers. One may construct the following basic topological invariant called the *Euler characteristic* of M:

$$\chi(M) := \sum_{k=0}^{n} (-1)^k b_k(M) \,. \tag{1.45}$$

Clearly, topologically equivalent manifolds have the same Euler characteristic. To see how the topology enters the game let us consider three topologically different two-dimensional spaces: the plane $\mathbb{R}^2$, the sphere S^2 and the torus T^2.

1. $M = \mathbb{R}^2$. Clearly $H^1(\mathbb{R}^2) = H^2(\mathbb{R}^2) = 0$. Closed zero-forms are nothing but constant functions and therefore $H^0(\mathbb{R}^2) \cong \mathbb{R}$ which leads to $\chi(\mathbb{R}^2) = 1$.

2. $M = S^2$. Now $H^0(S^2) = H^2(S^2) \cong \mathbb{R}$ and all other cohomology groups are trivial. One has $\chi(S^2) = 2$. Actually, it is easy to show that

$$\chi(S^n) = \begin{cases} 2, & \text{if } n \text{ is even} \\ 0, & \text{if } n \text{ is odd} \end{cases}.$$

3. $M = T^2$. Let θ_1 and θ_2 be coordinates on each of the two circles making the torus $T^2 = S^1 \times S^1$. The differential forms $d\theta_i$ are obviously closed but not exact, since the θ's are defined only modulo 2π and do not define global coordinates. Therefore, $b_1 = 2$ (the $d\theta$'s form a two-dimensional basis) and, as in the case of S^2, $H^0(T^2) = H^2(T^2) \cong \mathbb{R}$ which results in $\chi(T^2) = 0$.

Summarizing,

$$\chi(M) = \begin{cases} 0, & \text{for } M = T^2 \\ 1, & \text{for } M = \mathbb{R}^2 \\ 2, & \text{for } M = S^2 \end{cases}.$$

Proposition 1.1.6 *A contractible manifold M, i.e., a manifold that may be continuously contracted to a single point, has trivial de Rham cohomology groups $H^k(M)$ for all $k \geq 1$.*

Let us note that the manifold $M = \mathbb{R}^2 - \{0\}$ from Example 1.1.9 is not contractible. One easily finds that

$$H^1(\mathbb{R}^2 - \{0\}) \cong \mathbb{R},$$

i.e., its first cohomology group is not trivial. We close this section with the beautiful notion of *Poincaré duality*.

Proposition 1.1.7 *Let M be a compact, connected, orientable n-dimensional manifold. Then*

$$H^k(M) \cong H^{n-k}(M),$$

for any $k = 0, 1, \ldots, n$.

Note that Poincaré duality applies to S^2 and T^2 but not to $\mathbb{R}^2$ — both S^2 and T^2 are compact, connected, orientable two-dimensional manifolds and, therefore, $H^0(S^2) = H^2(S^2)$, and the same holds for a torus T^2. However $H^0(\mathbb{R}^2) \neq H^2(\mathbb{R}^2)$.

1.1.5 Lie derivative

Consider a vector field X on a manifold M. The *flow* of X is the collection of maps $F_t : M \longrightarrow M$ satisfying

$$\frac{d}{dt} F_t(x) = X(F_t(x)), \tag{1.46}$$

for each $x \in X$ and $t \in \mathbb{R}$. Fixing a point $x \in M$ one obtains a map

$$\mathbb{R} \ni t \longrightarrow F_t(x) \in M .$$

Clearly, this map defines a curve in M called an *integral curve* of X passing through a point x. Note, that a flow F_t satisfies the following property:

$$F_t \circ F_s = F_s \circ F_t = F_{t+s} . \tag{1.47}$$

Example 1.1.10 Consider a vector field in $\mathbb{R}^n$ defined by the linear operator $\hat{A} : \mathbb{R}^n \longrightarrow \mathbb{R}^n$:

$$\mathbb{R}^n \ni \mathbf{x} \longrightarrow \hat{A}\mathbf{x} \in T_{\mathbf{x}}\mathbb{R}^n \cong \mathbb{R}^n .$$

The corresponding flow, F_t, satisfies

$$\frac{d}{dt} F_t(\mathbf{x}) = \hat{A} \, F_t(\mathbf{x}) ,$$

and hence

$$F_t(\mathbf{x}) = e^{\hat{A}t} \, \mathbf{x} .$$

The flow property (1.47) immediately follows. $\diamond$

Definition 1.1.5 *Let $X \in \mathfrak{X}(M)$ and T be a tensor field on M. The Lie derivative of T with respect to the vector field X is defined by*

$$(\mathcal{L}_X T)(x) := \frac{d}{dt} (F_t^* \, T)(x)\Big|_{t=0} , \tag{1.48}$$

where F_t is the flow of X.

By the very definition of a pull-back, if T is a (k, l)-tensor, then so is $\mathcal{L}_X T$. If f is a function on M, then $\mathcal{L}_X f$, given by

$$\mathcal{L}_X f = \frac{d}{dt} (F_t^* f)\Big|_{t=0} = \frac{d}{dt} (f \circ F_t)\Big|_{t=0} = X(f) , \tag{1.49}$$

is the directional derivative of f along X. Consider now the action of $\mathcal{L}_X$ on vector fields. If $Y \in \mathfrak{X}(M)$, then

$$\mathcal{L}_X Y = [X, Y] , \tag{1.50}$$

where the *Lie bracket* (often called a *commutator*)

$$[\, , \,] : \mathfrak{X}(M) \times \mathfrak{X}(M) \longrightarrow \mathfrak{X}(M) ,$$

is defined by

$$[X, Y]^i := X^k \partial_k Y^i - Y^k \partial_k X^i . \tag{1.51}$$

Finally, let us turn to differential forms. One may show that if $\omega \in \Lambda(M)$, then

$$\mathcal{L}_X \omega = d(i_X \omega) + i_X \, d\omega \, , \tag{1.52}$$

where i_X stands for an interior product (contraction with X) — see (1.19). In particular, we have for any one-form α:

$$\begin{aligned} d\alpha(u, v) &= \mathcal{L}_v(\alpha(u)) - \mathcal{L}_u(\alpha(v)) - \alpha(\mathcal{L}_u v) \\ &= v(\alpha(u)) - u(\alpha(v)) - \alpha([u, v]) \, . \end{aligned} \tag{1.53}$$

This equation may be generalized to the celebrated Cartan formula, as follows:

$$\begin{aligned} d\omega(v_0, v_1, \ldots, v_k) &= \sum_{i=0}^{k} (-1)^i \, v_i \left[\omega(v_0, v_1, \ldots, \check{v}_i, \ldots, v_k) \right] \\ &\quad + \sum_{0 \le i < j \le k} \omega([v_i, v_j], v_0, \ldots, \check{v}_i, \ldots, \check{v}_j, \ldots, v_k) \, , \end{aligned} \tag{1.54}$$

where $\check{v}_i$ denotes the omission of v_i. The formula (1.54) may be regarded as a coordinate-free definition of an exterior derivative d.

1.2 Groups, Lie algebras and actions

1.2.1 Basic definitions

A Lie group G is a group that is also a differentiable manifold such that the differentiable structure is compatible with the group structure, i.e., the group operation $G \times G \longrightarrow G$, defined by $(g, h) \longrightarrow g \cdot h$, and the inversion $g \longrightarrow g^{-1}$ are smooth maps.

Example 1.2.1 The simplest examples of Lie groups are linear spaces. If V is a linear space, then the following operations

$$V \times V \ni (u, v) \longrightarrow u + v \in V \, , \quad \text{and} \quad V \ni x \longrightarrow -x \in V,$$

endow V with the structure of an abelian Lie group. Hence $\mathbb{R}^n$ defines an n-dimensional Lie group. $\diamond$

The most important examples of Lie groups are classical matrix groups.

Example 1.2.2 (General linear group) Denote by $M(n, \mathbb{R})$ the space of $n \times n$ real matrices. Then the general linear group is defined by

$$GL(n, \mathbb{R}) := \left\{ X \in M(n, \mathbb{R}) \,\middle|\, \det X \neq 0 \right\} \, .$$

One may show that $GL(n, \mathbb{R})$ is an n^2-dimensional differential manifold. The group operation is a composition of matrices:

$$GL(n, \mathbb{R}) \ni A, B \longrightarrow A \cdot B \in GL(n, \mathbb{R}) \, ,$$

and the inversion map is defined by $A \longrightarrow A^{-1}$. Both maps are smooth and so $GL(n, \mathbb{R})$ is a Lie group. For other matrix Lie groups see Appendix A. $\diamondsuit$

Example 1.2.3 Let M be a differentiable manifold. Then a set of diffeomorphisms

$$\varphi : M \longrightarrow M ,$$

defines an infinite-dimensional Lie group denoted by $\mathrm{Diff}(M)$. Clearly, the group operation is composition of differeomphisms. $\diamondsuit$

Recall now that a *Lie algebra L* is a vector space endowed with a bilinear operation,

$$[\, , \,] : L \times L \longrightarrow L ,$$

called a *Lie bracket,*[2] which satisfies the following two conditions:

1. $[x, y] = -[y, x]$,

2. $[[x, y], z] + [[z, x], y] + [[y, z], x] = 0$ (Jacobi identity) ,

for any $x, y, z \in L$. If $(e_1, \ldots, e_n)$ is a basis in L, then commutation relations

$$[e_i, e_j] = f^k_{ij} e_k , \tag{1.55}$$

uniquely determine the structure of the Lie algebra. The constants f^k_{ij} are called the *structure constants* of L.

Example 1.2.4 (Matrix Lie algebra) Evidently, a set of real $n \times n$ matrices together with

$$[A, B] := A \cdot B - B \cdot A ,$$

defines an n^2-dimensional Lie algebra, which we denote $gl(n, \mathbb{R})$. For other matrix Lie algebras see Appendix A. $\diamondsuit$

Example 1.2.5 If M is a differentiable manifold, then a set of vector fields $\mathfrak{X}(M)$ endowed with a Lie bracket as defined in (1.51), gives rise to an infinite-dimensional Lie algebra. $\diamondsuit$

Definition 1.2.1 *Consider two Lie algebras* $(L_1, [\, , \,]_1)$ *and* $(L_2, [\, , \,]_2)$. *A linear map* $\phi : L_1 \longrightarrow L_2$ *is a Lie algebra homomorphism or anti-homomorphism iff, respectively,*

$$[\phi(x), \phi(y)]_2 = \phi([x, y]_1) \ \ or \ \ [\phi(x), \phi(y)]_2 = -\phi([x, y]_1) ,$$

for any $x, y \in L_1$.

---

[2]Physicists often call it a commutator.

It turns out that every Lie group G has a Lie algebra $\mathfrak{g}$ associated with it, called the Lie algebra of G, which may be constructed as follows. Any element $g \in G$ gives rise to the following natural mappings:

$$L_g, R_g : G \longrightarrow G$$

defined by

$$L_g(h) := gh \quad \text{and} \quad R_g(h) := hg , \tag{1.56}$$

for any $h \in G$. They are called *left* (L_g) and *right* (R_g) *translations*. A vector field $X \in \mathfrak{X}(G)$ is called *left-invariant* if

$$(L_g)_* X = X , \tag{1.57}$$

for any $g \in G$, i.e., $[(L_g)_* X](h) = X(gh)$. One defines *right-invariant* vector fields analogously. Denote by $\mathfrak{X}_L(G)$ the space of left-invariant vector fields on G. Now, take $X, Y \in \mathfrak{X}_L(G)$ and compute $[X, Y]$ as a Lie bracket in $\mathfrak{X}(G)$ ($\mathfrak{X}_L(G) \subset \mathfrak{X}(G)$). It turns out that $[X, Y] \in \mathfrak{X}_L(G)$, for any $X, Y \in \mathfrak{X}_L(G)$. A pair $(\mathfrak{X}_L(G), [,])$ defines a Lie algebra called the Lie algebra of G. This algebra may be equivalently described as follows: any left-invariant vector field on G is uniquely determined by its value in the identity element $e \in G$:

$$\mathfrak{X}_L(G) \ni X \longleftrightarrow X(e) \in T_e G .$$

Take $X, Y \in \mathfrak{X}_L(G)$, and let $\xi = X(e)$ and $\eta = Y(e)$. Define the Lie bracket in $T_e G$ by

$$[\xi, \eta] := [X, Y](e) . \tag{1.58}$$

One usually denotes $T_e G = \mathfrak{g}$ and calls it the Lie algebra of G.

Example 1.2.6 The matrix algebra $gl(n, \mathbb{R})$ is the Lie algebra of $GL(n, \mathbb{R})$. $\diamond$

Example 1.2.7 If M is a differential manifold, then the set of vector fields $\mathfrak{X}(M)$ is the Lie algebra of $\mathrm{Diff}(M)$. $\diamond$

1.2.2 Actions of Lie groups

Let M be a smooth manifold and G be a Lie group.

Definition 1.2.2 *A left action of G on M is a smooth map* $\Phi : G \times M \longrightarrow M$ *such that*

1. $\Phi_e(x) = x$, *for all* $x \in M$,

2. $\Phi_{g_1} \circ \Phi_{g_2} = \Phi_{g_1 g_2}$, *for all* $g_1, g_2 \in G$,

where e denotes the identity element in G and $\Phi_g(x) := \Phi(g, x)$.

Sometimes one uses a simplified notation $g \cdot x := \Phi_g(x)$. A *right action* of G on M is defined in the same way with an obvious replacement: instead of the second condition one has $\Phi_{g_1} \circ \Phi_{g_2} = \Phi_{g_2 g_1}$. In the following section we shall consider only left actions but everything may be easily expressed in terms of right actions. Clearly, a left (right) action of a Lie group on a manifold defines a homomorphism (anti-homomorphism) from G to the group of diffeomorphisms of M:

$$G \ni g \longrightarrow \Phi_g \in \mathrm{Diff}(M) \, . \tag{1.59}$$

Example 1.2.8 A particular type of group action is defined by a *group representation*. This is a left action of G on a vector space V such that Φ_g is a linear operator in V. Whenever the dimension of V is finite, say n, each element $g \in G$ may be represented by an $n \times n$ matrix from $GL(n, \mathbb{R})$. $\diamond$

Having defined an action Φ of G on M one introduces an orbit passing through a point $x \in M$:

$$\mathcal{O}_x := \left\{ \, \Phi_g(x) \, \middle| \, g \in G \, \right\} \subset M \, , \tag{1.60}$$

and the isotropy subgroup of Φ at x by

$$G_x = \left\{ \, g \in G \, \middle| \, \Phi_g(x) = x \, \right\} \subset G \, . \tag{1.61}$$

It is evident that $\mathcal{O}_x \cong G/G_x$. We may define a natural relation between points of M:

$$x \sim y \iff \exists\, g \in G \, , \quad y = \Phi_g(x) \, ,$$

that is, x and y belong to the same orbit. An action is said to be

1. *Transitive* if there is only one orbit, i.e., for every two points $x, y \in M$ there is an element $g \in G$ such that $y = \Phi_g(x)$;

2. *Effective* (or *faithful*) if $\Phi_g = \mathrm{id}_M$ implies $g = e$, i.e., the map $g \longrightarrow \Phi_g$ is one-to-one;

3. *Free* if it has no fixed points, that is, $\Phi_g(x) = x$ implies $g = e$. Note that the action is free iff $G_x = e$ for all $x \in M$. Evidently, every free action is effective.

Let us observe that each element $\xi \in \mathfrak{g}$ defines a one-parameter subgroup of G. Indeed, let $X_\xi \in \mathfrak{X}_L(G)$ be a unique left-invariant field corresponding to ξ, i.e. $X_\xi(e) = \xi$. The flow property (1.47) implies that the integral curve $g_\xi(t)$ of X_ξ, passing at $t = 0$ through a point $e \in G$, satisfies

$$g_\xi(t_1) g_\xi(t_2) = g_\xi(t_1 + t_2) \, , \tag{1.62}$$

and hence $g_\xi(t)$ defines a one-parameter subgroup of G. A map

$$\mathfrak{g} \ni \xi \longrightarrow \exp(\xi) := g_\xi(1) \, ,$$

is called an *exponential map*.

Definition 1.2.3 *Suppose that the map* $\Phi : G \times M \longrightarrow M$ *is a left action of a group* G *on a manifold* M. *For any* $\xi \in \mathfrak{g}$ *define*

$$\mathbf{X}_\xi(x) := \frac{d}{dt} \Phi(g_\xi(t), x)\Big|_{t=0} . \tag{1.63}$$

$\mathbf{X}_\xi$ *is a vector field on* M *called the infinitesimal generator of the action* Φ *corresponding to* ξ.

Example 1.2.9 When $G = GL(n, \mathbb{R})$ acts by linear operations on $\mathbb{R}^n$, one has

$$\exp(tX) := \mathbb{1}_n + tX + \frac{1}{2}(tX)^2 + \dots , \tag{1.64}$$

for any $X \in gl(n, \mathbb{R})$. The flow property follows now from

$$\exp(t_1 X) \cdot \exp(t_2 X) = \exp((t_1 + t_2)X) .$$

Note that

$$\mathbf{X}_X := \frac{d}{dt} \exp(tX)\Big|_{t=0} = X , \tag{1.65}$$

that is, the infinitesimal generator $\mathbf{X}_X$ corresponding to X is X itself. $\diamond$

Let us observe that the tangent space to an orbit $\mathcal{O}_x$, passing through a point $x \in M$, is spanned by the corresponding infinitesimal generators:

$$T_y\mathcal{O}_x = \left\{ \mathbf{X}_\xi(y) \,\Big|\, \xi \in \mathfrak{g} \right\} \tag{1.66}$$

for any point $y \in \mathcal{O}_x$. Moreover, the following subspace of $\mathfrak{g}$:

$$\mathfrak{g}_x := \left\{ \xi \in \mathfrak{g} \,\Big|\, \mathbf{X}_\xi(x) = 0 \right\} , \tag{1.67}$$

defines a Lie algebra of the isotropy group G_x.

Proposition 1.2.1 *Let* Φ *be a left action of* G *on* M. *Then the map*

$$\mathfrak{g} \ni \xi \longrightarrow \mathbf{X}_\xi \in \mathfrak{X}(M) \tag{1.68}$$

is a Lie algebra anti-homomorphism:

$$[\mathbf{X}_\xi, \mathbf{X}_\eta] = -\mathbf{X}_{[\xi, \eta]} . \tag{1.69}$$

It is clear that left and right translations (L_g, R_g) define left and right actions, respectively, of the group on itself. The following composition:

$$I_g := L_g \circ R_{g^{-1}} : G \longrightarrow G , \tag{1.70}$$

is called an inner automorphism of G. Note that $I_g(e) = geg^{-1} = e$ and, therefore, the tangent map

$$T_e I_g : T_e G \longrightarrow T_e G \tag{1.71}$$

defines a linear operator on $T_e G$, i.e., on the Lie algebra $\mathfrak{g}$ of G. The map

$$G \ni g \longrightarrow T_e I_g =: \mathrm{Ad}_g \in \mathrm{Aut}(\mathfrak{g}) \tag{1.72}$$

is called the *adjoint* representation of G.

Example 1.2.10 For the matrix groups one finds the following well-known formula:

$$\mathrm{Ad}_A(X) = \frac{d}{dt}\left(A \cdot \exp(tX) \cdot A^{-1}\right)\Big|_{t=0} = A \cdot X \cdot A^{-1} , \tag{1.73}$$

where $A \in GL(n, \mathbb{R})$, and $X \in gl(n, \mathbb{R})$. $\Diamond$

Example 1.2.11 Consider the standard (left) action of the rotation group $SO(3)$ on $\mathbb{R}^3$ defined by

$$\Phi_A(\mathbf{x}) := A\mathbf{x} , \tag{1.74}$$

for any $\mathbf{x} \in \mathbb{R}^3$ and $A \in SO(3)$. Clearly, the rotation group $SO(3)$ is a Lie subgroup of $GL(3, \mathbb{R})$ and the corresponding Lie algebra $so(3)$ consists of three-dimensional antisymmetric matrices, and hence is isomorphic to $\mathbb{R}^3$. Define the following basis in $so(3)$:

$$s_1 = \begin{pmatrix} 0 & 0 & 0 \\ 0 & 0 & 1 \\ 0 & -1 & 0 \end{pmatrix} , \quad s_2 = \begin{pmatrix} 0 & 0 & 1 \\ 0 & 0 & 0 \\ -1 & 0 & 0 \end{pmatrix} , \quad s_3 = \begin{pmatrix} 0 & 1 & 0 \\ -1 & 0 & 0 \\ 0 & 0 & 0 \end{pmatrix} . \tag{1.75}$$

Each s_k generates the rotation about the corresponding kth axis. One easily recovers the standard commutation relations of $so(3)$:

$$[s_i, s_j] = \sum_{k=1}^{3} \epsilon_{ijk}\, s_k . \tag{1.76}$$

The isomorphism between $so(3)$ and $\mathbb{R}^3$ may be established as follows:

$$\mathbb{R}^3 \ni \mathbf{x} = (x_1, x_2, x_3) \longleftrightarrow \hat{\mathbf{x}} := \sum_{i=1}^{3} x_i s_i \in so(3) . \tag{1.77}$$

It is easy to prove that

$$\mathrm{Ad}_A \hat{\mathbf{x}} := A \cdot \hat{\mathbf{x}} \cdot A^{-1} = \hat{\mathbf{z}} , \tag{1.78}$$

with $\mathbf{z} = A\mathbf{x}$, and hence the following diagram commutes:

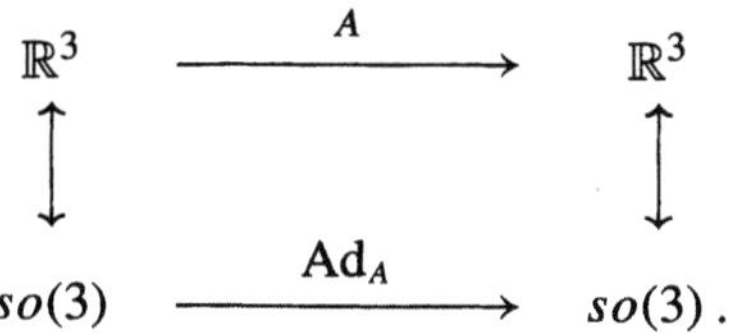

$$\diamond$$

1.2.3 Homogeneous spaces

Definition 1.2.4 *A manifold M on which a Lie group G acts transitively is called a homogeneous space of G.*

If M is a homogeneous space of G, then all isotropy groups G_x are isomorphic. Denote by H the common isotropy group. One may introduce the following relation between elements of G:

$$g_1 \sim g_2 \iff \exists\, h \in H, \quad g_2 = hg_1.$$

Clearly, it is an equivalence relation, and hence one may define a space of equivalence classes G/H, that is, the space of left cosets gH of H in G. Now, if M is a homogeneous space of G, then

$$M \cong G/H. \tag{1.79}$$

Example 1.2.12 Clearly, G is a homogeneous space of G, i.e., the left action $L_g : G \longrightarrow G$ is transitive. In particular, $\mathbb{R}^n$ is a homogeneous space (since $\mathbb{R}^n$ is an abelian group, cf. Example 1.2.1). $\quad\diamond$

Example 1.2.13 (Spheres) Consider a unit sphere S^n in $\mathbb{R}^{n+1}$. The orthogonal group $O(n+1)$ acts transitively on S^n: if $A \in O(n+1)$, then for any $\mathbf{x} \in \mathbb{R}^{n+1}$, $|A\mathbf{x}| = |\mathbf{x}|$. To find the common isotropy subgroup H, let us look for the isotropy group of $\mathbf{e}_1 := (1, 0, \dots, 0)$. It consists of all elements of the form

$$A = \left(\begin{array}{c|c} 1 & 0 \\ \hline 0 & B \end{array} \right),$$

where B represents an arbitrary $O(n)$-rotation. Thus $H \cong O(n)$ and, hence,

$$S^n \cong \frac{O(n+1)}{O(n)}. \tag{1.80}$$

The group $SO(n+1)$ also acts transitively on S^n, with $H \cong SO(n)$. Therefore, we may identify S^n with the quotient $SO(n+1)/SO(n)$.

One may apply the same procedure in the complex and quaternionic cases. In the complex case one has a transitive action of $U(n+1)$ on the unit sphere S^{2n+1} in $\mathbb{C}^{n+1}$, with an isotropy subgroup $H \cong U(n)$. Therefore,

$$S^{2n+1} \cong \frac{U(n+1)}{U(n)} \, , \tag{1.81}$$

or using $SU(n)$, one gets $S^{2n+1} \cong SU(n+1)/SU(n)$.

In the quaternionic case (cf. Appendix B) the symplectic group $Sp(n+1)$ acts transitively on the unit sphere S^{4n+3} in $\mathbb{H}^{n+1}$, with an isotropy subgroup $H \cong Sp(n)$, which gives the following representation for S^{4n+3}:

$$S^{4n+3} \cong \frac{Sp(n+1)}{Sp(n)} \, . \tag{1.82}$$

The above representations of spheres will play an essential role in what follows. $\diamond$

Example 1.2.14 (Projective spaces) Let $\mathbb{F}$ be a field ($\mathbb{R}$, $\mathbb{C}$ or $\mathbb{H}$). An $\mathbb{F}$-projective space $\mathbb{F}P^n$ is a space of $\mathbb{F}$-lines in $\mathbb{F}^{n+1}$, that is, $\mathbb{F}P^n$ is a set of equivalence classes with respect to the following equivalence relation: if $\mathbf{x}, \mathbf{y} \in \mathbb{F}^{n+1}$, then

$$\mathbf{x} \sim \mathbf{y} \iff \exists \lambda \in \mathbb{F}, \quad \mathbf{y} = \lambda \cdot \mathbf{x} \, .$$

Let us consider the complex case, which is the most important in physical applications. Let us restrict the above relation to the unit sphere in $\mathbb{C}^{n+1}$:

$$S^{2n+1} := \left\{ \mathbf{z} \in \mathbb{C}^{n+1} \,\middle|\, |\mathbf{z}| = 1 \right\} \, .$$

Now, if $\mathbf{z}_1$ and $\mathbf{z}_2$ are two points from S^{2n+1}, then

$$\mathbf{z}_1 \sim \mathbf{z}_2 \iff \exists \alpha \in \mathbb{R}, \quad \mathbf{z}_2 = e^{i\alpha}\mathbf{z}_1 \, .$$

The set of equivalence classes $S^{2n+1}/\sim$ is the complex projective space $\mathbb{C}P^n$. Note that we have a natural action of a Lie group $U(1)$ on S^{2n+1}:

$$U(1) \times S^{2n+1} \longrightarrow S^{2n+1} \, ,$$

defined by

$$(e^{i\alpha}, \mathbf{z}) \longrightarrow e^{i\alpha}\mathbf{z} \, ,$$

for any $\mathbf{z} \in S^{2n+1}$. Clearly, the complex projective space $\mathbb{C}P^n$ coincides with the space of orbits of the above $U(1)$-action, i.e.,

$$\mathbb{C}P^n \cong S^{2n+1}/U(1) \, . \tag{1.83}$$

Recalling (1.81), we have another representation:

$$\mathbb{C}P^n \cong \frac{U(n+1)}{U(n) \times U(1)} \, . \tag{1.84}$$

Let us note that there is a natural action of $U(n+1)$ on $\mathbb{C}P^n$. Let $\mathbf{z} \in \mathbb{C}^{n+1}$ and denote by $[\mathbf{z}]$ the corresponding equivalence class (i.e., complex projective line) in $\mathbb{C}P^n$. For any $U \in U(n+1)$ define

$$U[\mathbf{z}] := [U\mathbf{z}] .$$

Clearly, this action is transitive and has $H = U(n) \times U(1)$ as a common isotropy subgroup.

The above arguments may be easily repeated in the real and quaternionic cases. In the real case one has a transitive action of $O(n+1)$ on $\mathbb{R}P^n$ with an isotropy subgroup $O(n) \times O(1)$ and, therefore,

$$\mathbb{R}P^n \cong \frac{O(n+1)}{O(n) \times O(1)} , \tag{1.85}$$

or equivalently, using (1.80),

$$\mathbb{R}P^n \cong S^n/O(1) = S^n/\mathbb{Z}_2 , \tag{1.86}$$

since $O(1) = \mathbb{Z}_2 = \{1, -1\}$. In the quaternionic case

$$\mathbb{H}P^n \cong \frac{Sp(n+1)}{Sp(n) \times Sp(1)} , \tag{1.87}$$

or, using (1.82),

$$\mathbb{H}P^n \cong S^{3n+1}/Sp(1) = S^{3n+1}/SU(2) , \tag{1.88}$$

since $Sp(1) = SU(2)$. $\Diamond$

1.2.4 Lie algebras and differential forms

Having learned about differential forms and Lie algebras let us combine these two notions and consider the space $\Lambda(M) \otimes L$, i.e., a set of L-valued differential forms on a manifold M, with $(L, [\,,\,])$ being a Lie algebra. If $\alpha \in \Lambda^k(M) \otimes L$, then

$$\alpha_x(v_1, \ldots, v_k) \in L , \tag{1.89}$$

for any $v_1, \ldots, v_k \in T_x M$. Let $(e_1, \ldots, e_r)$ be a basis in L. Any L-valued form α may be written as

$$\alpha = \alpha^i \otimes e_i , \tag{1.90}$$

with $\alpha^i \in \Lambda(M)$. Thus $\alpha^1, \ldots, \alpha^r$ are 'ordinary' forms on M. Now, a Lie bracket in L may be extended to the following operation in $\Lambda(M) \otimes L$:

$$[\,,\,] : (\Lambda^k(M) \otimes L) \times (\Lambda^l(M) \otimes L) \longrightarrow \Lambda^{k+l}(M) \otimes L , \tag{1.91}$$

defined by

$$[\alpha, \beta] := \sum_{i,j=1}^{r} (\alpha^i \wedge \beta^j) \otimes [e_i, e_j] = \sum_{i,j=1}^{r} f_{ij}^k (\alpha^i \wedge \beta^j) \otimes e_k , \qquad (1.92)$$

for any $\alpha \in \Lambda^k(M) \otimes L$ and $\beta \in \Lambda^l(M) \otimes L$.

Proposition 1.2.2 *The bracket operation (1.92) satisfies the following properties:*

1. $[\alpha, \beta] = (-1)^{kl+1}[\beta, \alpha]$,

2. $(-1)^{km}[[\alpha, \beta], \gamma] + (-1)^{ml}[[\gamma, \alpha], \beta] + (-1)^{kl}[[\beta, \gamma], \alpha] = 0$,

3. $d[\alpha, \beta] = [d\alpha, \beta] + (-1)^k [\alpha, d\beta]$,

for any L-valued k-form α, l-form β and m-form γ on M.

If L is a matrix algebra $L = gl(n, \mathbb{R})$, then a wedge product operation may be extended from the space of forms $\Lambda(M)$ into L-valued forms:

$$\wedge : (\Lambda^k(M) \otimes L) \times (\Lambda^l(M) \otimes L) \longrightarrow \Lambda^{k+l}(M) \otimes L , \qquad (1.93)$$

using

$$\alpha \wedge \beta := \sum_{i,j=1}^{n^2} (\alpha^i \wedge \beta^j) \otimes (\lambda_i \cdot \lambda_j) , \qquad (1.94)$$

where $(\lambda_1, \ldots, \lambda_{n^2})$ is a basis in $gl(n, \mathbb{R})$, and $\lambda_i \cdot \lambda_j$ denotes a matrix multiplication. Note, that there is a direct relation between '[,]' defined in (1.92) and '$\wedge$' defined in (1.94):

$$[\alpha, \beta] = \alpha \wedge \beta - (-1)^{kl} \beta \wedge \alpha , \qquad (1.95)$$

for any k-form α and l-form β.

Consider now a Lie group G together with the space of differential forms $\Lambda(G)$. In analogy to left- (right-) invariant vector fields let us define left- (right-) invariant differential forms on G. A form $\alpha \in \Lambda(G)$ is left-invariant if

$$L_g^* \alpha = \alpha , \qquad (1.96)$$

for any $g \in G$. Denote the set of left-invariant forms by $\Lambda_L(G)$. Let us observe that, due to the basic properties of a wedge product and exterior derivative (cf. Propositions 1.1.1 and 1.1.2), we have

$$\alpha \in \Lambda_L(G) \implies d\alpha \in \Lambda_L(G) ,$$

and

$$\alpha, \beta \in \Lambda_L(G) \implies \alpha \wedge \beta \in \Lambda_L(G) .$$

Evidently, any left-invariant form on G is uniquely determined by its value in $e \in G$. Therefore, there is a one-to-one correspondence between left-invariant one-forms and cotangent space T_e^*G:

$$\Lambda_L^1(G) \ni \alpha \longleftrightarrow \alpha(e) \in T_e^*G .$$

Let $\{e_1, \ldots, e_n\}$ be a basis in $\mathfrak{g} \cong T_eG$. The structure of $\mathfrak{g}$ is entirely encoded in the set of commutation relations

$$[e_i, e_j] = f_{ij}^k e_k . \tag{1.97}$$

Let $\{\theta^1, \ldots, \theta^n\}$ be the dual basis in T_e^*G. Denote by θ_L^k the unique left-invariant one-form such that $\theta_L^k(e) = \theta^k$. Using the Cartan formula (1.54) it is easy to show that formula (1.97) implies the following equation for the dual forms θ_L's:

$$d\theta_L^k = -\frac{1}{2} f_{ij}^k \, \theta_L^i \wedge \theta_L^j . \tag{1.98}$$

The above formula is known under the name the *Maurer–Cartan equation*.

Definition 1.2.5 *The Maurer–Cartan or the canonical form on a Lie group G is a left-invariant one-form ω_0, taking values in the Lie algebra $\mathfrak{g}$, defined by*

$$[\omega_0(X)](g) := [(L_{g^{-1}})_* X](e) \in \mathfrak{g} , \tag{1.99}$$

for any $g \in G$ and $X \in \mathfrak{X}(G)$.

Let us observe that the action of ω_0 on a vector $X \in T_gG$ consists in pushing X forward from a point g to a point e. Since the formula (1.99) for the Maurer–Cartan form is rather complicated it is desirable to have a coordinate representation of ω_0. One can prove that

$$\omega_0 = \theta_L^k \otimes e_k , \tag{1.100}$$

which shows that ω_0 is manifestly left-invariant. The fundamental properties of the Maurer–Cartan form are summarized in the following

Proposition 1.2.3 *The Maurer–Cartan form satisfies*

$$d\omega_0 = -\frac{1}{2}[\omega_0, \omega_0] . \tag{1.101}$$

Moreover, under a right action of G it transforms according to

$$R_g^* \omega_0 = \mathrm{Ad}_{g^{-1}} \omega_0 , \tag{1.102}$$

for any $g \in G$.

Example 1.2.15 Let us examine the above abstract formulae in the case of a matrix algebra $gl(n, \mathbb{R})$. Note that the Maurer–Cartan form may be written as

$$\omega_0 = g^{-1}dg \,, \tag{1.103}$$

with $g \in GL(n, \mathbb{R})$. The left-invariance follows from

$$L_h^*\omega_0 = (hg)^{-1}d(hg) = g^{-1}h^{-1}hdg = g^{-1}dg = \omega_0 \,. \tag{1.104}$$

Similarly,

$$R_h^*\omega_0 = (gh)^{-1}d(gh) = h^{-1}g^{-1}(dg)h = h^{-1}\omega_0 h = \mathrm{Ad}_{h^{-1}}\omega_0 \,, \tag{1.105}$$

in agreement with (1.102). Moreover, since $dg^{-1} = -g^{-1}(dg)g^{-1}$, we have

$$\begin{aligned}
d\omega_0 &= d(g^{-1}dg) = dg^{-1} \wedge dg = -g^{-1}dg \wedge g^{-1}dg \\
&= -\omega_0 \wedge \omega_0 = -\frac{1}{2}[\omega_0, \omega_0] \,,
\end{aligned} \tag{1.106}$$

due to (1.95). $\diamond$

Example 1.2.16 Consider the abelian group $U(1)$. Any element $g \in U(1)$ may be represented by $g = e^{i\lambda}$, with $\lambda \in \mathbb{R}$. Now,

$$\omega_0 = g^{-1}dg = id\lambda \in u(1) \,, \tag{1.107}$$

where we used the identification $u(1) \cong i\mathbb{R}$. The group $U(1)$ and its corresponding Maurer–Cartan form (1.107) will play an important role in what follows. $\diamond$

Example 1.2.17 Another important group in this book is $SU(2)$. Any element $g \in SU(2)$ may be written in the following form:

$$g = \begin{pmatrix} \alpha & \beta \\ -\overline{\beta} & \overline{\alpha} \end{pmatrix} \,, \tag{1.108}$$

with $\alpha, \beta \in \mathbb{C}$ such that $|\alpha|^2 + |\beta|^2 = 1$, and $\overline{\alpha}$ denoting complex conjugation of α. The reader may easily show that:

$$g^{-1}dg = \begin{pmatrix} \overline{\alpha}d\alpha + \beta d\overline{\beta} & \overline{\alpha}d\beta - \beta d\overline{\alpha} \\ \overline{\beta}d\alpha - \alpha d\overline{\beta} & \alpha d\overline{\alpha} + \overline{\beta}d\beta \end{pmatrix} \,, \tag{1.109}$$

and, hence,

$$\mathrm{Tr}(g^{-1}dg) = 0 \,, \quad \text{and} \quad (g^{-1}dg)^* = -g^{-1}dg \,, \tag{1.110}$$

which shows that $g^{-1}dg$ does belong to the Lie algebra $su(2)$. $\diamond$

1.3   Bundles and connections

1.3.1   Fibre bundle

Fibre bundles frequently appear in differential geometry. It turns out that many important concepts in modern physics may be interpreted in terms of the geometry of fibre bundles. Maxwell's theory of electromagnetism and Yang–Mills gauge theories are essentially theories of connections in the appropriate fibre bundles over the space-time manifold. General Relativity may be interpreted in terms of the geometry of so-called frame bundles. As we shall see, the geometric phase also finds its natural description in terms of certain fibre bundles.

Roughly speaking, a fibre bundle is a manifold that looks locally like a Cartesian product of two spaces but may have nontrivial global geometry. To define a fibre bundle one needs five elements:

- Manifolds: E — the bundle (or total) space; M — the base space; and F — the standard (or typical) fibre.

- A structure Lie group G which acts effectively on F, i.e., there exists a map

$$\Phi \ : \ G \times F \ \longrightarrow \ F \,,$$

  such that if $\Phi_g(f) = f$, then $g = e$.

- A bundle projection:

$$\pi \ : \ E \ \longrightarrow \ M \,,$$

  such that each space $F_x := \pi^{-1}(x)$, called a fibre at $x \in M$, is homeomorphic to F.

These elements are not independent and we demand that

1. The bundle is locally trivial, i.e., it is locally homeomorphic to a cartesian product of two spaces. More precisely, for any covering of M by a family of open sets $\{U_j\}$ there exists a set of homeomorphisms

$$\varphi_j \ : \ \pi^{-1}(U_j) \ \longrightarrow \ U_j \times F \tag{1.111}$$

   of the form

$$\varphi_j(p) = (\pi(p), \phi_j(p)) \,, \tag{1.112}$$

   where

$$\phi_j \ : \ \pi^{-1}(U_j) \ \longrightarrow \ F \,,$$

   such that the following diagram commutes:

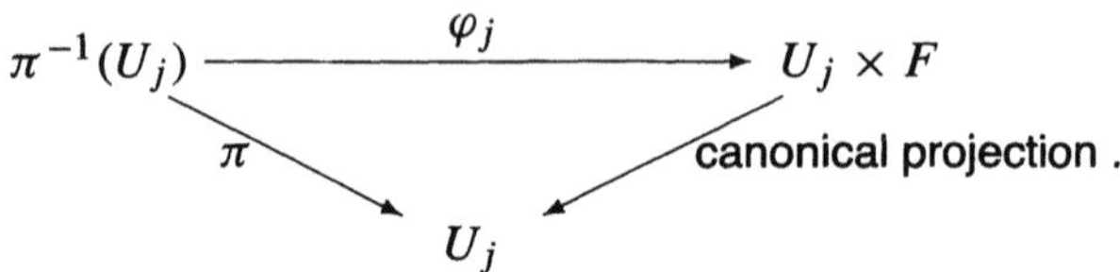

The set $\{(U_j, \varphi_j)\}$ is called a family of local trivializations. This way $\pi^{-1}(U_j)$ — a "portion" of the bundle over U_j — may be identified with the cartesian product $U_j \times F$:

$$\pi^{-1}(U_j) \cong U_j \times F \ .$$

2. The restriction of the map ϕ_j to the fibre over $x \in U_j$ defines the diffeomorphism

$$\phi_{j,x} := \phi_j|_{F_x} : F_x \longrightarrow F \ . \tag{1.113}$$

Now, for $x \in U_j \cap U_k$, the induced diffeomorphism

$$\phi_{k,x} \circ \phi_{j,x}^{-1} : F \longrightarrow F \ , \tag{1.114}$$

corresponds to an element of the structure group G, i.e., there exists $\gamma_{kj}(x) \in G$ such that

$$\Phi(\gamma_{kj}(x), \ \cdot \ ) = \phi_{k,x} \circ \phi_{j,x}^{-1} \ .$$

The maps

$$U_j \cap U_k \ni x \longrightarrow \gamma_{kj}(x) \in G \ ,$$

are called *transition functions* (cf. Fig. 1.2).

Instead of listing all elements (E, M, π, G, F) one often speaks about a G-bundle over M, or G-bundle $E \longrightarrow M$, or simply a bundle $\pi : E \longrightarrow M$. If the structure group G is not specified then one takes the entire group of all diffeomorphisms of a typical fibre, i.e. $G = \mathrm{Diff}(F)$. Transition functions encode the entire information about how the local pieces $U_j \times F$ are glued together.

Proposition 1.3.1 *Transition functions satisfy the following conditions:*

1. $\gamma_{ii}(x) = e \ , \quad x \in U_i \ ,$

2. $\gamma_{ij}(x) = (\gamma_{ji}(x))^{-1} \ , \quad x \in U_i \cap U_j \ ,$

3. $\gamma_{jk}(x)\gamma_{kl}(x) = \gamma_{jl}(x) \ , \quad x \in U_j \cap U_k \cap U_l \qquad$ (*cocycle condition*).

A fibre bundle is (up to an isomorphism) completely determined by the set of transition functions satisfying the above conditions.

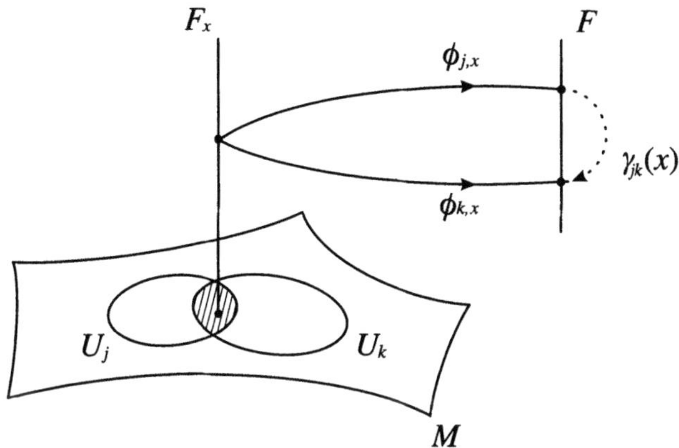

Figure 1.2: Transition functions

Definition 1.3.1 *A fibre bundle is trivial if*

$$E \cong M \times F \,,$$

that is, there is a diffeomorphism $h : M \times F \longrightarrow E$ such that

$$\pi(h(x, f)) = x \,,$$

for any $x \in M$ and $f \in F$.

It turns out that any fibre bundle over a contractible base space is trivial. Therefore, all fibre bundles over a base which is topologically equivalent to a ball in $\mathbb{R}^n$ are trivial. Nontrivial bundles can only be constructed when the global topology of the base is nontrivial (for example, over a sphere S^n).

Example 1.3.1 (Möbius strip) The classical example of a nontrivial fibre bundle is provided by the *Möbius strip*. It may be constructed as follows: take as a base manifold the unit circle S^1 parametrized by the angle $\theta \in [0, 2\pi)$. We cover S^1 by two coordinate patches:

$$U_+ := \{\, \theta \mid -\epsilon < \theta < \pi + \epsilon \,\} \,, \qquad U_- := \{\, \theta \mid \pi - \epsilon < \theta < \epsilon \,\} \,,$$

with a "small" $\epsilon > 0$. Let the typical fibre be an interval of the real line

$$F := [-1, 1] \subset \mathbb{R} \,,$$

parameterized by $t \in [-1, 1]$. To construct a bundle over S^1 with a typical fibre $[-1, 1]$, we have to glue together two pieces, namely,

$$U_+ \times F, \;\; \text{parametrized by } (\theta, t_+) \,,$$

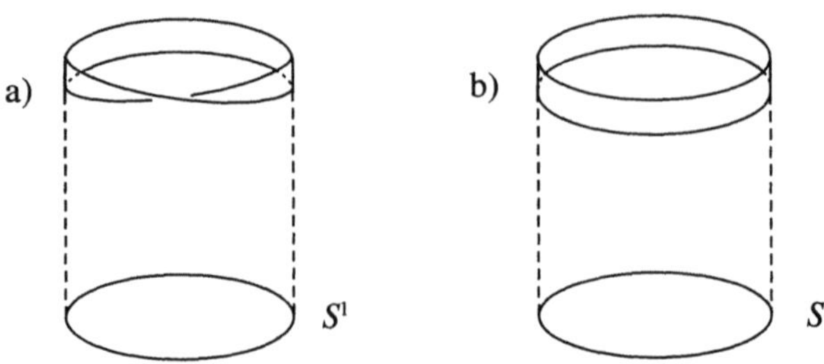

Figure 1.3: Möbius strip a) is a nontrivial (twisted) bundle over S^1, whereas a cylinder b) defines a trivial (nontwisted) bundle

and

$$U_- \times F, \quad \text{parametrized by } (\theta, t_-) .$$

Take as a structure group $G = \mathbb{Z}_2$, i.e., the two element group $\{e, -e\}$. We know that the information about gluing is encoded into the transition function $\gamma_{+-}(x)$ with $x \in U_+ \cap U_-$. Note that

$$U_+ \cap U_- = A \cup B ,$$

where

$$A = \{\, \theta \mid -\epsilon < \theta < \epsilon \,\}, \qquad B = \{\, \theta \mid \pi - \epsilon < \theta < \pi + \epsilon \,\} .$$

We define the transition function as follows:

$$\gamma_{+-}(x) := \begin{cases} e, & \text{for } x \in A \\ -e, & \text{for } x \in B \end{cases} .$$

That is, it describes a twist $t_+ = -t_-$ in the region B, giving rise to the nontrivial topology of the Möbius strip, as shown in Fig. 1.3. Gluing without twisting, i.e., taking $\gamma_{+-}(x) = e$ for any x, we simply obtain a cylinder $S^1 \times [-1, 1]$, which exemplifies a trivial bundle over S^1. $\diamond$

Definition 1.3.2 *A local section of the bundle* $\pi : E \longrightarrow M$ *is a mapping*

$$f : U \longrightarrow E ,$$

with $U \subset M$, such that $\pi \circ f = \mathrm{id}_U$.

A section is called *global* if it is defined over the entire base manifold M. It turns out that the existence of global sections depends on the global geometry of the bundle E. In physical applications the most important are vector bundles and principal bundles:

- a *vector bundle* is a fibre bundle with a typical fibre being a k-dimensional vector space, i.e., $F = \mathbb{R}^k$ (or $\mathbb{C}^k$ in the case of a complex bundle), and with the structure group G being a subgroup of $GL(k, \mathbb{R})$ (or $GL(k, \mathbb{C})$);

- a *principal bundle* is a fibre bundle whose typical fibre F coincides with the structure group G, which acts on itself by the left translation L_g.

As we shall see, principle fibre bundles lie in the heart of the theory, since any other bundle may be constructed from some principle fibre bundle.

Proposition 1.3.2 *For any principal bundle* (P, M, π, G) *there is a natural smooth, free right action of G on the total space P,*

$$\widetilde{R} : G \times P \longrightarrow P,$$

with the property that the corresponding orbits coincide with the fibres, that is, if $p \in P$, then the G-orbit passing through p defines a fibre over $x = \pi(p)$, i.e., $F_x = \mathcal{O}_p$.

Moreover, one may prove the following

Proposition 1.3.3 *Every principal bundle is obtained from the free right action of a Lie group G on some manifold P.*

The structure of the fibre bundle with base space M, structure group G and a typical fibre F is determined essentially by the transition functions

$$\gamma_{kl} : U_k \cap U_l \longrightarrow G,$$

satisfying conditions from Proposition 1.3.1. Therefore, having an action of G on a manifold F', we may construct another fibre bundle simply changing F to F'. Such bundle is said to be *associated* with the original one. It is therefore evident that every fibre bundle can be obtained as a fibre bundle associated with some principal fibre bundle. In particular, the problem of the classification of fibre bundles reduces to the classification of principal fibre bundles. There is also a simple criterion for triviality of principal fibre bundles.

Theorem 1.3.4 *A principal fibre bundle* (P, M, π, G) *is trivial if and only if it admits a global section.*

Proof. If the bundle is trivial then according to Definition 1.3.1 there is a diffeomorphism h:

$$M \times G \xrightarrow{\;h\;} P.$$

Therefore, we may take as a global section $f : M \longrightarrow P$:

$$f(x) := h(x, e),$$

with e being the unit element in G. Conversely, if there is global section $f : M \longrightarrow P$, then we may define a diffeomorphism h as follows:

$$h(x, g) := \widetilde{R}_g f(x).$$

Since the right action $\widetilde{R}_g$ is free and effective, h indeed defines a diffeomorphism. $\square$

1.3.2 Examples of fibre bundles

Let us list the most important examples of fibre bundles.

Example 1.3.2 (Tangent and cotangent bundles) Let M be an n-dimensional manifold and $T_x M$ denote the tangent space at x. Define a fibre bundle with a base space M and a total space

$$TM := \bigcup_{x \in M} T_x M \ . \tag{1.115}$$

Clearly, a typical fibre $F = \mathbb{R}^n$ and a fibre over $x \in M$ is $\pi^{-1}(x) = T_x M \cong \mathbb{R}^n$. The structure group $G = GL(n, \mathbb{R})$ acts on the typical fibre by matrix multiplication:

$$v \longrightarrow g \cdot v \ , \tag{1.116}$$

for any $g \in GL(n, \mathbb{R})$ and $v \in \mathbb{R}^n$. To show that this construction does indeed define a fibre bundle we proceed as follows: Let (U_k, ψ_k) be a family of local charts on M. To define a bundle we have to specify the set of transition functions

$$\gamma_{kl} \ : \ U_k \cap U_l \ \longrightarrow \ GL(n, \mathbb{R}) \ ,$$

satisfying the conditions of Proposition 1.3.1. We define these functions as derivatives of the corresponding overlap maps:

$$\gamma_{kl}(x) := T_x \, \psi_{kl} \ , \tag{1.117}$$

with $\psi_{kl} = \psi_k \circ \psi_l^{-1}$. To see that this really does the job, let $\psi_i = (x_{(i)}^1, \dots, x_{(i)}^n)$ be a local coordinate system in a patch (U_i, ψ_i), and let $x \in U_i \cap U_j$. For $x \in U_i$ any tangent vector $v \in T_x M$ may be represented as follows:

$$v = \sum_{k=1}^{n} X_{(i)}^k \, \frac{\partial}{\partial x_{(i)}^k} \ . \tag{1.118}$$

Now, define $\phi_{i,x} : T_x M \longrightarrow \mathbb{R}^n$ by

$$\phi_{i,x}(v) := (X_{(i)}^1, \dots, X_{(i)}^n) \ . \tag{1.119}$$

Thus, the corresponding transition function,

$$\gamma_{ij}(x) = \phi_{i,x} \circ \phi_{j,x}^{-1} \in GL(n, \mathbb{R}) \ , \tag{1.120}$$

is defined by

$$\left[\gamma_{ij}(x) \right]_l^k = \frac{\partial x_{(i)}^k}{\partial x_{(j)}^l}(x) \ , \tag{1.121}$$

i.e., as a derivative of the corresponding overlap map in M. The above formula reproduces the well-known transformation law

$$X^k_{(i)} = \sum_{l=1}^{n} \frac{\partial x^k_{(i)}}{\partial x^l_{(j)}} X^l_{(j)} \ .$$ (1.122)

In the same way, one defines a cotangent bundle of M as a union of cotangent spaces:

$$T^*M := \bigcup_{x \in M} T^*_x M \ .$$ (1.123)

The construction of the corresponding transition function we leave as an exercise to the reader. $\Diamond$

Example 1.3.3 (Frame bundle) Let M be an n-dimensional differential manifold and denote by $F_x M$ the set of all n-frames[3] in $T_x M \cong \mathbb{R}^n$. Let $(e_1, e_2, \ldots , e_n)$ be a standard frame in $\mathbb{R}^n$. Therefore, any frame $(v_1, v_2, \ldots , v_n)$ may be written

$$v_i = \sum_{k=1}^{n} A^k_i e_k \ ,$$ (1.124)

with $A \in GL(n, \mathbb{R})$. Hence, a typical fibre F may be identified with the general linear group:

$$F \cong GL(n, \mathbb{R}) = G \ .$$ (1.125)

This way one obtains a principal bundle

$$FM := \bigcup_{x \in M} F_x M \ ,$$ (1.126)

with the structure group $GL(n, \mathbb{R})$. It is called a *frame bundle* over M.

Let us construct a local trivialization of FM. As in the previous example let (U_i, ψ_i) be a family of local coordinate systems on M, with $\psi_{(i)} = (x^1_{(i)}, \ldots , x^n_{(i)})$. Take any point $x \in U_i$ and define

$$\phi_{i,x} : F_x \longrightarrow G$$ (1.127)

by

$$[\phi_{i,x}(v_1, \ldots , v_n)]^k{}_l := dx^k_{(i)}(v_l) \ ,$$ (1.128)

where $(v_1, \ldots , v_n)$ is a frame at the point $x \in M$. Now, in the region $U_i \cap U_j$ we have

$$x^k_{(i)} = x^k_{(i)}(x^1_{(j)}, \ldots , x^n_{(j)}) \ , \quad k = 1, \ldots , n \ .$$

---

[3] An n-frame is a basis in $T_x M$, i.e., a collection of n independent tangent vectors.

Therefore

$$dx_{(i)}^k(v_l) = \sum_{m=1}^{n} \frac{\partial x_{(i)}^k}{\partial x_{(j)}^m}\, dx_{(j)}^m(v_l)\,, \qquad (1.129)$$

and hence $\gamma_{ij}(x)$ is given by (1.121). Finally, let us look for the canonical right action of $GL(n, \mathbb{R})$ on FM. Let

$$p = (x; \xi_1, \ldots, \xi_n) \in \pi^{-1}(x) \qquad (1.130)$$

be a frame at $x \in M$. Then for any $g \in GL(n, \mathbb{R})$ we have

$$\widetilde{R}_g p = p \cdot g = (x; \widetilde{\xi}_1, \ldots, \widetilde{\xi}_n)\,, \qquad (1.131)$$

with

$$\widetilde{\xi}_k = \sum_{i=1}^{n} [g]^i{}_k \xi_i\,. \qquad (1.132)$$

Thus the action of $GL(n, \mathbb{R})$ on FM consists of transforming each frame ξ attached at $x \in M$ by $\xi \longrightarrow g \cdot \xi$. $\qquad\qquad \diamond$

Remark 1.3.1 Taking as a structure group $G = SO(n)$ instead of $GL(n, \mathbb{R})$, we obtain a principal fibre bundle of orthonormal frames over M that plays a crucial role in General Relativity. $\qquad\qquad \diamond$

It turns out that there is a general scheme to construct principal fibre bundles.

Theorem 1.3.5 *Let H be a closed subgroup of G. Then $(G, G/H, \pi, H)$ with a canonical projection*

$$\pi : G \to G/H$$

defines a principal bundle, with H as the structure group.

Example 1.3.4 (Classical bundles over spheres) Example 1.2.13 gives rise to the following natural principal bundles over spheres:

  1. $O(n)$-bundle

$$O(n+1) \longrightarrow S^n \cong \frac{O(n+1)}{O(n)}\,, \qquad (1.133)$$

  2. $U(n)$-bundle

$$U(n+1) \longrightarrow S^{2n+1} \cong \frac{U(n+1)}{U(n)}\,, \qquad (1.134)$$

3. $Sp(n)$-bundle

$$Sp(n+1) \longrightarrow S^{4n+3} \cong \frac{Sp(n+1)}{Sp(n)} \,. \tag{1.135}$$

Thus spheres serve as base spaces for principal fibre bundles.   ◇

The next example shows that spheres may also serve as total spaces of fibre bundles.

Example 1.3.5 (Hopf fibrations) Using the results of Example 1.2.14 it is easy to construct the following principal bundles over projective spaces:

1. $O(1) \cong \mathbb{Z}_2$-bundle

$$S^n \longrightarrow \mathbb{R}P^n \cong S^n/\mathbb{Z}_2 \,, \tag{1.136}$$

2. $U(1) \cong SO(2)$-bundle

$$S^{2n+1} \longrightarrow \mathbb{C}P^n \cong S^{2n+1}/U(1) \,, \tag{1.137}$$

3. $Sp(1) \cong SU(2)$-bundle

$$S^{4n+3} \longrightarrow \mathbb{H}P^n \cong S^{3n+1}/Sp(1) \,. \tag{1.138}$$

The above bundles are usually called *Hopf bundles* or *Hopf fibrations*. For $n = 1$ we obtain the celebrated Hopf bundles:

$$U(1)\text{-bundle:} \qquad S^3 \longrightarrow \mathbb{C}P^1 \cong S^2$$

and

$$SU(2)\text{-bundle:} \qquad S^7 \longrightarrow \mathbb{H}P^1 \cong S^4 \,.$$

These two bundles, also called monopole and instanton bundles by physicists, respectively, will play a fundamental role throughout this book.   ◇

1.3.3 Connections — general theory

In this section we introduce a geometric object which allows one to compare different fibres of the bundle and to transport elements from one fibre to another. This object, called a *connection*, plays crucial role in the theory of geometric phases.

Consider an arbitrary fibre bundle (E, M, π, G, F). To begin with we shall disregard the structure group, which means that $G = \mathrm{Diff}(F)$. A fibre bundle equipped with a connection may be intuitively represented as follows: One has a family of fibres F_x whose union gives the total space

$$E = \bigcup_{x \in M} F_x \,.$$

Now, consider a curve

$$[0, 1] \ni t \longrightarrow \gamma(t) \in M .$$

A connection provides us with a rule of parallel transporting the fibre F along the path γ from one end to the other, i.e., it defines a map

$$\mathbf{T}_\gamma : F_{x_0} \longrightarrow F_{x_1} , \qquad x_0 = \gamma(0) , \quad x_1 = \gamma(1) , \tag{1.139}$$

satisfying the following conditions:

1. $\mathbf{T}_\gamma$ depends continuously on the path γ ,

2. $\mathbf{T}_{\gamma_1 * \gamma_2} = \mathbf{T}_{\gamma_1} \circ \mathbf{T}_{\gamma_2}$,

3. $\mathbf{T}_{\gamma^{-1}} = (\mathbf{T}_\gamma)^{-1}$,

where the operation of multiplying curves $\gamma_1 * \gamma_2$ is defined as follows: If γ_1 and γ_2 are two paths such that $\gamma_1(1) = \gamma_2(0)$, i.e., the end of γ_1 is the beginning of γ_2, then $\gamma_1 * \gamma_2$ is a new curve defined by

$$(\gamma_1 * \gamma_2)(t) := \begin{cases} \gamma_1(2t) & \text{for } 0 \le t \le 1/2 , \\ \gamma_2(2t - 1) & \text{for } 1/2 \le t \le 1 . \end{cases} \tag{1.140}$$

Moreover, the inversion γ^{-1} of the curve γ is defined by

$$\gamma^{-1}(t) := \gamma(1 - t) , \tag{1.141}$$

i.e., γ^{-1} goes "backwards in time."

Equivalently, the connection may be defined as follows: Take any vector v from $T_p E$. We shall call it a *vertical* vector if $v \in T_p F_x$, with $x = \pi(p)$, i.e., v is vertical at p if it is tangent to the fibre passing through p. Denote by V_p the space of vertical vectors at p. One obviously has

$$V_p := \left\{ v \in T_p E \,\middle|\, T_p \pi(v) = 0 \right\} , \tag{1.142}$$

and calls V_p a *vertical subspace*. Let $\mathfrak{X}_{ver}(E)$ denote the space of vertical vector fields on E:

$$v \in \mathfrak{X}_{ver}(E) \iff v(p) \in V_p , \tag{1.143}$$

for any point $p \in E$.

Definition 1.3.3 *A connection of a general type,*[4] *or, simply, a connection, is a smooth assignment*

$$E \ni p \longrightarrow H_p \subset T_p E ,$$

such that H_p is transverse to V_p, and

$$T_p E = V_p \oplus H_p . \tag{1.144}$$

---

[4]Mathematicians often call it the Ehresmann connection.

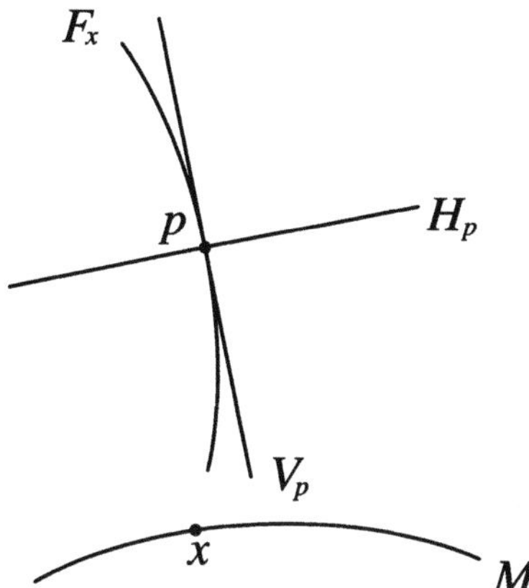

Figure 1.4: Decomposition of $T_p E$ into the horizontal subspace H_p and the vertical subspace V_p

H_p is called a *horizontal subspace* and a vector $v \in H_p$ is called a *horizontal vector*, see Fig. 1.4. Due to the decomposition (1.144) any vector $v \in T_p P$ may be uniquely decomposed as follows:

$$v = \text{hor } v + \text{ver } v , \tag{1.145}$$

where hor $v \in H_p$ is called the horizontal part of v, and ver $v \in V_p$ the vertical part of v.

Definition 1.3.4 *A curve*

$$[0, 1] \ni t \longrightarrow c(t) \in E$$

is said to be horizontal if its velocity vector dc/dt is horizontal.

Let $\gamma(t)$ be a curve in M. We call a curve $\widetilde{\gamma}$ a *lift* of γ if

$$\pi(\widetilde{\gamma}(t)) = \gamma(t) .$$

Moreover, $\widetilde{\gamma}$ is called a *horizontal lift* if $\widetilde{\gamma}$ is a horizontal curve.

Equipped with this abstract definition let us see how one can define the notion of parallel transport shown in formula (1.139). Let $\gamma(t)$ be a curve in M such that $\gamma(0) = x_0$ and $\gamma(1) = x_1$. Define the map $\mathbf{T}_\gamma$ as follows: Let $p_0 \in F_{x_0}$ and denote by $\widetilde{\gamma}(t)$ a horizontal lift of γ such that $\widetilde{\gamma}(0) = p_0$. Then

$$\mathbf{T}_\gamma(p_0) := \widetilde{\gamma}(1) \in F_{x_1} . \tag{1.146}$$

It is evident that $\mathbf{T}_\gamma$ fulfills all the natural requirements of parallel transport. $\mathbf{T}_\gamma$ is usually called a map (or an operator) of parallel transport determined by the connection. Clearly, an assignment $\gamma \longrightarrow \mathbf{T}_\gamma$ is equivalent to endowing a bundle with a connection.

Note that there is an equivalent way to introduce a connection into a bundle (E, M, π, F). Instead of describing a connection in terms of vectors from TE we may equivalently use a dual description and define it in the language of forms. The mapping

$$T_p E \ni u \longrightarrow \operatorname{ver} u \in V_p$$

allows us to introduce a one-form $\mathcal{A}$ on E with values in a vertical subspace. That is,

$$\mathcal{A}_p(u) := \operatorname{ver} u \ \in \ V_p \tag{1.147}$$

for any $u \in T_p E$. It is evident that the horizontal subspace H_p may be defined as follows:

$$H_p := \left\{ u \in T_p E \ \middle| \ \mathcal{A}_p(u) = 0 \right\} . \tag{1.148}$$

We call $\mathcal{A}$ a *connection form*.

Consider the space of V–valued differential forms on E.

Definition 1.3.5 *The covariant exterior derivative*

$$D \ : \ \Lambda^k(E) \otimes V \longrightarrow \Lambda^{k+1}(E) \otimes V$$

is defined by

$$D\alpha(u_1, \dots, u_{k+1}) := d\alpha(\operatorname{hor} u_1, \dots, \operatorname{hor} u_{k+1}) , \tag{1.149}$$

for any $\alpha \in \Lambda^k(E)$ and $u_1, \dots, u_{k+1} \in \mathfrak{X}(E)$.

Using Cartan's formula (1.54) we may rewrite $D\alpha$ as follows:

$$D\alpha(u_1, \dots, u_{k+1}) := \sum_{i=1}^{k+1}(-1)^i \operatorname{ver}[\operatorname{hor} u_i, \alpha(\operatorname{hor} u_1, \dots, \check{u}_i, \dots, \operatorname{hor} u_{k+1})] \tag{1.150}$$

$$+ \sum_{1 \le i < j \le k+1} (-1)^{i+j}\alpha([\operatorname{hor} u_i, \operatorname{hor} u_j], \operatorname{hor} u_1, \dots, \check{u}_i, \dots, \check{u}_j, \dots, \operatorname{hor} u_{k+1}) ,$$

where $\check{u}_i$ denotes that u_i is omitted.

Definition 1.3.6 *The two-form $\mathcal{F} \in \Lambda^2(E) \otimes V$ defined by*

$$\mathcal{F} := D\mathcal{A} \tag{1.151}$$

is called the curvature form of the connection $\mathcal{A}$.

Now, since $\mathcal{A}$ vanishes on horizontal vectors the formula (1.150) implies that, for any $u, v \in \mathfrak{X}(E)$,

$$\mathcal{F}(u, v) = -\mathcal{A}([\operatorname{hor} u, \operatorname{hor} v]) . \tag{1.152}$$

Hence, using defining equation (1.147), we obtain

$$\mathcal{F}(u, v) = -\operatorname{ver}([\operatorname{hor} u, \operatorname{hor} v]) . \tag{1.153}$$

Theorem 1.3.6 (Bianchi identity) *The curvature two-form $\mathcal{F}$ satisfies the following Bianchi identity:*

$$D\mathcal{F} = 0 \,. \tag{1.154}$$

Proof. Formula (1.150) implies

$$D\mathcal{F}(u_1, u_2, u_3)$$

$$= -\sum_{i,j,k=1}^{3} \epsilon_{ijk} \left(\text{ver } [\text{hor } u_i, \mathcal{F}(\text{hor } u_j, \text{hor } u_k)] + \mathcal{F}([\text{hor } u_i, \text{hor } u_j], \text{hor } u_k) \right)$$

$$= \sum_{i,j,k=1}^{3} \epsilon_{ijk} \text{ ver } \left([\text{hor } u_i, \text{ver } [\text{hor } u_j, \text{hor } u_k]] + [\text{hor } [\text{hor } u_i, \text{hor } u_j], \text{hor } u_k] \right) ,$$

where we have used (1.153). Now, by the very definition of verticality, we have

$$\text{ver } [\text{hor } u_j, \text{hor } u_k] = [\text{hor } u_j, \text{hor } u_k] - \text{hor } [\text{hor } u_j, \text{hor } u_k] \,, \tag{1.155}$$

and hence we are left with

$$D\mathcal{F}(u_1, u_2, u_3) = \sum_{i,j,k=1}^{3} \epsilon_{ijk} \text{ ver } [\text{hor } u_i, [\text{hor } u_j, \text{hor } u_k]] \equiv 0 \,, \tag{1.156}$$

due to the Jacobi identity. $\qquad\square$

1.3.4 Connection in a principal bundle

Now, we are going to apply the general theory of connections to the case of principal fibre bundles. Any principal bundle is endowed with a canonical right action of the structure group G on the total bundle space P (cf. Proposition 1.3.2): $\widetilde{R}_g : P \longrightarrow P$. We shall require that the assignment of horizontal subspaces H_p is compatible with that action.

Definition 1.3.7 *A connection on a principle bundle (P, M, π, G) is a smooth assignment of linear subspaces H_p of $T_p P$*

$$P \ni p \longrightarrow H_p \subset T_p P \,, \tag{1.157}$$

such that the following conditions hold:

1. *The linear map*

$$T_p\pi : H_p \longrightarrow T_{\pi(p)}M \tag{1.158}$$

   is an isomorphism for any $p \in P$.

2. *The mapping (1.157) is invariant under the right action of G, i.e.,*

$$T_p \widetilde{R}_g(H_p) = H_{p \cdot g} \tag{1.159}$$

for any $p \in P$, and $g \in G$.

Let us turn to the connection form $\mathcal{A}$. In a general case $\mathcal{A}$ takes its value in the vertical space. However, in the case of a principal bundle there exists a canonical isomorphism between the vertical subspace V_p and the Lie algebra $\mathfrak{g}$ of G. That V_p and $\mathfrak{g}$ are isomorphic is evident; any fibre $F_x \cong G$, and hence the vertical space V_p, being a tangent space to a fibre, is isomorphic to $\mathfrak{g}$. This isomorphism is defined as follows: Take any element $\xi \in \mathfrak{g}$ and let $\mathbf{X}_\xi$ denote the infinitesimal generator of the canonical right action of G on P corresponding to ξ (cf. Definition 1.2.3). Since $\widetilde{R}_g$ acts along the fibres, each infinitesimal generator has to be tangent to the corresponding fibre, and hence

$$\mathbf{X}_\xi(p) \in V_p \, ,$$

for any $p \in P$. Thus

$$\mathfrak{g} \ni \xi \longrightarrow \mathbf{X}_\xi(p) \in V_p \, , \tag{1.160}$$

defines an isomorphism between $\mathfrak{g}$ and V_p. This isomorphism may be used to define a connection form. The vector field

$$P \ni p \longrightarrow \mathbf{X}_\xi(p) \in V_p \, ,$$

is usually called a *fundamental vector field*. For any $v \in V_p$ let $\widehat{v}$ be a unique element in $\mathfrak{g}$ such that

$$\mathbf{X}_{\widehat{v}}(p) = v \, . \tag{1.161}$$

Definition 1.3.8 *A connection form $\mathcal{A}$ on a principal fibre bundle (P, M, π, G) is a $\mathfrak{g}$-valued one-form on P defined by*

$$\mathcal{A}(u) := \widehat{\mathrm{ver}\, u} \, , \tag{1.162}$$

for any $u \in \mathfrak{X}(P)$.

Let us recall that each fibre $F_x = \pi^{-1}(x) \cong G$ and hence the restriction of $\mathcal{A}$ to $F_x \cong G$ defines a $\mathfrak{g}$-valued one-form on the Lie group G. Moreover, the restricted form $\mathcal{A}|_G$ satisfies the following property:

$$\mathcal{A}|_G(v) = \widehat{v} \, , \tag{1.163}$$

which is the defining property of the canonical Maurer–Cartan form on G (see Section 1.2.4), that is, $\mathcal{A}|_G = \omega_0$. This observation implies the following

Proposition 1.3.7 *The canonical right action of G on P induces the following transformation law for the connection form $\mathcal{A}$:*

$$\widetilde{R}_g^* \mathcal{A} = \mathrm{Ad}_{g^{-1}} \mathcal{A} \,, \tag{1.164}$$

for any $g \in G$.

If G is a matrix group, then

$$\widetilde{R}_g^* \mathcal{A} = g^{-1} \cdot \mathcal{A} \cdot g \,. \tag{1.165}$$

Note, that formulae (1.159) and (1.164) are dual to each other. They both state that the connection is compatible with the right action $\widetilde{R}_g$. Now, following Definition 1.3.6 we define the curvature of the connection $\mathcal{A}$,

$$\mathcal{F} := D\mathcal{A} \,, \tag{1.166}$$

to be a $\mathfrak{g}$-valued two-form on P. In analogy with Proposition 1.3.7, we have

$$\widetilde{R}_g^* \mathcal{F} = \mathrm{Ad}_{g^{-1}} \mathcal{F} \,, \tag{1.167}$$

or in the case of a matrix group,

$$\widetilde{R}_g^* \mathcal{F} = g^{-1} \cdot \mathcal{F} \cdot g \,. \tag{1.168}$$

Using Cartan's formula (1.54) one may prove the following

Proposition 1.3.8 *The curvature two-form $\mathcal{F} = D\mathcal{A}$ satisfies the following Cartan structural equation:*

$$\mathcal{F} = d\mathcal{A} + \frac{1}{2}[\mathcal{A}, \mathcal{A}] \,, \tag{1.169}$$

where $[\ ,\ ]$ denotes a Lie bracket in $\mathfrak{g}$. If G is a matrix group, then the above formula is equivalent to

$$\mathcal{F} = d\mathcal{A} + \mathcal{A} \wedge \mathcal{A} \,. \tag{1.170}$$

Note that the Bianchi identity $D\mathcal{F} = 0$ follows easily from (1.169). Indeed, one has

$$
\begin{aligned}
D\mathcal{F} &= d\mathcal{F} + [\mathcal{A}, \mathcal{F}] = d\left(d\mathcal{A} + \frac{1}{2}[\mathcal{A}, \mathcal{A}]\right) + \left[\mathcal{A}, d\mathcal{A} + \frac{1}{2}[\mathcal{A}, \mathcal{A}]\right] \\
&= d(d\mathcal{A}) + \frac{1}{2}[d\mathcal{A}, \mathcal{A}] - \frac{1}{2}[\mathcal{A}, d\mathcal{A}] + [\mathcal{A}, d\mathcal{A}] + \frac{1}{2}[\mathcal{A}, [\mathcal{A}, \mathcal{A}]] \\
&= \frac{1}{2}[\mathcal{A}, [\mathcal{A}, \mathcal{A}]] = 0 \,,
\end{aligned}
\tag{1.171}
$$

due to Proposition 1.2.2.

Let $P \longrightarrow M$ be a principal G-bundle. Suppose that the bundle is endowed with a connection form $\mathcal{A}$. Recall that if γ is a curve in M, e.g.,

$$[0, 1] \ni t \longrightarrow \gamma(t) \in M , \qquad \gamma(0) = x_0 , \quad \gamma(1) = x_1 ,$$

then one defines a parallel transport from $\pi^{-1}(x_0)$ to $\pi^{-1}(x_1)$ along γ as follows:

$$\pi^{-1}(x_0) \ni p_0 \longrightarrow \mathbf{T}_\gamma(p_0) := \widetilde{\gamma}(1) \in \pi^{-1}(x_1) , \tag{1.172}$$

where $\widetilde{\gamma}$ is the unique horizontal lift of γ such that $\widetilde{\gamma}(0) = p_0$. It is easy to see that a parallel transport commutes with a right action of G on P:

$$\widetilde{R}_g \circ \mathbf{T}_\gamma = \mathbf{T}_\gamma \circ \widetilde{R}_g , \tag{1.173}$$

that is,

$$p_0' = p_0 \cdot g \quad \Longrightarrow \quad \mathbf{T}_\gamma(p_0') = \mathbf{T}_\gamma(p_0) \cdot g , \tag{1.174}$$

for any $g \in G$. If γ is a closed curve beginning and ending at $x_0 \in M$, then a parallel transport along γ leads back to the fibre $\pi^{-1}(x_0)$. Consequently, it uniquely determines an element $\Phi[\widetilde{\gamma}]$ from G,

$$\mathbf{T}_\gamma(p_0) =: p_0 \cdot \Phi[\widetilde{\gamma}] , \tag{1.175}$$

which is called the *holonomy* of a curve $\widetilde{\gamma}$, with respect to $\mathcal{A}$. Note that a horizontal lift of a closed curve γ needs not be closed. Hence, in general, $\widetilde{\gamma}$ is open ($\widetilde{\gamma}$ is closed iff $\Phi[\widetilde{\gamma}] = e$). As γ varies over all closed paths based at $x \in M$, the corresponding $\Phi[\widetilde{\gamma}]$'s form a subgroup of G,

$$\mathrm{Hol}(p_0) := \left\{ \Phi[\widetilde{\gamma}] \;\middle|\; \gamma - \text{closed} , \, \gamma(0) = \gamma(1) = x_0 , \, \widetilde{\gamma}(0) = p_0 \right\} ,$$

which is called the holonomy group of the connection $\mathcal{A}$ with reference point p_0. Now, let $p_0' = p_0 \cdot g$, and let $\widetilde{\gamma}'$ be a horizontal lift of a closed curve γ such that $\widetilde{\gamma}'(0) = p_0'$. On the one hand, one has

$$\mathbf{T}_\gamma(p_0') = p_0' \cdot \Phi[\widetilde{\gamma}'] = p_0 \cdot g \cdot \Phi[\widetilde{\gamma}'] . \tag{1.176}$$

On the other hand, using (1.174) one finds

$$\mathbf{T}_\gamma(p_0') = \mathbf{T}_\gamma(p_0) \cdot g = p_0 \cdot \Phi[\widetilde{\gamma}] \cdot g . \tag{1.177}$$

Hence

$$\Phi[\widetilde{\gamma}'] = g^{-1} \cdot \Phi[\widetilde{\gamma}] \cdot g . \tag{1.178}$$

This implies the following relation between the corresponding holonomy groups $\mathrm{Hol}(p_0)$ and $\mathrm{Hol}(p_0 \cdot g)$:

$$\mathrm{Hol}(p_0 \cdot g) = g^{-1} \cdot \mathrm{Hol}(p_0) \cdot g , \tag{1.179}$$

that is, $\mathrm{Hol}(p_0 \cdot g)$ is the subgroup of G conjugate to $\mathrm{Hol}(p_0)$ and, hence, $\mathrm{Hol}(p_0 \cdot g)$ and $\mathrm{Hol}(p_0)$ are isomorphic.

1.3.5   Gauge transformations — paving the way to physics

Consider a principal G-bundle $P \longrightarrow M$ endowed with a connection form $\mathcal{A}$. Recall that $\mathcal{A}$ and the corresponding curvature $\mathcal{F}$ are $\mathfrak{g}$-valued forms on the bundle space P. Now we are going to introduce the local connection A and the local curvature F to be $\mathfrak{g}$-valued forms living on the base space M. Usually, in physical applications, M serves as a model for a physical space-time or a space of states for some physical system. Therefore, it is of great importance to show how the geometric objects defined on the total space P may be projected down to the base manifold M. Note that there is an obvious recipe for such a projection: If f is a local section of a bundle,

$$f : U \longrightarrow P ,$$

with $U \subset M$, then any $\mathfrak{g}$-valued form α on P may be projected to M by performing a pull-back *via* f, as follows:

$$\Lambda(P) \otimes \mathfrak{g} \ni \alpha \longrightarrow f^*\alpha \in \Lambda(M) \otimes \mathfrak{g} .$$

Evidently the projected form $f^*\alpha$ depends upon a chosen section f. Let us define local sections

$$f_i : U_i \longrightarrow P ,$$

which are canonically associated with a family of local trivializations (U_i, φ_i). For any $x \in U_i$ define

$$f_i(x) := \phi_{i,x}^{-1}(e) , \tag{1.180}$$

where e denotes a unit element in G. We call

$$A_{(i)} := f_i^* \mathcal{A} \tag{1.181}$$

the *connection form in the local trivialization φ_i.*[5] Suppose that we have another trivialization (U_j, φ_j), which gives rise to another local connection form $A_{(j)}$. What is the relation between $A_{(i)}$ and $A_{(j)}$ in the intersection $U_i \cap U_j$? The answer to this question is given by the following

Theorem 1.3.9 *For any point $x \in U_i \cap U_j$ the local connections forms $A_{(i)}$ and $A_{(j)}$ are related by*

$$A_{(i)}(x) = \mathrm{Ad}(\gamma_{ji}^{-1}(x))A_{(j)}(x) + (\gamma_{ji}^*\omega_0)(x) , \tag{1.182}$$

where $\gamma_{ij}^*\omega_0$ denotes the pull-back of the canonical Maurer–Cartan form on G.

---

[5]The index "(i)" should remind the reader that it does *not* denote the ith component of A but points to the ith trivialization φ_i used to define $A_{(i)}$.

If G is a matrix group, then the formula (1.182) may be rewritten in a more "friendly" form, as follows:

$$A_{(i)}(x)(v) = \gamma_{ji}^{-1}(x) \cdot A_{(j)}(x)(v) \cdot \gamma_{ji}(x) + \gamma_{ji}^{-1}(x) \cdot d\gamma_{ji}(x)(v) , \qquad (1.183)$$

for any $v \in T_x M$. The same procedure may be applied for the curvature form $\mathcal{F}$. One defines

$$F_{(i)} := f_i^* \mathcal{F} , \qquad (1.184)$$

and calls $F_{(i)}$ the curvature form in the local trivialization φ_i. In analogy to Theorem 1.3.9 one may prove that

$$F_{(i)}(x) = \mathrm{Ad}(\gamma_{ji}^{-1}(x)) F_{(j)}(x) , \qquad (1.185)$$

or, if G is matrix group, that

$$F_{(i)}(x)(u, v) = \gamma_{ji}^{-1}(x) \cdot F_{(j)}(x)(u, v) \cdot \gamma_{ji}(x) , \qquad (1.186)$$

for any $u, v \in T_x M$.

In physical applications one usually uses slightly different terminology. Instead of a change of trivializations one speaks about gauge transformations. By a local *gauge transformation* we mean a map:

$$g : U \longrightarrow G , \qquad (1.187)$$

with $U \subset M$. Any local section

$$f : U \longrightarrow P , \qquad (1.188)$$

is called a *local gauge*. It is evident that any two local gauges $f, f' : U \longrightarrow P$ differ by a gauge transformation, i.e.,

$$f'(x) = g(x) f(x) , \quad x \in U , \qquad (1.189)$$

for some g. Each local gauge gives rise to a local connection form

$$A := f^* \mathcal{A} , \qquad (1.190)$$

and local curvature

$$F := f^* \mathcal{F} . \qquad (1.191)$$

A local gauge transformation induces the following transformations of local connections:

$$A' = g^{-1} \cdot A \cdot g + g^{-1} \cdot dg , \qquad (1.192)$$

and curvatures:

$$F' = g^{-1} \cdot F \cdot g \,, \qquad (1.193)$$

where A' and F' are the corresponding objects in a local gauge f'. Physicists usually call A — gauge potential (in a local gauge f), and F — gauge field (in a local gauge f). Two connections (gauge potentials) A and A' are said to be *gauge equivalent* in a region $U \subset M$, if there exists a gauge transformation $g : U \longrightarrow G$ such that (1.192) holds.

Now, let γ be a closed curve contained in a patch U such that $\gamma(0) = \gamma(1)$. Choosing a local gauge $f : U \longrightarrow P$, we may lift γ to a curve $f(\gamma)$. Clearly, the lifted curve is also closed: $f(\gamma(0)) = f(\gamma(1)) = p_0 \in P$. Recall, that we have defined the holonomy of the horizontal lift $\widetilde{\gamma}$ to be an element $\Phi[\widetilde{\gamma}]$ such that

$$\widetilde{\gamma}(1) = p_0 \cdot \Phi[\widetilde{\gamma}] \,,$$

with $\widetilde{\gamma}(0) = p_0$. One can show that

$$\Phi[\widetilde{\gamma}] = \mathrm{P} \exp\left(\int_{f(\gamma)} \mathcal{A} \right) = \mathrm{P} \exp\left(\int_{\gamma} A \right) =: \Phi_f[\gamma] \,, \qquad (1.194)$$

where 'P' denotes a path ordering, and $A = f^*\mathcal{A}$ is the local connection form in the gauge f. Hence, any closed curve γ gives rise to a holonomy $\Phi_f[\gamma]$. Clearly, this definition is gauge-dependent. If we perform a local gauge transformation (1.187) and choose a local section $f'(x) = g(x)f(x)$, then we obtain

$$\Phi_{f'}[\gamma] = g(x_0)^{-1} \cdot \Phi_f[\gamma] \cdot g(x_0) \,, \qquad (1.195)$$

in perfect agreement with the formula (1.178). Note, however, that the trace of $\Phi_f[\gamma]$ is gauge invariant:

$$\mathrm{Tr}\, \Phi_f[\gamma] = \mathrm{Tr}\, \Phi_{f'}[\gamma] \,. \qquad (1.196)$$

Physicists call $\Phi_f[\gamma]$ a *Wilson loop* in the local gauge f. Let us illustrate the above discussion with the following important physical examples.

Example 1.3.6 (Electrodynamics) Let M denote a physical space-time. It is usually modelled as a four-dimensional pseudo-Riemannian manifold (e.g., Minkowski space-time). Consider an open region $U \in M$ and let A_μ denote the electromagnetic four-potential on U. Denote by $F_{\mu\nu}$ the corresponding electromagnetic field tensor,

$$F_{\mu\nu} = \partial_\mu A_\nu - \partial_\nu A_\mu \,. \qquad (1.197)$$

Let us observe that this scheme corresponds to a $U(1)$-principal bundle over M. The electromagnetic potential gives rise to a $u(1)$-valued one-form

$$A = i A_\mu dx^\mu \,.$$

A local gauge transformation

$$g : U \longrightarrow G = U(1)$$

may be written as

$$g(x) = e^{i\lambda(x)} , \quad x \in U ,$$

with $\lambda : U \longrightarrow \mathbb{R}$, and hence formula (1.192) implies

$$A' = A + id\lambda , \tag{1.198}$$

which reproduces the well-known rule

$$A'_\mu = A_\mu + \partial_\mu \lambda . \tag{1.199}$$

Now, since the structure group $U(1)$ is abelian, the general formula for the gauge field F (the curvature of the connection) simplifies to

$$F = DA = dA + \frac{1}{2}[A, A] = dA ,$$

that is,

$$F = \frac{i}{2} F_{\mu\nu} \, dx^\mu \wedge dx^\nu .$$

Clearly, a field tensor is gauge invariant:

$$F' = F .$$

Moreover, the identity $dF = d^2A = 0$ may be written as the following equation for F:

$$\partial_\mu F_{\nu\lambda} + \partial_\nu F_{\lambda\mu} + \partial_\lambda F_{\mu\nu} = 0 , \tag{1.200}$$

and it corresponds to the Bianchi identity $DF = 0$. For a detailed exposition of electrodynamics in terms of differential forms see, e.g., Ingarden and Jamiołkowski 1985.

◇

Example 1.3.7 (Yang–Mills theory) Now, instead of an abelian group $U(1)$ take a non-abelian group $SU(N)$ — the corresponding theory is called a Yang–Mills theory. Let $L_1, \ldots , L_{N^2-1}$ form a basis in the Lie algebra $su(N)$. Denote by f^a_{bc} the corresponding structure constants, i.e., $[L_a, L_b] = f^c_{ab} L_c$. Clearly, the generators L_a may be represented as anti-hermitian $N \times N$ matrices.[6] The local components of A and F are given by

$$A = A_\mu \, dx^\mu \quad \text{and} \quad F = \frac{1}{2} F_{\mu\nu} \, dx^\mu \wedge dx^\mu , \tag{1.201}$$

---

[6]Physicists prefer to work with hermitian quantities and, hence, instead of L's one often uses hermitian matrices $\lambda_a = -iL_a$. In terms of λ's, one finds $[\lambda_a, \lambda_b] = if^c_{ab}\lambda_c$.

where

$$A_\mu = A_\mu^a \, L_a \quad \text{and} \quad F_{\mu\nu} = F_{\mu\nu}^a \, L_a \, . \tag{1.202}$$

Clearly, the gauge potentials A_μ^a and gauge fields $F_{\mu\nu}^a$ are real quantities and A_μ and $F_{\mu\nu}$ are represented by anti-hermitian matrices. The definition of F,

$$F = dA + \frac{1}{2}[A, A] \, , \tag{1.203}$$

implies the following relations between the corresponding components:

$$F_{\mu\nu} = \partial_\mu A_\nu - \partial_\nu A_\mu + [A_\mu, A_\nu] \, , \tag{1.204}$$

and

$$F_{\mu\nu}^a = \partial_\mu A_\nu^a - \partial_\nu A_\mu^a + f_{bc}^a A_\mu^b A_\nu^c \, . \tag{1.205}$$

The gauge potentials A_μ^a, one for each generator of $SU(N)$, are analogs of the electromagnetic potential A_μ. In the quantized theory they correspond to spin-1 massless particles (so-called *gauge bosons*). Note that the Bianchi identity $DF = 0$ may be written as the following equation:

$$D_\mu F_{\nu\lambda} + D_\nu F_{\lambda\mu} + D_\lambda F_{\mu\nu} = 0 \, , \tag{1.206}$$

where we introduce the covariant derivative

$$D_\mu := \mathbb{1} \cdot \partial_\mu - A_\mu \, . \tag{1.207}$$

The Yang–Mills theory may be derived from a variational principle based on the following Yang–Mills action:

$$S_{\text{YM}}[A] := \int_M \text{Tr}(F \wedge \star F) = \int_M \mathcal{L}_{\text{YM}} \sqrt{|g|} \, d^4x \, . \tag{1.208}$$

In the case of $SU(N)$ it is possible to choose L's such that

$$\text{Tr}(L_a \cdot L_b) = -\frac{1}{2} \delta_{ab} \, . \tag{1.209}$$

With such a choice one finds for the Yang–Mills Lagrangian

$$\mathcal{L}_{\text{YM}} = -\frac{1}{4} \sum_{a=1}^{N^2-1} F_{\mu\nu}^a F^{a\mu\nu} \, , \tag{1.210}$$

which generalizes the well-known Maxwell Lagrangian $\mathcal{L}_{\text{M}} = -\frac{1}{4} F_{\mu\nu} F^{\mu\nu}$. $\diamond$

1.3.6   Characteristic classes

In this section we shall briefly discuss the problem of the classification of principal
fibre bundles. Let G be a Lie group and $\mathfrak{g}$ its Lie algebra. A symmetric, k-linear mapping

$$f : \overbrace{\mathfrak{g} \times \mathfrak{g} \times \ldots \times \mathfrak{g}}^{k} \longrightarrow \mathbb{R}$$

is called G-invariant (or simply invariant) if, for any $g \in G$ and any $\xi_1, \ldots, \xi_k \in \mathfrak{g}$,
we have

$$f(\mathrm{Ad}_g \xi_1, \ldots, \mathrm{Ad}_g \xi_k) = f(\xi_1, \ldots, \xi_k) \,. \tag{1.211}$$

If G is a matrix group, then G-invariance means

$$f(g \cdot \xi_1 \cdot g^{-1}, \ldots, g \cdot \xi_k \cdot g^{-1}) = f(\xi_1, \ldots, \xi_k) \,. \tag{1.212}$$

Any G-invariant, symmetric k-linear function f gives rise to an invariant polynomial
of order k, defined by

$$\mathfrak{f}_k(\xi) := f(\xi, \ldots, \xi) \,. \tag{1.213}$$

Denote by $I_k(G)$ the space of such polynomials and let

$$I(G) := \bigoplus_k I_k(G) \,.$$

Consider now a principal fibre bundle (P, M, π, G) together with a connection $\mathcal{A}$ and
curvature $\mathcal{F} = D\mathcal{A}$. Recall that any local section $s : U \longrightarrow P$, with U being an open
patch in M, gives rise to a local curvature $F = s^*\mathcal{F}$, which is a two-form on U. Now,
let us take a symmetric function f and define $\mathfrak{f}_k(F) \in \Lambda^{2k}(U)$ by

$$\mathfrak{f}_k(F) = f(F, \ldots, F) = f(e_{\alpha_1}, \ldots, e_{\alpha_k}) \, F^{\alpha_1} \wedge \ldots \wedge F^{\alpha_k} \,, \tag{1.214}$$

where $F = F^\alpha \otimes e_\alpha$, with $\{e_1, e_2, \ldots, e_r\}$ being a basis in $\mathfrak{g}$. Equivalently, $\mathfrak{f}_k(F)$
may be defined as follows:

$$\mathfrak{f}_k(F)(v_1, \ldots, v_{2k}) = \frac{1}{(2k)!} \sum_\sigma (-1)^\sigma \, f\Big(F(v_{\sigma(1)}, v_{\sigma(2)}), \ldots, F(v_{\sigma(2k-1)}, v_{\sigma(2k)}) \Big) \,,$$

$$\tag{1.215}$$

for any $v_1, \ldots, v_{2k} \in T_p P$. We sum over all permutations σ of the set $\{1, 2, \ldots, 2k\}$
and, as usual,

$$(-1)^\sigma = \begin{cases} +1 \,, & \sigma \ \text{is even} \\ -1 \,, & \sigma \ \text{is odd} \end{cases} \,.$$

One then may prove the following

Theorem 1.3.10 *The 2k-form $\mathfrak{f}_k(F)$ enjoys the following properties:*

1. *It may be globally defined on M, i.e., various locally defined 2k-forms $\mathfrak{f}_k(F)$ fit together to produce a well-defined form on M.*

2. *It is closed and hence defines a cohomology class $[\mathfrak{f}_k(F)] \in H^{2k}(M)$.*

3. *The cohomology class $[\mathfrak{f}_k(F)]$ does not depend on the particular choice of the connection form $\mathcal{A}$, that is, if $\mathcal{A}'$ is another connection one-form and $\mathcal{F}' = D\mathcal{A}'$, the corresponding curvature, then $[\mathfrak{f}_k(F)] = [\mathfrak{f}_k(F')]$.*

The above theorem implies the existence of a map

$$w \,:\, I(G) \longrightarrow H^*(M) := \bigoplus_k H^k(M) \,,$$

defined by

$$I_k(G) \ni \mathfrak{f}_k \longrightarrow [\mathfrak{f}_k(F)] \in H^{2k}(M) \,. \tag{1.216}$$

The elements from the image $w(I(G)) \subset H^*(M)$ are called *characteristic classes* of the G-bundle $P \longrightarrow M$.

Suppose now that we have two bundles over M with the same structure group G, namely, (P_1, π_1, M, G) and (P_2, π_2, M, G). We call them equivalent iff there exists a map $\phi : P_1 \longrightarrow P_2$ preserving the bundle structures, i.e., it sends a fibre over x in one bundle into a fibre over x in the other:

$$\phi(\pi_1^{-1}(x)) = \pi_2^{-1}(x) \,,$$

for all $x \in M$. Any invariant polynomial $\mathfrak{f} \in I(G)$ gives rise to two characteristic classes, $[\mathfrak{f}(F_1)]$ and $[\mathfrak{f}(F_2)]$, where F_1 and F_2 are local curvatures in $P_1 \longrightarrow M$ and $P_2 \longrightarrow M$ bundles, respectively.

Proposition 1.3.11 *If the bundles P_1 and P_2 are equivalent, then*

$$[\mathfrak{f}(F_1)] = [\mathfrak{f}(F_2)] \,,$$

for any invariant polynomial $\mathfrak{f}$.

This means that equivalent bundles have the same characteristic classes.

Example 1.3.8 (Chern classes) Consider a bundle with the structure group $G = GL(n, \mathbb{C})$ (or one of its subgroups). We define the invariant polynomials $c_k \in I_k(G)$ by

$$\det\left(\mathbb{1} + \frac{i}{2\pi} F \right) =: \sum_{k=0}^{n} c_k(F) \,. \tag{1.217}$$

One calls $c_k(F)$ a kth *Chern form*. The kth Chern class $C_k(P)$ of the bundle P is defined by

$$C_k(P) := [c_k(F)] \,. \tag{1.218}$$

It is easy to show that

$$c_k(F) = \frac{(-1)^k}{(2\pi i)^k} \, \epsilon^{j_1 \ldots j_k}_{i_1 \ldots i_k} \, F^{i_1}{}_{j_1} \wedge \ldots \wedge F^{i_k}{}_{j_k} \,, \tag{1.219}$$

where

$$\epsilon^{j_1 \ldots j_k}_{i_1 \ldots i_k} := \delta^{j_1}_{[i_1} \ldots \delta^{j_k}_{i_k]} \,. \tag{1.220}$$

One easily finds that

$$\begin{aligned}
C_0(P) &= 1 \,, \\
C_1(P) &= \frac{i}{2\pi} \operatorname{Tr} F \,, \\
C_2(P) &= \frac{1}{2} \left(\frac{i}{2\pi} \right)^2 \left[\operatorname{Tr} F \wedge \operatorname{Tr} F - \operatorname{Tr}(F \wedge F) \right] \,, \\
&\vdots \\
C_n(P) &= \left(\frac{i}{2\pi} \right)^n \det F \,.
\end{aligned} \tag{1.221}$$

For example, $U(1)$-bundles are characterized by the first Chern class $C_1(P)$:

$$C_1(P) = \frac{i}{2\pi} \operatorname{Tr} F = \frac{i}{2\pi} F \,. \tag{1.222}$$

Note that in this case the local curvature F is actually globally defined. For an $SU(2)$-bundle the curvature F is given by:

$$F = F^a L_a = F^a \frac{\sigma_a}{2i} \,, \tag{1.223}$$

where σ_a are the Pauli matrices:

$$\sigma_1 = \begin{pmatrix} 0 & 1 \\ 1 & 0 \end{pmatrix} \,, \quad \sigma_2 = \begin{pmatrix} 0 & -i \\ i & 0 \end{pmatrix} \,, \quad \sigma_3 = \begin{pmatrix} 1 & 0 \\ 0 & -1 \end{pmatrix} \,. \tag{1.224}$$

One finds that $C_1(P) = 0$, and, since $\operatorname{Tr}(\sigma_a \sigma_b) = \delta_{ab}$, that

$$C_2(P) = -\frac{1}{16\pi^2} F^a \wedge F_a \tag{1.225}$$

defines a real four-form on the base manifold M.

Let $P \longrightarrow M$ be a principal G-bundle and suppose that the base M is an oriented compact manifold of dimension $2n$. Then the value of the integral

$$\int_M C_n(P) = \int_M c_n(F) \tag{1.226}$$

is called the *Chern number* of the bundle. As we shall see in due course, this abstract mathematical concept plays an important role in physics. $\diamond$

1.4 Topology, bundles and physics

1.4.1 Elements from homotopy theory

Consider two topological spaces X and Y together with two maps

$$f_0, f_1 : X \longrightarrow Y .$$

One says that f_1 and f_2 are *homotopic* if there exists a continuous family of maps

$$F : X \times [0, 1] \longrightarrow Y ,$$

such that

$$F(x, 0) = f_0(x) \quad \text{and} \quad F(x, 1) = f_1(x) ,$$

for $x \in X$. A map F is called a *homotopy* between f_0 and f_1. Intuitively, it means that f_0 and f_1 can be continuously deformed one into another. It is evident that *to be homotopic* defines an equivalence relation in the space of maps from X to Y. One denotes by $[f]$ a class of maps which are homotopic to f and calls $[f]$ a *homotopy class* of f. Consider now the following space:

$$\Omega(X, x_0) = \{ \text{ space of loops in } X \text{ with a base point } x_0 \} ,$$

that is, $\gamma \in \Omega(X, x_0)$, iff

$$\gamma : [0, 1] \longrightarrow X , \qquad \gamma(0) = \gamma(1) = x_0 .$$

Recall that $\Omega(X, x_0)$, endowed with an operation of multiplication (1.140) and an inverse (1.141), defines a group called a *loop group* at $x_0 \in X$. Two loops $\gamma_1, \gamma_2 \in \Omega(x_0, X)$ are said to be (based) homotopic if there exists a homotopy F between γ_1 and γ_2 such that $F(0, t) = F(1, t) = x_0$, i.e., we may deform γ_1 into γ_2 keeping a base point x_0 fixed. A loop is called *null-homotopic* if it is homotopic to a trivial loop $\gamma(t) = x_0$. Moreover, one easily shows that

$$\gamma_1 \sim \gamma_2 \quad \text{and} \quad \zeta_1 \sim \zeta_2 \quad \Longrightarrow \quad \gamma_1 * \zeta_1 \sim \gamma_2 * \zeta_2 , \tag{1.227}$$

and

$$\gamma * \gamma^{-1} \sim \gamma^{-1} * \gamma \sim \text{trivial loop at } x_0 . \tag{1.228}$$

Hence, both operations are well defined on homotopy classes in $\Omega(X, x_0)$. Therefore, we may define a quotient space

$$\pi_1(X, x_0) := \Omega(X, x_0)/\sim \tag{1.229}$$

called a *fundamental group* (or a *first homotopy group*) of X at a point x_0. Our construction of $\pi_1(X, x_0)$ depends on the base point x_0. Assume now, that a topological space X is path-connected, that is, for any two points $x, y \in X$ there is path α in X joining x and y.

Proposition 1.4.1 *If X is path-connected then the fundamental groups $\pi_1(X, x_0)$ and $\pi_1(X, x_1)$ are isomorphic for any $x_0, x_1 \in X$.*

Actually, it is easy to construct the above isomorphism. Let $\gamma \in \Omega(X, x_1)$ and take any path α starting at x_0 and ending at x_1. Clearly,

$$\alpha * \gamma * \alpha^{-1} \in \Omega(X, x_0) ,$$

and hence the path α gives rise to a map $\Omega(X, x_1) \longrightarrow \Omega(X, x_0)$. One may show that this map induces an isomorphism between the corresponding fundamental groups (see, e.g., Schwarz 1996, Morandi 1992). Therefore, one usually does not indicate a base point but simply writes $\pi_1(X)$. A topological space with a trivial fundamental group is called *simply connected*.

Example 1.4.1 It is evident that

$$\pi_1(\mathbb{R}^2) = 0 ,$$

i.e., that any loop in $\mathbb{R}^2$ may be continuously shrunk to a point, or, equivalently, any loop is null-homotopic. It is no longer true if we remove one point from $\mathbb{R}^2$. One easily shows, for example, that

$$\pi_1(\mathbb{R}^2 - \{0\}) = \mathbb{Z} .$$

Hence, each class of loops in $\mathbb{R}^2 - \{0\}$ is characterized by an integer $n \in \mathbb{Z}$ which says how many times any loop in this class winds around '0'. Obviously, we obtain the same result by removing a single line from $\mathbb{R}^3$:

$$\pi_1(\mathbb{R}^3 - \text{Line}) = \mathbb{Z} .$$

Note, however, that

$$\pi_1(\mathbb{R}^3 - \{0\}) = 0 ,$$

and hence punctured $\mathbb{R}^3$ is simply connected. The same is true for $\mathbb{R}^n - \{0\}$, with $n \geq 3$. $\diamond$

A loop in X may be regarded as a map from a circle S^1 to X. Indeed, if we parametrize a circle by an angle $\varphi \in [0, 2\pi]$, then a map

$$\gamma : [0, 2\pi] \longrightarrow X$$

can be seen as a map from S^1 to X provided $\gamma(0) = \gamma(2\pi)$. Replacing S^1 by S^k one may introduce higher homotopy groups. Consider a map

$$f : S^k \longrightarrow X ,$$

taking one fixed point on S^k, say the south pole s, into the fixed point x_0 of X. We call such a map *k-loop* (some authors, e.g., Schwarz (1996), call it a *k*-dimensional spheroid). To define a multiplication of k-loops let us observe that we may equivalently represent any k-loop as a map from a k-dimensional unit cube $I^k := [0, 1]^k$ into X, such that the boundary ∂I^k is mapped to x_0, i.e.,

$$f : I^k \longrightarrow X , \quad \text{and} \quad f(\partial I^k) = x_0 .$$

For $k = 1$ this construction reproduces a unit interval $I^1 = [0, 1]$. A loop is a map $\gamma : I^1 \longrightarrow X$, such that $\gamma(0) = \gamma(1) = x_0$, that is, $\gamma(\partial I^1) = x_0$. We call k-loops f_0 and f_1 (based) homotopic if there is a homotopy $F : I^k \times I \longrightarrow X$ between f_0 and f_1 such that

$$F(t_1, \ldots, t_k, t) = x_0 \quad \text{if} \quad (t_1, \ldots, t_k) \in \partial I^k ,$$

where $(t_1, \ldots, t_k)$ are coordinates in I^k. Denote by $\pi_k(X, x_0)$ the space of homotopy classes of k-loops based on x_0. Note that any two k-loops f and g may be multiplied:

$$(f * g)(t_1, \ldots, t_k) := \begin{cases} f(2t_1, t_2, \ldots, t_k) , & \text{for } 0 \le t_1 \le 1/2 \\ g(2t_1 - 1, t_2, \ldots, t_k) , & \text{for } 1/2 \le t_1 \le 1 ; \end{cases} \tag{1.230}$$

and that for any k-loop f we may define an inverse:

$$f^{-1}(t_1, t_2, \ldots, t_k) := f(1 - t_1, t_2, \ldots, t_k) . \tag{1.231}$$

It is easy to show that both the multiplication (1.230) and the inverse (1.231) are well defined on $\pi_k(X, x_0)$, and, therefore, they endow $\pi_k(X, x_0)$ with a group structure. One may show that $\pi_k(X, x_0)$ is abelian for $k > 1$. For a path-connected X the k-dimensional homotopy groups $\pi_k(X, x_0)$ and $\pi_k(X, x_1)$ are isomorphic for any points $x_0, x_1 \in X$. Hence, one usually writes $\pi_k(X)$.

Example 1.4.2 Recall that the punctured $\mathbb{R}^3$ is simply connected. However,

$$\pi_2(\mathbb{R}^3 - \{0\}) = \mathbb{Z} ,$$

and

$$\pi_2(\mathbb{R}^n - \{0\}) = 0 , \quad \text{for } n > 3 .$$

Analogously,

$$\pi_m(\mathbb{R}^{m+1} - \{0\}) = \mathbb{Z} ,$$

and

$$\pi_m(\mathbb{R}^n - \{0\}) = 0 , \quad \text{for } n > m + 1 .$$

$$\diamondsuit$$

It turns out that homotopy groups are topological invariants. If X and Y are path-connected, topologically equivalent spaces, then the corresponding homotopy groups $\pi_k(X)$ and $\pi_k(Y)$ are isomorphic.

Example 1.4.3 (Brouwer degree) Suppose X and Y are two compact, connected, orientable n-dimensional manifolds, and $f : X \longrightarrow Y$ is a smooth map. Let $\omega \in \Lambda^n(Y)$ be a volume form on Y. The *Brouwer degree* of f, denoted by $\deg(f)$, is defined by

$$\int_X f^*\omega =: \deg(f) \int_Y \omega \, . \tag{1.232}$$

It turns out that $\deg(f) \in \mathbb{Z}$, and that it does not depend upon the choice of ω. Roughly speaking, it measures how many times the image of X wraps around Y. Now, it turns out that if f and g are homotopic, then

$$\deg(f) = \deg(g) \, ,$$

that is, the degree of a map is a homotopic invariant. In particular, the degree of a map $f : S^k \longrightarrow S^k$ is called the *winding number* of f. To illustrate a concept of a winding number, choose any positive integer n and consider two maps $f, g : S^1 \longrightarrow S^1$ defined by $f(z) = z^n$ and $g(z) = \bar{z}^n$, where we identify S^1 with the set of complex numbers $z \in \mathbb{C}$ with $|z| = 1$. Then it is easy to show that

$$\deg(f) = n \quad \text{and} \quad \deg(g) = -n \, ,$$

and hence f and g are not homotopic. $\Diamond$

Example 1.4.4 (Homotopy groups of spheres) The previous example shows that

$$\pi_1(S^1) = \mathbb{Z} \, .$$

Actually, one can show that

$$\pi_n(S^n) = \mathbb{Z} \, ,$$

that is, maps from S^n to S^n may be classified according to their winding numbers. What about maps between spheres with different dimensions $f : S^n \longrightarrow S^m$? It is quite easy to show that

$$\pi_n(S^m) = 0 \, , \quad \text{for } n < m \, .$$

However, the opposite case, i.e., $n > m$, is much more subtle. It turns out that, for example,

$$\pi_n(S^1) = 0 \, , \quad \text{for } n > 1 \, ,$$

but this is not true for S^2. Actually, studying maps $f : S^3 \longrightarrow S^2$, Hopf discovered celebrated *Hopf fibration*: $S^3 \longrightarrow S^2$. One shows that

$$\pi_3(S^2) = \mathbb{Z} \, ,$$

that is, a homotopy class $[f]$ is characterized by an integer number. It turns out, that the integer $[f]$ is equal to the Chern number of the Hopf bundle $S^3 \longrightarrow S^2$. $\Diamond$

Example 1.4.5 (Classification of bundles) It turns out that homotopy theory plays a significant role in the classification of fibre bundles. Let us consider a principal fibre bundles over S^n (such bundles will play an important role in our book). One can prove the following

Theorem 1.4.2 (Classification theorem) *If the structure group G is path connected, then G-bundles over S^n are classified (up to equivalence) by elements of the homotopy group $\pi_{n-1}(G)$.*

Thus for $U(1)$-bundles over S^n one has

$$\pi_{n-1}(U(1)) = \pi_{n-1}(S^1) = \begin{cases} \mathbb{Z}, & \text{for } n = 2 \\ 0, & \text{for } n > 2 \end{cases}.$$

This means that $U(1)$-bundles over S^n are necessarily trivial if $n > 2$. For $n = 2$ we recover Hopf $U(1)$-bundles $f : S^3 \longrightarrow S^2$, which are classified by the homotopy classes $[f]$. The same is true for another important Hopf $SU(2)$-bundle: $S^7 \longrightarrow S^4$, i.e.,

$$\pi_4(SU(2)) \cong \pi_4(S^3) \cong \mathbb{Z}.$$

As we shall see, these integers have clear physical interpretation. $\diamond$

1.4.2   Monopole bundle

Let us first briefly recall Dirac's famous idea of a magnetic pole (Dirac 1931, 1948). Suppose that a magnetic charge 'g' is placed at the origin of $\mathbb{R}^3$. It produces a Coulomb-like magnetic field

$$\mathbf{B}(\mathbf{r}) = g\,\frac{\mathbf{r}}{r^3}\,, \tag{1.233}$$

with $\mathbf{r} = (x, y, z) \in \mathbb{R}^3$ and $r = |\mathbf{r}|$. Obviously, the origin $\mathbf{r} = 0$ has to be excluded and, therefore, we are dealing with the punctured manifold $M := \mathbb{R}^3 - \{0\}$. Evidently $\nabla \cdot \mathbf{B} = 0$ on M. However, as is well known, $\mathbf{B}$ does not admit a globally defined smooth vector potential $\mathbf{A}$, such that $\mathbf{B} = \nabla \times \mathbf{A}$. Let us note that $\mathbf{B}$ admits singular potentials, however. Take, for example,

$$\mathbf{A}_+(x, y, z) = \frac{g}{r}\,\frac{1}{z+r}\,(-y, x, 0)\,; \tag{1.234}$$

one easily finds that

$$\nabla \times \mathbf{A}_+ = g\,\frac{\mathbf{r}}{r^3}\,. \tag{1.235}$$

Using spherical coordinates one has

$$A_+^r = A_+^\theta = 0\,, \quad A_+^\varphi = g\,\frac{1 - \cos\theta}{r\sin\theta}\,. \tag{1.236}$$

The above formulae show that $\mathbf{A}_+$ is singular along the line $(x = 0, y = 0, z \leq 0)$ or, equivalently for $\theta = \pi$. Such a singular line is called a *Dirac string*. Thus, any attempt to define a global vector potential leads to the appearance of a fictitious string singularity ending on a pole. If we take another potential $\mathbf{A}_-$:

$$\mathbf{A}_- = \frac{g}{r}\frac{1}{z-r}(-y, x, 0) \tag{1.237}$$

or, in spherical coordinates,

$$A_-^r = A_-^\theta = 0 , \quad A_-^\varphi = -g\,\frac{1 + \cos\theta}{r\sin\theta} , \tag{1.238}$$

we obtain a Dirac string along $(x = 0, y = 0, z \geq 0)$ or, equivalently for $\theta = 0$.

Let us translate the problem into the language of differential forms. Observe that the following one-forms:

$$A_+ = \frac{g}{r}\frac{1}{z+r}(-ydx + xdy) , \qquad A_- = \frac{g}{r}\frac{1}{z-r}(xdy - ydx) , \tag{1.239}$$

have the property that

$$dA_+ = dA_- =: B = \frac{g}{r^3}\left(xdy \wedge dz + ydz \wedge dx + zdx \wedge dy\right) . \tag{1.240}$$

The reader can easily check that B is closed. Is B exact? Using standard spherical coordinates in $\mathbb{R}^3$ we find

$$B = g\,\sin\theta\,d\theta \wedge d\varphi , \tag{1.241}$$

which means that B is proportional to the standard volume form on a unit sphere. Hence,

$$\int_{S^2} B = 4\pi g , \tag{1.242}$$

which shows that B is not exact (otherwise the Stokes theorem implies the vanishing of $\int_{S^2} B$). Note that rewriting covectors $A_\pm$ in spherical coordinates gives

$$A_+ = g(1 - \cos\theta)d\varphi , \tag{1.243}$$

and

$$A_- = -g(1 + \cos\theta)d\varphi . \tag{1.244}$$

They are related by the following gauge transformation:

$$A_+ = A_- + df , \tag{1.245}$$

where $f : \mathbb{R}^3 \longrightarrow S^1 \cong U(1)$ depends only upon φ and is defined by

$$f(\varphi) = 2g\varphi \, .$$

Periodicity of f, i.e., that

$$f(2\pi) = f(0) \, , \qquad \mathrm{mod} \ 2\pi n \, , \tag{1.246}$$

implies that

$$g = \frac{n}{2} \, , \qquad n \in \mathbb{Z} \, , \tag{1.247}$$

which is the famous *Dirac quantization condition*;[7] The integer n is called a *magnetic number*. We refer the interested reader to the review article by Goddard and Olive (1978) for a detailed discussion of the Dirac magnetic monopole and its generalizations.

The above construction of the magnetic pole gives rise to a principal $U(1)$-bundle over S^2 called by physicists a *monopole bundle*. Let us note that the quantities B and $A_\pm$ depend only upon the angles θ and φ. The radial coordinate is irrelevant and, therefore, we may reduce the problem from $\mathbb{R}^3$ to the unit sphere S^2. Observe, now, that $\mathbf{A}_+$ is singular at the south pole of S^2 only, whereas $\mathbf{A}_-$ has its sole singularity at the north pole. Following Wu and Yang (1975) we divide S^2 into two coordinate patches and define the corresponding fields on each patch separately. This way we avoid the use of a singular vector potential. Let U_N and U_S be open subsets in S^2 such that (cf. Example 1.1.1)

1. U_N (U_S) contains the north (south) pole,

2. $U_N \cup U_S = S^2$, and $U_N \cap U_S \neq \emptyset$.

Now, a $U(1)$-bundle over S^2 is uniquely determined by a transition function

$$\gamma_{NS} : U_N \cap U_S \longrightarrow U(1) \, .$$

Let us take

$$\gamma_{NS}(\theta, \varphi) := e^{in\varphi} \, . \tag{1.248}$$

Clearly, the local connection forms are related by

$$A_S = A_N + \gamma_{NS}^{-1} d\gamma_{NS} = A_N + ind\varphi \, , \tag{1.249}$$

---

[7]Dirac showed that the quantum mechanics of a magnetic pole g implies the following quantization condition

$$g = \frac{n}{2} \frac{\hbar}{e} \, ,$$

where e denotes the elementary electric charge.

and

$$F_N = F_S .\tag{1.250}$$

Let us compute the corresponding Chern number of the magnetic bundle:

$$\text{Chern number} = \int_{S^2} C_1(P).$$

To do this, let us define *up*, S^2_+, and *down*, S^2_-, hemispheres, as follows:

$$S^2_+ := \left\{ (x^1, x^2, x^3) \in S^2 \,\middle|\, x^3 \geq 0 \right\},$$
$$S^2_- := \left\{ (x^1, x^2, x^3) \in S^2 \,\middle|\, x^3 \leq 0 \right\}.$$

Evidently, $S^2_+ \cup S^2_+ = S^2$ and $S^2_+ \cap S^2_- = S^1$, where S^1 is an "equatorial" circle that we provide with the orientation it inherits from S^2_+. One has

$$\begin{aligned} F_N = dA_N \quad &\text{on} \quad S^2_+ \subset U_N , \\ F_S = dA_S \quad &\text{on} \quad S^2_- \subset U_S , \end{aligned}$$

and, since $\operatorname{Tr} F = F$, the Stokes theorem (pay attention to the orientations) implies that

$$\int_{S^2_+} F_N = \int_{S^2_+} dA_N = \int_{S^1} A_N ,\tag{1.251}$$

and

$$\int_{S^2_-} F_S = \int_{S^2_-} dA_S = - \int_{S^1} A_S .\tag{1.252}$$

Thus we obtain the following formula for the Chern number:

$$\begin{aligned} \int_{S^2} C_1(P) &= \frac{i}{2\pi} \int_{S^2} F = \frac{i}{2\pi} \int_{S^2_+} F_N + \frac{i}{2\pi} \int_{S^2_-} F_S \\ &= \frac{i}{2\pi} \int_{S^1} (A_N - A_S) = \frac{i}{2\pi} \int_{S^1} (-in\,d\varphi) = n . \end{aligned}\tag{1.253}$$

Note, that the Chern number does not depend on a particular choice of A but only on the transition function γ_{NS}, which uniquely defines the monopole bundle. Any two bundles with the same

$$\text{magnetic number} = \text{Chern number} = n$$

are equivalent. In particular, we may take

$$A_N = -\frac{in}{2} (1 - \cos\theta) \, d\varphi$$

on U_N, and

$$A_S = \frac{in}{2}(1 + \cos\theta)\, d\varphi$$

on U_S. One then obtains

$$F_N = F_S = -\frac{in}{2}\sin\theta\, d\theta \wedge d\varphi\, ,$$

i.e., quantities corresponding to the field of a magnetic pole with a quantized strength $g = n/2$. For that reason, the above $U(1)$-bundle is called a monopole bundle.

1.4.3   Instanton bundle

Following our example of a monopole bundle — $U(1)$-bundle over S^2 — we are going to construct a principal $SU(2)$-bundle over S^4. Let us cover S^4 with two patches U_N and U_S. To define the bundle we show how to construct the transition function γ_{NS}. Let

$$\gamma_{NS} : U_N \cap U_S \longrightarrow SU(2)\, , \tag{1.254}$$

be a smooth function such that on the equatorial three-sphere

$$S^3 := S_+^4 \cap S_-^4 \subset U_N \cap U_S,$$

it is defined by

$$\gamma_{NS}(x) = U(x) \in SU(2)\, , \tag{1.255}$$

where

$$U(x) = x^\alpha \tau_\alpha = \begin{pmatrix} x^0 - ix^3 & -x^2 - ix^1 \\ x^2 - ix^1 & x^0 + ix^3 \end{pmatrix}. \tag{1.256}$$

In the above formula

$$\tau_0 = \mathbb{1}_2\, , \quad \text{and} \quad \tau_k = i\sigma_k\, , \quad k = 1, 2, 3\, ,$$

and $x = (x^0, x^1, x^2, x^3) \in \mathbb{R}^4$. Clearly, we have two $su(2)$-valued local connection forms A_N and A_S related by

$$A_S = \gamma_{NS}^{-1} \cdot A_N \cdot \gamma_{NS} + \gamma_{NS}^{-1} \cdot d\gamma_{NS}\, , \tag{1.257}$$

and two local curvatures,

$$F_N = dA_N + A_N \wedge A_N \quad \text{and} \quad F_S = dA_S + A_S \wedge A_S\, . \tag{1.258}$$

Let us calculate the corresponding Chern number,

$$\int_{S^4} C_2(P) = \int_{S^4} [c_2(F)] ,$$

where

$$c_2(F) = \frac{1}{8\pi^2} \, \mathrm{Tr}\,(F \wedge F) .$$

Note that on $U_N \cap U_S$, we have

$$F_S = \gamma_{NS}^{-1} \cdot F_N \cdot \gamma_{NS} , \tag{1.259}$$

yet

$$\mathrm{Tr}\,(F_S \wedge F_S) = \mathrm{Tr}\,(F_N \wedge F_N) .$$

Now, the four-form $\mathrm{Tr}\,(F \wedge F)$ is exact, and can be written

$$\begin{aligned}
\mathrm{Tr}(F \wedge F) &= d\left\{ \mathrm{Tr}\left(dA \wedge A + \frac{2}{3} A \wedge A \wedge A \right) \right\} \\
&= d\left\{ \mathrm{Tr}\left(F \wedge A - \frac{1}{3} A \wedge A \wedge A \right) \right\} .
\end{aligned} \tag{1.260}$$

Thus using the Stokes theorem (pay attention to the orientation), one obtains

$$\begin{aligned}
\int_{S^4_+} \mathrm{Tr}\,(F_N \wedge F_N) &= \int_{S^4_+} d\left\{ \mathrm{Tr}\left(dA_N \wedge A_N + \frac{2}{3} A_N \wedge A_N \wedge A_N \right) \right\} \\
&= \int_{S^3} \mathrm{Tr}\left(dA_N \wedge A_N + \frac{2}{3} A_N \wedge A_N \wedge A_N \right) , \tag{1.261} \\
\int_{S^4_-} \mathrm{Tr}\,(F_S \wedge F_S) &= \int_{S^4_-} d\left\{ \mathrm{Tr}\left(dA_S \wedge A_S + \frac{2}{3} A_S \wedge A_S \wedge A_S \right) \right\} \\
&= -\int_{S^3} \mathrm{Tr}\left(dA_S \wedge A_S + \frac{2}{3} A_S \wedge A_S \wedge A_S \right) , \tag{1.262}
\end{aligned}$$

which implies that

$$\begin{aligned}
\int_{S^4} C_2(P) &= \frac{1}{8\pi^2} \int_{S^4_+} \mathrm{Tr}\,(F_N \wedge F_N) + \frac{1}{8\pi^2} \int_{S^4_-} \mathrm{Tr}\,(F_S \wedge F_S) \\
&= \frac{1}{8\pi^2} \int_{S^3} \mathrm{Tr}\left(dA_N \wedge A_N + \frac{2}{3} A_N \wedge A_N \wedge A_N - dA_S \wedge A_S - \frac{2}{3} A_S \wedge A_S \wedge A_S \right) .
\end{aligned}$$

This fairly complicated integral should not depend on a particular choice of A_N and A_S (related according to (1.257)), but may only depend upon the transition function

γ_{NS}, which entirely defines the bundle. Indeed, for any A_N and A_S related by (1.257) one finds that

$$dA_N \wedge A_N + \frac{2}{3} A_N \wedge A_N \wedge A_N - dA_S \wedge A_S - \frac{2}{3} A_S \wedge A_S \wedge A_S$$
$$= -\frac{1}{3} \left(\gamma_{NS}^{-1} \cdot d\gamma_{NS} \wedge \gamma_{NS}^{-1} \cdot d\gamma_{NS} \wedge \gamma_{NS}^{-1} \cdot d\gamma_{NS} \right) . \qquad (1.263)$$

A purely algebraic proof we leave as an exercise to the reader. Now, on the equatorial three-sphere $\gamma_{NS}(x) = U(x)$, and hence

$$\int_{S^4} C_2(P) = -\frac{1}{24\pi^2} \int_{S^3} \mathrm{Tr} \left(U^{-1} \cdot dU \wedge U^{-1} \cdot dU \wedge U^{-1} \cdot dU \right) . \qquad (1.264)$$

This integral is computed in several books — see, e.g., Rajaramaran 1982. Actually, it represents a Euclidean action of $SU(2)$ Yang–Mills theory in $\mathbb{R}^4$ corresponding to a special solution of this theory called an *instanton*. For that reason the above $SU(2)$-bundle over S^4 is usually called an *instanton bundle*. One finds that

$$\int_{S^4} C_2(P) = -1 . \qquad (1.265)$$

Physicists call the value of $\int_{S^4} C_2(P)$ an *instanton number*, in analogy to the monopole number $\int_{S^2} C_1(P)$ of the $U(1)$-bundle over S^2. Thus, for the instanton bundle,

$$\text{instanton number} = \text{Chern number} = -1 .$$

The reader can easily show that if we modify the transition function as follows:

$$\gamma_{NS} = U \longrightarrow \gamma_{NS} = U^k ,$$

then

$$\text{instanton number} = \text{Chern number} = -k .$$

As we shall see, the instanton bundle, which has already been applied in Yang–Mills theory, finds new application in nonrelativistic quantum mechanics.

1.4.4   Hopf fibration $S^3 \longrightarrow S^2$

Interestingly, almost at the same time as Dirac discovered magnetic poles, Heinz Hopf investigated the properties of maps from S^3 into S^2 (Hopf 1931, see also Hopf 1964). Define a map $\pi : \mathbb{C}^2 \longrightarrow \mathbb{R}^3$ by:

$$\pi(z_1, z_2) := \left(\bar{z}_1 z_2 + z_1 \bar{z}_2, \, i(\bar{z}_1 z_2 - z_1 \bar{z}_2), \, |z_1|^2 - |z_2|^2 \right) . \qquad (1.266)$$

Note that any point on the unit three-sphere

$$S^3 = \left\{ (z_1, z_2) \in \mathbb{C}^2 \,\middle|\, |z_1|^2 + |z_2|^2 = 1 \right\}$$

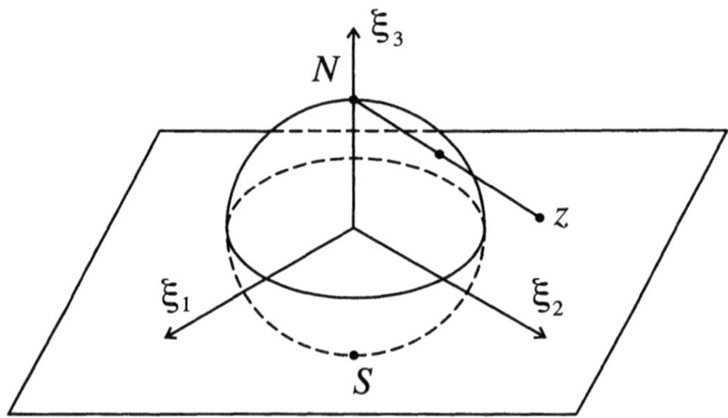

Figure 1.5: Stereographic projection

is mapped via π to a unit two-sphere in $\mathbb{R}^3$, that is, $\pi(S^3) \subset S^2$. One calls the restriction of π to S^3 a *Hopf map*. The crucial property of π is $U(1)$-invariance:

$$\pi(e^{i\lambda}z_1, e^{i\lambda}z_2) = \pi(z_1, z_2) \,. \tag{1.267}$$

This enables us to construct a principal $U(1)$-bundle $S^3 \longrightarrow S^2$ with π as a projection. One calls $(S^3, S^2, \pi, U(1))$ a *Hopf bundle* or a *Hopf fibration*. Introducing real coordinates

$$z_1 = x_1 + i x_2 \ \text{ and } \ z_2 = x_3 + i x_4 \,, \tag{1.268}$$

we have

$$\pi(z_1, z_2) = (\xi_1, \xi_2, \xi_3) \,, \tag{1.269}$$

with

$$\begin{aligned}
\xi_1 &= 2(x_1 x_3 + x_2 x_4) \,, \\
\xi_2 &= 2(x_2 x_3 - x_1 x_4) \,, \\
\xi_3 &= x_1^2 + x_2^2 - x_3^2 - x_4^2 \,.
\end{aligned} \tag{1.270}$$

The sphere S^2 may alternatively be parametrized by the coordinates on the equatorial plane via the stereographic projection (see Fig. 1.5). Introducing the planar coordinates $z = x + iy$, one finds that

$$z = \frac{\xi_1 + i\xi_2}{1 - \xi_3} = \frac{x_1 + i x_2}{x_3 + i x_4} = \frac{z_1}{z_2} \,. \tag{1.271}$$

It is evident that (z_1, z_2) and $(e^{i\lambda}z_1, e^{i\lambda}z_2)$ give rise to the same point 'z' on the equatorial plane. Introducing three angle variables (θ, ϕ, ψ) on S^3, as follows:

$$z_1 = \cos\frac{\theta}{2}\exp\left(i\frac{\psi+\phi}{2}\right), \tag{1.272}$$

$$z_2 = \sin\frac{\theta}{2}\exp\left(i\frac{\psi-\phi}{2}\right), \tag{1.273}$$

one obtains

$$\begin{aligned}
\xi_1 &= \sin\theta\cos\phi, \\
\xi_2 &= \sin\theta\sin\phi, \\
\xi_3 &= \cos\theta,
\end{aligned} \tag{1.274}$$

which defines the standard parametrization of S^2.

Having described the bundle projection π, let us describe in more detail the structure of the Hopf bundle $S^3 \longrightarrow S^2$. Covering the base manifold S^2 with two coordinate patches U_N and U_S (cf. section 1.4.2), we define

$$\phi_N : \pi^{-1}(U_N) \longrightarrow U(1),$$

and

$$\phi_S : \pi^{-1}(U_S) \longrightarrow U(1),$$

by

$$\phi_N(z_1, z_2) := \frac{z_1}{|z_1|}, \qquad \phi_S(z_1, z_2) := \frac{z_2}{|z_2|}. \tag{1.275}$$

The corresponding transition function γ_{NS} is therefore given by

$$\gamma_{NS} = \phi_N \circ \phi_S^{-1} = \frac{z_1}{|z_1|}\frac{|z_2|}{z_2} = e^{i\phi}, \tag{1.276}$$

where we have used the angle parametrization (1.272)–(1.273).

Our next step is to define a connection. Let $\langle\ ,\ \rangle : \mathbb{C}^2 \longrightarrow \mathbb{C}$ be the standard hermitian inner product in $\mathbb{C}^2$:

$$\langle (z_1, z_2), (w_1, w_2)\rangle := \bar{z}_1 w_1 + \bar{z}_2 w_2. \tag{1.277}$$

Let $p \in S^3$ and take any vector $v \in T_p S^3$. Using the canonical embedding $S^3 \hookrightarrow \mathbb{C}^2$, both p and v may be identified with points in $\mathbb{C}^2$. If $p \in S^3$ corresponds to $(z_1, z_2) \in \mathbb{C}^2$, then the vertical space V_p reads

$$V_p = \{(\alpha z_1, \alpha z_2) \,|\, \alpha \in \mathbb{C}\}. \tag{1.278}$$

As the horizontal space we take

$$H_p = \{(w_1, w_2) \in \mathbb{C}^2 \,|\, \langle (z_1, z_2), (w_1, w_2)\rangle = 0\}, \tag{1.279}$$

that is, H_p consists of all vectors tangent to S^3 at p which are orthogonal to p viewed as vectors in $\mathbb{C}^2$. We shall call a connection defined this way a *natural* or *canonical connection*. It is therefore evident that the connection one-form $\mathcal{A}$ on S^3 is defined by[8]

$$\begin{aligned} \mathcal{A} &= 2g\,(\bar{z}_1 dz_1 + \bar{z}_2 dz_2) \\ &= 2g\,\mathrm{Re}(\bar{z}_1 dz_1 + \bar{z}_2 dz_2) + 2ig\,\mathrm{Im}(\bar{z}_1 dz_1 + \bar{z}_2 dz_2)\,. \end{aligned} \qquad (1.280)$$

However, since $(z_1, z_2) \in S^3$, we have

$$2\,\mathrm{Re}(\bar{z}_1 dz_1 + \bar{z}_2 dz_2) = d(|z_1|^2 + |z_2|^2) = 0\,, \qquad (1.281)$$

and, hence,

$$\mathcal{A} = 2ig\,\mathrm{Im}(\bar{z}_1 dz_1 + \bar{z}_2 dz_2)\,. \qquad (1.282)$$

Finally, taking into account (1.268) one finds

$$\mathcal{A} = 2g(x_1 dx_2 - x_2 dx_1 + x_3 dx_4 - x_4 dx_3)\,, \qquad (1.283)$$

or, in terms of angle variables (θ, ϕ, ψ),

$$\mathcal{A} = ig\,(d\psi + \cos\theta\, d\phi)\,. \qquad (1.284)$$

The corresponding curvature $\mathcal{F}$ reads

$$\mathcal{F} = -ig\,\sin\theta\, d\theta \wedge d\phi\,. \qquad (1.285)$$

The canonical local sections (cf. paragraph 1.3.5) are defined as follows:

$$f_N(\theta, \phi) = (\theta, \phi, \psi = -\phi) \qquad (1.286)$$

and

$$f_S(\theta, \phi) = (\theta, \phi, \psi = \phi)\,. \qquad (1.287)$$

Hence the pulled back forms A_N and A_S are given by

$$A_N = f_N^* \mathcal{A} = -ig\,(1 - \cos\theta)d\phi\,, \qquad (1.288)$$

and

$$A_S = f_S^* \mathcal{A} = ig\,(1 + \cos\theta)d\phi\,. \qquad (1.289)$$

Interestingly, the construction of the Hopf bundle reproduces the Dirac quantization condition. Indeed, the transformation law

$$A_S = A_N + \gamma_{NS}^{-1} \cdot d\gamma_{NS}\,, \qquad (1.290)$$

---

[8] A connection one-form is defined only up to a numerical factor. To have a direct correspondence to the physical formulae we put this factor equal to '$2g$.'

with $\gamma_{NS} = e^{i\phi}$, implies that

$$g = \frac{1}{2}. \tag{1.291}$$

This proves that the Hopf bundle corresponds to the magnetic bundle with magnetic charge, or, equivalently, Chern number, $n = 1$. This example nicely illustrates how the beautiful mathematical construction fits the deep physical idea.

Finally, we shall present the formula for the connection one-form in terms of local coordinates defined by the stereographic projection (1.271). Taking into account (1.291) one finds

$$A = 2i \, \text{Im}\left(\bar{z}_2 dz_2(1 + |z|^2) + |z_2|^2 \bar{z} dz\right), \tag{1.292}$$

with $z = z_1/z_2$. Taking $z_2 = re^{i\chi}$ one easily finds

$$A = id\chi + \frac{1}{2} \frac{\bar{z}dz - zd\bar{z}}{1 + |z|^2}, \tag{1.293}$$

and, therefore,

$$A = \frac{1}{2} \frac{\bar{z}dz - zd\bar{z}}{1 + |z|^2} = i \, \text{Im}\left(\frac{\bar{z}dz}{1 + |z|^2}\right) \tag{1.294}$$

for the local connection form, and

$$F = dA = \frac{d\bar{z} \wedge dz}{(1 + |z|^2)^2} \tag{1.295}$$

for the corresponding curvature two-form. Later on, we shall meet these formulae frequently.

1.4.5   Hopf fibration $S^7 \longrightarrow S^4$

The Hopf $SU(2)$-fibration $S^7 \longrightarrow S^4$ is defined in perfect analogy to the $S^3 \longrightarrow S^2$ case by replacing complex numbers, $\mathbb{C}$, by quaternionic ones, $\mathbb{H}$ (see Appendix B for the brief introduction to quaternions). Let us define a unit sphere in $\mathbb{H}^2$:

$$S^7 = \left\{(q_1, q_2) \in \mathbb{H}^2 \,\middle|\, |q_1|^2 + |q_2|^2 = 1\right\}.$$

For any $(q_1, q_2) \in S^7$ define

$$q := q_1 q_2^{-1},$$

and identify q with a point on S^4 via a generalized stereographic projection: $S^4 \longleftrightarrow \mathbb{R}^4 \cong \mathbb{H}$, cf. Example 1.1.1. In this way we have constructed a map $S^7 \longrightarrow S^4$:

$$S^7 \ni (q_1, q_2) \longrightarrow q = q_1 q_2^{-1} \in \mathbb{H} \cong \mathbb{R}^4 \xrightarrow{\text{stereographic projection}} S^4$$

Now, observe that if $(q_1, q_2) \in S^7$, then the same is true of $(q_1 u, q_2 u)$, where $u \in \mathbb{H}$ is a unit quaternion. However, the set of unit quaternions is isomorphic with the unitary group $SU(2)$. Hence, the above map is $SU(2)$-invariant and defines a canonical projection in the Hopf principal fibre bundle $S^7 \longrightarrow S^4$.

By analogy with the Hopf bundle $S^3 \longrightarrow S^2$, we endow this bundle with the natural connection defined by the following connection one-form (cf. formula (1.282)):

$$\mathcal{A} = \mathrm{Im}(\overline{q}_1 dq_1 + \overline{q}_2 dq_2) \,. \tag{1.296}$$

Let $q_2 = qu$, with $u \in \mathbb{H}$ being a unit quaternion. Inserting q_2 into (1.296), one obtains

$$\mathcal{A} = \mathrm{Im}\left(u^{-1} du + u^{-1} \frac{\overline{q} dq}{1 + |q|^2} u \right) , \tag{1.297}$$

which is the quaternionic analog of formula (1.293). The corresponding curvature two-form $\mathcal{F}$ reads

$$\mathcal{F} = \mathrm{Im}\left(u^{-1} \frac{d\overline{q} \wedge dq}{(1 + |q|^2)^2} u \right) \,. \tag{1.298}$$

It is evident from (1.297) and (1.298) that the corresponding local forms A and F on a base manifold S^4 are given by

$$A = \mathrm{Im}\left(\frac{\overline{q} dq}{1 + |q|^2} \right) \tag{1.299}$$

and

$$F = \frac{d\overline{q} \wedge dq}{(1 + |q|^2)^2} \,, \tag{1.300}$$

respectively. It turns out that the above formulae are well-known in $SU(2)$ Yang–Mills theory. Actually, they correspond to the instanton configuration discussed in section 1.4.3. Here again we witness the elegant interrelation between mathematics and physics; the Hopf bundle $S^7 \longrightarrow S^4$ is equivalent to the instanton bundle with an instanton number, or, equivalently, Chern number, $k = -1$. We shall discuss this bundle in more detail in section 6.6.

Further reading

Section 1.1. There are several standard references devoted to classical differential geometry and topology and their applications in physics, see for example the following: classic mathematics books: Kobayashi and Nomizu 1969; Spivak 1999; Dubrovin, Fomenko and Novikov 1984; and those addressed to the physics-oriented audience: Choquet-Bruhat and DeWitt-Morette 1982; Abraham, Marsden and Ratiu 1983; Flanders 1963; Felsager 1998; Trautman 1984; Nash and Sen 1983; Nakahara 1990;

Göckeler and Schücker 1997; Schwarz 1996; and Isham 1999. See also the review article by Eguchi, Gilkey and Hanson (1980).

Section 1.2. The reader may consult Wigner 1959; Wybourne 1974; Gilmore 1974; Helgason 1978; Abraham and Marsden 1978; Marsden and Ratiu 1999; and Barut and Rączka 1980.

Section 1.3. The reader interested in a more detailed presentation is referred to the classic mathematics books by Steenrod (1999) and Husemoller (1966). We recommend also the following books written for physicists: Hermann 1970; Trautman 1984; Nash and Sen 1983; Choquet-Bruhat, DeWitt-Morette and Dillard-Bleick 1982; Göckeler and Schücker 1999; Isham 1999. See also the useful reviews by Eguchi, Gilkey and Hanson (1980) and Thomas (1980). Characteristic classes are treated in Chern 1967 and Milnor and Stasheff 1974. For a popular introduction to the theory of bundles and connections we refer the reader to the article by Bernstein and Phillips (1981).

For the geometric approach to gauge theory see, e.g., Lubkin 1963; Trautman 1970; 1979; Drechsler and Mayer 1977; Mayer 1977; Daniel and Viallet 1980; Marathe and Martucci 1989; Naber 1997, 2000; and Felsager 1998.

Section 1.4. To see how topology enters modern theoretical physics see, e.g., Schwarz 1996 and Morandi 1992. The monopole bundle is discussed in many books see, e.g., Balachandran et al. 1983, 1991; Trautman 1984; and Naber 2000. For a detailed discussion of applications of the Hopf map to monopole theory see also Ryder 1980; Minami 1979, 1980; and Aitchison 1987. We recommend also Urbantke 1991 for instructive figures. The reader interested in instanton solutions in gauge theory is referred to Coleman 1977 and Rajamaran 1982.

Problems

1.1. Verify the formula for an overlap map φ_{NS} from Example 1.1.1.

1.2. Show that if $\varphi : M \longrightarrow N$ and $\psi : N \longrightarrow P$, then

$$(\psi \circ \varphi)_* = \psi_* \circ \varphi_* ,$$

and

$$(\psi \circ \varphi)^* = \varphi^* \circ \psi^* .$$

1.3. Prove that the dimension of $\Lambda^k(\mathbb{R}^n)$ equals to $\binom{n}{k}$ by showing that a basis in $\Lambda^k(\mathbb{R}^n)$ is defined by

$$\left\{ e^{i_1} \wedge \ldots \wedge e^{i_k} \right\},$$

where $\{e_1, \ldots, e_n\}$ is a basis in $\mathbb{R}^n$, and $\{e^1, \ldots, e^n\}$ is its dual basis.

1.4. Prove that the wedge product satisfies

$$\alpha \wedge \beta = (-1)^{kl}\beta \wedge \alpha \, ,$$

where α and β are k- and l-forms, respectively.

1.5. Check the following property of the exterior derivative:

$$d(\alpha \wedge \beta) = d\alpha \wedge \beta + (-1)^{k}\alpha \wedge d\beta \, ,$$

where α is a k-form and β is of arbitrary order.

1.6. Consider $\mathbb{R}^4$ with a standard euclidean metric $g(e_i, e_j) = \delta_{ij}$. Find $\star(e^i \wedge e^j)$, the Hodge dual of $e^i \wedge e^j$, for $i, j = 1, 2, 3, 4$.

1.7. Show that there is one and only one map $d : \Lambda(M) \longrightarrow \Lambda(M)$ satisfying the following properties:

    (1) d is linear,

    (2) $d(\Lambda^k(M)) \subset \Lambda^{k+1}(M)$,

    (3) if $f \in \Lambda^0(M)$, then $df = \frac{\partial f}{\partial x^i} dx^i$,

    (4) Leibnitz rule: $d(\alpha \wedge \beta) = d\alpha \wedge \beta + (-1)^{k}\alpha \wedge d\beta \, ,$ for $\alpha \in \Lambda^k(M)$,

    (5) $d^2 = 0$.

1.8. Show that for $[\omega_1] \in H^k(M)$ and $[\omega_2] \in H^l(M)$ one has

$$[\omega_1 \wedge \omega_2] = [\omega_1] \wedge [\omega_2] \in H^{k+l}(M) \, .$$

1.9. Show that for any $v \in \mathfrak{X}(M)$, $\alpha \in \Lambda^k(M)$ and $\beta \in \Lambda^l(M)$:

$$i_v (\alpha \wedge \beta) = (i_v \alpha) \wedge \beta + (-1)^{k}\alpha \wedge (i_v \beta) \, .$$

1.10. Derive the flow property

$$F_t \circ F_s = F_s \circ F_t = F_{t+s} \, .$$

1.11. Prove the following properties of the Lie derivative:

    (1) $\mathcal{L}_X (T_1 + T_2) = \mathcal{L}_X T_1 + \mathcal{L}_X T_2 \, ,$

    (2) $\mathcal{L}_X (T_1 \otimes T_2) = (\mathcal{L}_X T_1) \otimes T_2 + T_1 \otimes \mathcal{L}_X T_2 \, ,$

for any tensor fields T_1 and T_2.

1.12. Show that if φ is a smooth map $\varphi : M \longrightarrow N$, then a Lie bracket $[\, , \,]$ satisfies

$$\varphi_*[X, Y] = [\varphi_* X, \varphi_* Y] \, ,$$

for any $X, Y \in \mathfrak{X}(M)$.

1.13. Prove that the set of vector fields on a differential manifold M endowed with the Lie bracket introduced in (1.51), defines a Lie algebra.

1.14. Let Φ be a right action of a Lie group G on a differential manifold M. Show that instead of (1.69) one obtains

$$[\mathbf{X}_\xi, \mathbf{X}_\eta] = \mathbf{X}_{[\xi,\eta]} \,, \tag{1.301}$$

i.e., the map (1.68) is a Lie algebra homomorphism.

1.15. Using a coordinate representation of the Maurer–Cartan form (1.100) show that

$$d\omega_0 = -\frac{1}{2}[\omega_0, \omega_0] \,.$$

1.16. Construct transition functions in the cotangent bundle T^*M.

1.17. Following the examples of TM and T^*M given in the text, construct a vector bundle $T^{k,l}(M)$ of (k, l)-tensors over a manifold M.

1.18. Let (E, M, π, F) be an arbitrary fibre bundle. Show that if $v \in \mathcal{X}_{ver}(E)$, then

$$[v, u] \in \mathcal{X}_{ver}(E) \,,$$

for any $u \in \mathcal{X}(E)$.

1.19. Show that the curvature two-form $\mathcal{F}$ of a general (Ehresmann) connection in the bundle $E \longrightarrow M$ satisfies

$$\mathcal{F}(u, v) = \mathrm{hor}\,[u, v] - [\mathrm{hor}\,u, \mathrm{hor}\,v] \,,$$

for any vector fields $u, v \in \mathcal{X}(E)$.

1.20. Derive Cartan's structural equation, i.e.,

$$\mathcal{F} = d\mathcal{A} + \frac{1}{2}[\mathcal{A}, \mathcal{A}] \,. \tag{1.302}$$

1.21. Show that a parallel transport operator $\mathbf{T}_\gamma$ commutes with a canonical right action of the structure group, i.e.,

$$\widetilde{R}_g \circ \mathbf{T}_\gamma = \mathbf{T}_\gamma \circ \widetilde{R}_g \,, \qquad g \in G.$$

1.22. Let D_μ denote the covariant derivative (cf. formula (1.207)) and $F_{\mu\nu}$ be components of local curvature (Yang–Mills field strength). Verify the following formulae:

(1) $[D_\mu, D_\nu] = F_{\mu\nu}$,
(2) $D_\mu F_{\nu\lambda} = \partial_\mu F_{\nu\lambda} + [A_\mu, F_{\nu\lambda}]$.

1.23. Derive the Yang–Mills equations from the variational principle, i.e., show that

$$\frac{\delta S_{\mathrm{YM}}}{\delta A_\mu} = 0 \quad \Longrightarrow \quad D_\nu F^{\nu\mu} = 0 .$$

1.24. Verify the following formula for the 3rd Chern class:

$$C_3(P) = \frac{1}{3!} \left(\frac{i}{2\pi}\right)^3 \left\{ 2\,\mathrm{Tr}\,(F \wedge F \wedge F) - 3\,\mathrm{Tr}\,(F \wedge F) \wedge \mathrm{Tr}\,F + \mathrm{Tr}\,F \wedge \mathrm{Tr}\,F \wedge \mathrm{Tr}\,F \right\} .$$

1.25. Show that a two-form B representing the magnetic field of a magnetic pole is closed on $\mathbb{R}^3 - \{0\}$.

1.26. Show that

$$\begin{aligned}
\mathrm{Tr}(F \wedge F) &= d\left\{ \mathrm{Tr}\left(dA \wedge A + \frac{2}{3} A \wedge A \wedge A \right) \right\} \\
&= d\left\{ \mathrm{Tr}\left(F \wedge A - \frac{1}{3} A \wedge A \wedge A \right) \right\} ,
\end{aligned}$$

where $F = dA + A \wedge A$ is a local curvature two-form. Note, that for an abelian group, e.g., $U(1)$, this identity is trivially satisfied.

1.27. Consider the two-dimensional Hilbert space $\mathcal{H} = \mathbb{C}^2$. Any normalized vector $\psi \in \mathbb{C}^2$ may be represented by

$$\psi = z_1|+\rangle + z_2|-\rangle ,$$

where $|\pm\rangle$ are eigenvectors of σ_3, i.e. $\sigma_3|\pm\rangle = \pm|\pm\rangle$, and $|z_1|^2 + |z_2|^2 = 1$. Prove that the Hopf map introduced in (1.270) may be defined as follows:

$$\xi_k := \langle \psi|\sigma_k|\psi\rangle , \qquad k = 1, 2, 3 ,$$

where σ_k are Pauli matrices. This shows a direct relation between the Hopf map and quantum physics.

1.28. Using formulae (1.295) and (1.300) for the local curvature two-forms, compute the Chern numbers of

(1) the Hopf $U(1)$-bundle $S^3 \longrightarrow S^2$,

(2) the Hopf $SU(2)$-bundle $S^7 \longrightarrow S^4$.

2
Adiabatic Phases in Quantum Mechanics

2.1 Adiabatic evolution in quantum mechanics

2.1.1 Adiabatic approach of Born and Fock

The notion of *adiabaticity* has played important role in the history of physics. Roughly speaking, it lies on the border of statics and dynamics, taking into account dynamical effects but in the limit of infinitely slow changes. That is, the system is no longer static but its evolution is "infinitely slow." A typical situation where one applies adiabatic ideas is when a physical system may be divided into two subsystems with completely different time scales: a so-called *slow subsystem* and *fast subsystem*.

The adiabatic theorem in nonrelativistic quantum mechanics describes the long time behavior of solutions of the Schrödinger equation with the Hamiltonian slowly evolving in time. Even though the theorem itself is rather old (it goes back to M. Born and V. Fock (1928)), its proper formulation was found years later by T. Kato (1958) in the context of perturbation theory of linear operators.

Consider a time-dependent self-adjoint Hamiltonian $H = H(t)$. To simplify our discussion let us assume that the spectrum of H is discrete and nondegenerate for all t. We then write

$$H(t)|n(t)\rangle = E_n(t)|n(t)\rangle , \tag{2.1}$$

and we choose the time-dependent eigenvectors $|n(t)\rangle$ such that

$$\langle n(t)|m(t)\rangle = \delta_{nm} . \tag{2.2}$$

Clearly, the eigenvectors $|n(t)\rangle$ are not uniquely defined. We may perform the following time-dependent phase transformations:

$$|n(t)\rangle \longrightarrow |n'(t)\rangle = e^{i\lambda_n(t)}|n(t)\rangle ,\tag{2.3}$$

with arbitrary functions $\lambda_n : \mathbb{R} \longrightarrow \mathbb{R}$. For an obvious reason (cf. section 1.3.5) one calls (2.49) a gauge transformation. It is clear that physics should be gauge invariant, i.e., physics does not depend upon the particular choice of the eigenvectors $|n(t)\rangle$. Suppose that the initial state of the system is an eigenvector of $H(0)$, i.e., $\psi(0) = |n(0)\rangle$ for some n. Then the adiabatic theorem states that if the Hamiltonian $H(t)$ changes in time "slowly enough" then to a "good approximation" $\psi(t) = e^{i\alpha(t)}|n(t)\rangle$, that is, during the adiabatic evolution $\psi(t)$ stays in the nth eigenspace of $H(t)$. In order to have a rigorous mathematical theory we show in the next section how to define "slow enough" changes and what a "good approximation" means.

The state of the system at time t may be expanded in the orthonormal base $|n(t)\rangle$:

$$\psi(t) = \sum_m c_m(t) \exp\left(-\frac{i}{\hbar}\int_0^t E_m(\tau)d\tau\right)|m(t)\rangle ,\tag{2.4}$$

where we have separated out a standard dynamical phase factor $\exp(-i\int_0^t E_n(\tau)d\tau/\hbar)$. The Schrödinger equation implies the following equations for the coefficients $c_m(t)$:

$$\begin{aligned}
\dot{c}_m(t) &= -c_m(t)\left\langle m(t)\left|\frac{d}{dt}\right|m(t)\right\rangle \\
&\quad - \sum_{k\neq m} c_k(t)\left\langle m(t)\left|\frac{d}{dt}\right|k(t)\right\rangle \exp\left[-\frac{i}{\hbar}\int_0^t (E_k(\tau)-E_m(\tau))d\tau\right] .
\end{aligned}\tag{2.5}$$

Let us observe that by differentiating the formula (2.1) with respect to time, i.e.,

$$\dot{H}|k\rangle + H|\dot{k}\rangle = \dot{E}_k|k\rangle + E_k|\dot{k}\rangle ,\tag{2.6}$$

and then multiplying by $\langle m|$, we obtain

$$\langle m|\dot{k}\rangle = \frac{1}{E_k - E_m}\langle m|\dot{H}|k\rangle , \qquad \text{for } m \neq k .\tag{2.7}$$

where, for simplicity, the explicit time dependence of H and the corresponding eigenvectors has been suppressed. Now, it is time for the adiabatic approximation: The evolution generated by $H(t)$ is considered adiabatic if

$$|\langle m|\dot{H}|k\rangle| \ll \frac{|E_k - E_m|}{\Delta T_{km}} ,\tag{2.8}$$

where ΔT_{km} is the characteristic time of transition between states k and m. It means that the changes of H are slow compared to the natural time scale of our system, as defined by the transition between energy eigenstates. Clearly, in the adiabatic limit $\Delta T_{km} \to \infty$, the changes of H are infinitely slow

$$|\langle m|\dot{H}|k\rangle| \longrightarrow 0 ,\tag{2.9}$$

and hence

$$\langle m | \dot{k} \rangle \longrightarrow 0 , \qquad \text{for } m \neq k . \tag{2.10}$$

Therefore, in the adiabatic limit the system of equations (2.5) simplifies to

$$\dot{c}_m = -c_m \langle m | \dot{m} \rangle , \tag{2.11}$$

together with an initial condition $c_m(0) = \delta_{nm}$. Now the adiabatic theorem immediately follows: $c_m(t) = 0$ for $m \neq n$ and the formula (2.4) implies

$$\psi(t) = c_n(t) \exp\left(-\frac{i}{\hbar} \int_0^t E_n(\tau) d\tau \right) | n(t) \rangle , \tag{2.12}$$

i.e., during the adiabatic evolution $\psi(t)$ remains in the nth eigenspace of $H(t)$. The final step is to calculate $c_n(t)$. It is clear from (2.11) that $c_n(t)$ is a pure phase factor

$$c_n(t) = e^{i\phi_n(t)} , \tag{2.13}$$

where the phase $\phi_n(t)$ satisfies

$$\dot{\phi}_n = i \langle n | \dot{n} \rangle . \tag{2.14}$$

For almost 50 years this additional phase $\phi_n(t)$ was completely ignored. The argument was the following: Using the gauge freedom (2.3) in choosing $| n(t) \rangle$, we may take another eigenvector $| \tilde{n}(t) \rangle$ defined by:

$$| \tilde{n}(t) \rangle = e^{i\phi_n(t)} | n(t) \rangle . \tag{2.15}$$

The transformed eigenstate satisfies

$$\left\langle \tilde{n} \left| \frac{d}{dt} \right| \tilde{n} \right\rangle = 0 . \tag{2.16}$$

The eigenvector $| \tilde{n}(t) \rangle$ satisfying the above equation is said to be in the *Born–Fock gauge*. Using $| \tilde{n} \rangle$ instead of $| n \rangle$ the formula (2.12) for $\psi(t)$ may be rewritten as follows:

$$\psi(t) = \exp\left(-\frac{i}{\hbar} \int_0^t E_n(\tau) d\tau \right) | \tilde{n}(t) \rangle , \tag{2.17}$$

i.e., without an additional phase. This way the additional phase is completely removed. However, as we shall see in Section 2.2, there are important physical situations where the procedure of removing the additional phase $\phi_n(t)$ fails. In such cases the additional phase can not be ignored and it acquires an actual physical meaning.

2.1.2 Adiabatic theorem

In the previous section we considered a special situation where the time-dependent Hamiltonian $H(t)$ possesses a discrete and nondegenerate spectrum $E_n(t)$. However, the adiabatic theorem works in much more general situations. Traditionally, this theorem is stated for Hamiltonians which fulfill the so-called gap condition, i.e., they have a part of the spectrum (a band) separated by an energy gap from the other parts. The gap condition has clear physical implications. To speak about slow changes in time one needs an intrinsic time scale to determine what slow and fast mean. In quantum mechanics the intrinsic time scale is usually determined by the energy gaps in the spectrum. If the spectrum is nondegenerate then the gap condition is automatically satisfied.

To formulate our discussion of the adiabatic theorem more precisely it is convenient to replace the physical time t by the scaled time $s = t/T$, where T is the time scale associated with the gap. In the rescaled variables, the Schrödinger equation takes the following form:

$$i\hbar\partial_s\psi_T(s) = TH(s)\psi_T(s) \, . \tag{2.18}$$

The adiabatic limit, or the limit of "infinitely slow" changes of the Hamiltonian, corresponds to the limit $T \to \infty$. Let $P(s)$ denote the finite rank projector onto one part of the spectrum of $H(s)$, separated from the rest by a gap. If we start the time evolution with a state within this part of the spectrum, i.e., $\psi_T(0) \in \text{Range } P(0)$, the adiabatic theorem tells us that the state $\psi_T(t)$, at a later time t of order T, is still within this part of the spectrum, up to a small error term, which is controlled by the time scale T and the width of the gap.

Let us construct the unitary evolution operator U_{AD} for the adiabatic evolution. From its definition, U_{AD} would have to map Range $P(0)$ onto Range $P(s)$, i.e.,

$$P(s) = U_{\text{AD}}(s)P(0)U_{\text{AD}}^{-1}(s) \, . \tag{2.19}$$

Following Kato (1958) we introduce the so-called Kato Hamiltonian

$$H_{\text{Kato}}(s, P) := \frac{i\hbar}{T} \, [\partial_s P(s), P(s)] \, . \tag{2.20}$$

Consider now the following Schrödinger equation known as the *Kato equation*:

$$\dot\psi_{\text{AD}}(s) = [\partial_s P(s), P(s)]\psi_{\text{AD}}(s) \, , \tag{2.21}$$

together with the initial condition $\psi_{\text{AD}}(0) \in \text{Range } P(0)$.

Proposition 2.1.1 *The adiabatic evolution operator U_{AD} determined by the Kato equation*

$$i\hbar\partial_s U_{\text{AD}}(s) = TH_{\text{Kato}}(s, P)U_{\text{AD}}(s) \, , \tag{2.22}$$

satisfies the intertwining condition (2.19):

$$P(s)U_{\text{AD}}(s) = U_{\text{AD}}(s)P(0) \, . \tag{2.23}$$

Proof: Following Kato (1958) (see also Avron, Seiler and Yaffe 1989) we show that both sides of (2.23) solve the same initial value problem. Since $U_{\mathrm{AD}}(0) = 1$, equation (2.23) is an identity for $s = 0$. Now, for the r.h.s. of (2.23) one has

$$\partial_s(U_{\mathrm{AD}}(s)P(0)) = [\partial_s P(s),\, P(s)](U_{\mathrm{AD}}(s)P(0)) , \tag{2.24}$$

while, for the l.h.s.,

$$\begin{aligned}
\partial_s(P(s)U_{\mathrm{AD}}(s)) &= (\partial_s P(s))U_{\mathrm{AD}}(s) + P(s)[\partial_s P(s),\, P(s)]U_{\mathrm{AD}}(s) \\
&= \left[\partial_s P(s) + P(s)(\partial_s P(s))P(s) - P^2(s)\partial_s P(s)\right]U_{\mathrm{AD}}(s) .
\end{aligned} \tag{2.25}$$

Now, since $P^2 = P$, one finds

$$\partial_s P(s) = P(s)\partial_s P(s) + (\partial_s P(s))P(s) . \tag{2.26}$$

Therefore, multiplying from the right by $P(s)$, one gets

$$(\partial_s P(s))P(s) = P(s)(\partial_s P(s))P(s) + (\partial_s P(s))P(s) , \tag{2.27}$$

and, hence,

$$P(s)(\partial_s P(s))P(s) = 0 . \tag{2.28}$$

Therefore, formula (2.25) reduces to

$$\partial_s(P(s)U_{\mathrm{AD}}(s)) = \left[\partial_s P(s) - P(s)\partial_s P(s)\right]U_{\mathrm{AD}}(s) . \tag{2.29}$$

Now, using equation (2.26) once more, the above formula finally gives

$$\partial_s(P(s)U_{AD}(s)) = [\partial_s P(s),\, P(s)](P(s)U_{AD}(s)) , \tag{2.30}$$

which together with (2.24) proves (2.23). $\qquad\qquad\Box$

For the real evolution the intertwining condition no longer holds. The adiabatic theorem tells us how much this condition is violated.

Theorem 2.1.2 (Adiabatic theorem) *Let $H(s)$ be a smooth one-parameter family of Hamiltonians and let $U_T(s)$ be the physical evolution parametrized in the rescaled time $s = t/T$, i.e. $U_T(s)$ is a solution of*

$$i\hbar \dot{U}_T(s) = T H(s)U_T(s) .$$

Let $P(s)$ be a family of finite rank projectors onto the band of the spectrum separated by the gap. Then

$$U_T(s)P(0)U_T^{-1}(s) = P(s) + O(T^{-1}) . \tag{2.31}$$

The size of the error term depends on the size of the gap and the time scale T.

Clearly, in the adiabatic $T \to \infty$ limit, $U_T \longrightarrow U_{\mathrm{AD}}$.

2.1.3 Adiabaticity and geometry

In physical applications the time-dependence usually enters into the physical Hamiltonian via time-dependence of some external parameters (potentials, forces, electromagnetic fields, etc.). Suppose that these external parameters parametrize some manifold M, i.e., we have a family of systems parametrized by points of M:

$$M \ni x \longrightarrow H(x) .$$

Let $P(x)$ denote a finite rank projector onto part of the spectrum of $H(x)$ (as before, we assume that this part is separated by a gap). We define the following vector bundle. Denote by $\mathcal{H}$ the Hilbert space of the system in question. Take a trivial bundle $\mathcal{H} \times M$ and project out the sub-bundle $\mathcal{H}_P$ whose fibres are the vector subspaces Range $P(x)$ for $x \in M$, i.e., the fibre at x is defined by

$$F_x := \text{Range } P(x) \subset \mathcal{H} . \tag{2.32}$$

A fibre bundle

$$\mathcal{H}_P := \bigcup_{x \in M} F_x \tag{2.33}$$

is called a *spectral bundle*. Let us observe that the adiabatic evolution generated by the Kato Hamiltonian has a nice geometric interpretation. Indeed, the formula (2.28) implies that

$$(\partial_s P)P = P^\perp (\partial_s P) P , \tag{2.34}$$

$$P \partial_s P = P(\partial_s P) P^\perp , \tag{2.35}$$

where $P^\perp = \mathbb{1} - P$, and for simplicity we omit the argument of P. Hence, the Kato equation for ψ may be transformed as follows:

$$\begin{aligned} \partial_s \psi &= ((\partial_s P)P - P \partial_s P)\psi \\ &= (P^\perp(\partial_s P)P - P(\partial_s P)P^\perp)\psi . \end{aligned} \tag{2.36}$$

Therefore, if initially ψ belongs to the range of P, then according to the adiabatic theorem it stays in the range of P for later times, or, equivalently, $P^\perp \psi = 0$. Hence

$$\partial_s \psi = P^\perp(\partial_s P)P\psi = P^\perp(\partial_s P)\psi , \tag{2.37}$$

since $P\psi = \psi$. This means that $\partial_s \psi \in \text{Range } P^\perp$, or, equivalently, $P\partial_s \psi = 0$. This condition may be rewritten as

$$P d\psi = 0 , \tag{2.38}$$

where d denotes exterior differentiation on M, i.e.,

$$d\psi := \partial_s \psi ds = \sum_{k=1}^{n} \frac{\partial \psi}{\partial x^k} dx^k ,$$

where $(x^1, \ldots, x^n)$ are local coordinates on M. The operator

$$\nabla := P d \tag{2.39}$$

defines the covariant derivative in the vector bundle $\mathcal{H}_P$. It is evident that ∇ is $\mathbb{C}$-linear, i.e.,

$$\nabla(\lambda_1 \psi_1 + \lambda_2 \psi_2) = \lambda_1 \nabla \psi_1 + \lambda_2 \nabla \psi_2 , \tag{2.40}$$

for any $\lambda_1, \lambda_2 \in \mathbb{C}$, and it fulfills the Leibnitz rule,

$$\nabla(f \psi) = (df)\psi + f \nabla \psi , \tag{2.41}$$

for any $f \in C^\infty(M)$. Indeed, one has

$$
\begin{aligned}
\nabla(f\psi) &= P d(f\psi) = P(df \cdot \psi) + P(f d\psi) \\
&= (df)P\psi + f P d\psi = (df)\psi + f\nabla\psi ,
\end{aligned}
\tag{2.42}
$$

Since $P\psi = \psi$. Therefore, the adiabatic evolution $\nabla \psi = 0$ (cf. equation (2.38)) defines a parallel transport of the vector $\psi(s)$ in the bundle $\mathcal{H}_P$ along a curve in the base manifold M. This parallel transport is realized via the adiabatic time evolution operator $U_{AD}(s)$.

2.2   Berry's phase

2.2.1   Phase in quantum mechanics

In the standard approach to quantum mechanics, pure quantum states are represented by vectors in a complex Hilbert space $\mathcal{H}$. Each vector $\psi \in \mathcal{H}$ describes a state by the collection of expectation values

$$A \longrightarrow \frac{\langle \psi | A | \psi \rangle}{\langle \psi | \psi \rangle} , \tag{2.43}$$

where A is a self-adjoint operator in $\mathcal{H}$ representing some physical quantity. For this reason, two vectors ψ and φ describe the same physical state if and only if they are linearly dependent, i.e., $\psi = \lambda\varphi$, with $\lambda \in \mathbb{C}$. If we normalize the state vector, e.g., choose $\langle \psi | \psi \rangle = 1$, there is still a freedom to choose an overall phase factor $e^{i\alpha}$. Two normalized state vectors ψ and φ are physically equivalent:

$$\psi \sim \varphi \iff \psi = e^{i\alpha}\varphi .$$

Hence, one usually says that the above phase factor, $e^{i\alpha}$, has no physical meaning. For this reason we may equivalently represent (pure) quantum states as one-dimensional projectors in $\mathcal{H}$:

$$\psi \longrightarrow P_\psi := |\psi\rangle\langle\psi| ;$$

clearly,

$$\psi \sim \varphi \iff P_\psi = P_\varphi \,.$$

However, we know that it is the phase that controls the key effect of quantum mechanics — quantum interference. This effect is governed by the *relative phase*. If we have two (normalized) state vectors ψ and φ such that $\psi = e^{i\alpha}\varphi$, then one calls α a relative phase between ψ and φ. The relative phase, or, equivalently, the phase difference, does have physical meaning, and hence may be measured. Superposition of two states $\psi \sim \varphi$ differing in phase by α leads to the following interference formula:

$$I \propto |1 + e^{i\alpha}|^2 = 2(1 + \cos\alpha) = 4\cos^2(\alpha/2) \,, \tag{2.44}$$

which enables one to measure α. We stress the crucial difference between overall and relative phases. In the interference experiment the overall phases of ψ and φ are still unknown and are not important. It is evident that $e^{i\lambda}\psi$ and $e^{i\lambda}\varphi$ will produce the same interference as ψ and φ. Only the relative phase counts.

2.2.2 Standard derivation

Consider a curve C on a manifold of external parameters M:

$$t \longrightarrow x_t \in M,$$

and the adiabatic evolution of the quantum system described by the parameter dependent Hamiltonian $H = H(x)$ along the curve C. Then the Hamiltonian depends on time solely via the time-dependence of the external parameters: $H(t) = H(x_t)$. Suppose that for any $x \in M$ the Hamiltonian $H(x)$ has a purely discrete spectrum, i.e.,

$$H(x)|n(x)\rangle = E_n(x)|n(x)\rangle \,, \tag{2.45}$$

with

$$\langle n(x)|m(x)\rangle = \delta_{nm} \,. \tag{2.46}$$

Moreover, let us assume that the eigenvectors $|n(x)\rangle$ are single-valued (as functions of $x \in M$), that is, we assume the existence of the map

$$M \ni x \longrightarrow |n(x)\rangle \in \mathcal{H} \,, \tag{2.47}$$

with $\mathcal{H}$ being the system's Hilbert space. Obviously, this map need not be defined globally on M. Therefore, we assume its existence only locally. To find the adiabatic evolution we may apply the "adiabatic machinery" of the previous section. Let us assume that the nth eigenvalue $E_n(x)$ is nondegenerate and let $P_n(x) := |n(x)\rangle\langle n(x)|$ be the corresponding one-dimensional projector onto the nth eigenspace $\mathcal{H}_n(x)$, which we write

$$\mathcal{H}_n(x) := \text{Range } P_n(x) = \{\, \alpha\,|n(x)\rangle \mid \alpha \in \mathbb{C} \,\} \,. \tag{2.48}$$

The eigenvectors $|n(x)\rangle$ are not uniquely defined by (2.45) and (2.46); one may arbitrarily change the phase of $|n(x)\rangle$:

$$|n(x)\rangle \longrightarrow |n'(x)\rangle = e^{i\alpha_n(x)}|n(x)\rangle \,, \tag{2.49}$$

where $\alpha_n : M \longrightarrow \mathbb{R}$. Obviously, the phase transformation (2.49) does not change $P_n(x)$. Suppose that $\psi(0) = |n(x_0)\rangle$. Due to the adiabatic theorem, $\psi(t)$ stays in nth eigenspace of $H(x_t)$ during the adiabatic evolution, i.e.,

$$\psi(t) \in \mathcal{H}_n(x_t) \,. \tag{2.50}$$

Therefore, if the evolution is cyclic, i.e., a curve C is closed ($x_0 = x_T$ for some $T > 0$), then $\psi(0)$ and $\psi(T)$ both belong to $\mathcal{H}_n(x_0)$ and hence they may differ only by a phase factor:

$$\psi(T) = e^{i\gamma}\,\psi(0) \,. \tag{2.51}$$

The obvious guess for the phase γ would be

$$\gamma = -\frac{1}{\hbar}\int_0^T E_n(t)\,dt \,, \tag{2.52}$$

but, as was shown by Berry (Berry 1984), it is wrong! There is an additional component that has a purely geometric origin. It depends upon the geometry of the manifold M and the circuit C itself. To find it let us note that, due to (2.12) and (2.13), $\psi(t)$ and $|n(x_t)\rangle$ differ by a time-dependent phase factor:

$$\psi(t) = \exp\left(-\frac{i}{\hbar}\int_0^t E_n(\tau)d\tau\right) e^{i\phi_n(t)}|n(x_t)\rangle \,, \tag{2.53}$$

where the Schrödinger equation implies the following equation for the function ϕ_n:

$$\dot{\phi}_n = i\langle n|\dot{n}\rangle \,, \tag{2.54}$$

(for simplicity we have omitted the argument of $|n\rangle$). The last formula defines the following one-form on M:

$$A^{(n)} := i\langle n|dn\rangle \,, \tag{2.55}$$

or, in local coordinates $(x^1,\dots,x^n)$, $A^{(n)} = A_k^{(n)}dx^k$, with

$$A_k^{(n)} := i\langle n|\partial_k n\rangle \,.$$

Note that since $\langle n|dn\rangle$ is purely imaginary, the formula (2.55) may be rewritten as follows:[1]

$$A^{(n)} = -\text{Im}\,\langle n|dn\rangle \,. \tag{2.56}$$

---

[1] Our definition of $A^{(n)}$ agrees with that of Berry (1984). There is another convention in which A is not real but purely imaginary, i.e., $A \in u(1) \cong i\mathbb{R}$. However, most authors define A as a real quantity.

One solves (2.54) by simple integration:

$$\phi_n(t) = i \int_0^t \langle n(\tau) | \dot{n}(\tau) \rangle \, d\tau = \int_C A^{(n)} \, , \tag{2.57}$$

where one integrates the one-form $A^{(n)}$ along the curve C between x_0 and x_t. This shows that the formula (2.52) has to be supplemented by the following geometric quantity:

$$\gamma_n(C) := \phi_n(T) = \oint_C A^{(n)}, \tag{2.58}$$

i.e., the total phase shift γ splits into two parts:

$$\gamma = \underbrace{-\frac{1}{\hbar} \int_0^T E_n(\tau) \, d\tau}_{\text{dynamical phase}} + \underbrace{\gamma_n(C)}_{\text{geometric phase}} \, . \tag{2.59}$$

This geometric quantity $\gamma_n(C)$ defines the celebrated *Berry phase*, corresponding to the cyclic adiabatic evolution along C. Using the Stokes theorem one may rewrite Berry's phase as

$$\gamma_n(C) = \int_\Sigma F^{(n)} \, , \tag{2.60}$$

where Σ is an arbitrary two-dimensional submanifold in M such that $\partial \Sigma = C$, and

$$F^{(n)} = d A^{(n)} = -\mathrm{Im} \, \langle \, dn | \wedge | dn \, \rangle \, . \tag{2.61}$$

In local coordinates $(x^1, \ldots, x^n)$ on M one finds that

$$F^{(n)} = \frac{1}{2} F^{(n)}_{ij} dx^i \wedge dx^j \, ,$$

with

$$F^{(n)}_{ij} = -\mathrm{Im} \left(\langle \, \partial_i n | \partial_j n \, \rangle - \langle \, \partial_j n | \partial_i n \, \rangle \right) \, . \tag{2.62}$$

Note that in performing a phase transformation (2.49), the quantity $A^{(n)}$ transforms according to

$$A^{(n)} \longrightarrow A'^{(n)} = A^{(n)} - d\alpha_n \, , \tag{2.63}$$

or, using component notation,

$$A'^{(n)}_k = A^{(n)}_k - \partial_k \alpha_n \, , \tag{2.64}$$

i.e., in the same way as a vector potential in classical electrodynamics. For this reason it is usually called the *Berry vector potential* (or rather Berry's potential one-form). Now, since $d^2\alpha_n = 0$, the two-form $F^{(n)}$ is perfectly gauge invariant, and due to (2.60), so is the Berry phase $\gamma_n(C)$. The quantity $F^{(n)}$ plays a role of a magnetic field for the potential $A^{(n)}$, and equation (2.60) shows that the Berry phase $\gamma_n(C)$ is an analog of the magnetic flux in the electromagnetic theory:

$$\text{Berry's phase } \gamma_n(\partial\Sigma) = \text{flux of } F^{(n)} \text{ through } \Sigma .$$

Finally, let us express $F^{(n)}$ in terms of the energy eigenvalues E_k. Inserting into (2.61) the complete basis system

$$\mathbb{1} = \sum_m |m\rangle\langle m| , \tag{2.65}$$

one obtains

$$\begin{aligned}
F^{(n)} &= -\text{Im} \sum_m \langle dn|m\rangle \wedge \langle m|dn\rangle \\
&= -\text{Im} \sum_{m\neq n} \langle dn|m\rangle \wedge \langle m|dn\rangle \tag{2.66} \\
&= -\text{Im} \sum_{m\neq n} \overline{\langle m|d|n\rangle} \wedge \langle m|d|n\rangle .
\end{aligned}$$

The term with $m = n$ drops out from (2.66) since $\langle n|dn\rangle$ is purely imaginary. Now, the eigenvalue equation (2.45) implies that for $m \neq n$,

$$\langle m|d|n\rangle = \frac{\langle m|dH|n\rangle}{E_n - E_m} , \tag{2.67}$$

which, together with (2.66), gives

$$F^{(n)} = -\text{Im} \sum_{m\neq n} \frac{\langle n|dH|m\rangle \wedge \langle m|dH|n\rangle}{(E_m - E_n)^2} . \tag{2.68}$$

The above formula shows that $F^{(n)}$ is singular at all points $x^* \in M$ where the spectrum of $H(x^*)$ has accidental degeneracies, i.e., $E_n(x^*) = E_{n\pm1}(x^*)$. This observation will play an important role in what follows.

Proposition 2.2.1 *The two-form $F^{(n)}$ satisfies the following property:*

$$\sum_n F^{(n)} = 0 . \tag{2.69}$$

Proof. Using formula (2.66) one finds that

$$\begin{aligned}
F_{ij}^{(n)} &= -\text{Im} \sum_{m\neq n} \left(\langle \partial_i n|m\rangle\langle m|\partial_j n\rangle - (i \rightleftharpoons j)\right) \\
&= -\frac{1}{2} \text{Im} \sum_{m\neq n} \left(\langle \partial_i n|m\rangle\langle m|\partial_j n\rangle + \langle \partial_j m|n\rangle\langle n|\partial_i m\rangle - (i \rightleftharpoons j)\right) ,
\end{aligned}$$

since

$$\langle \partial_i n | m \rangle = -\langle n | \partial_i m \rangle .$$

It is clear, therefore, that

$$\sum_n F_{ij}^{(n)} = 0 , \tag{2.70}$$

since the formula $\sum_n F_{ij}^{(n)}$ is both symmetric and antisymmetric in i and j. $\qquad\square$

2.2.3 How to measure the Berry phase

If the quantum evolution is cyclic, then one can measure the relative phase between $\psi(0)$ and $\psi(T)$. Suppose that the system in question, prepared in the state $|n\rangle$, is split at $t = 0$ into two subsystems. One of them is adiabatically cycled and the other is not. During the evolution both subsystems will acquire dynamical phases φ_1 and φ_2, respectively. However, the cycled system, in addition, will gain a geometric Berry phase $\gamma_n(C)$ depending on the cycle, C, followed in the parameter space M. Now, if the subsystems are recombined at $t = T$ (the period of the cycle) then the intensity of their superposition is given by

$$\begin{aligned} I^2 \quad &\propto \quad \left| \exp[i(\varphi_1 + \gamma_n(C)] + \exp(i\varphi_2) \right|^2 \\ &= \quad 4\cos^2\left(\frac{1}{2}\left[\varphi_1 - \varphi_2 + \gamma_n(C)\right]\right) . \end{aligned} \tag{2.71}$$

Therefore, by knowing the dynamical phases φ_1 and φ_2 we may detect the geometric phase $\gamma_n(C)$ as a shift in the interference pattern.

This kind of interferometric experiment involves the same (initial) state and two Hamiltonians (one for each subsystem). A second class of experiments involves two (or more) states and the same Hamiltonian. Suppose that at $t = 0$ the initial state is a superposition

$$\psi(0) = a_n |n\rangle + a_m |m\rangle , \tag{2.72}$$

where $|n\rangle$ and $|m\rangle$ are eigenvectors of $H(0)$. Now, if H changes adiabatically and $H(T) = H(0)$, then, using an obvious notation, one obtains

$$\psi(T) = a_m \exp[i(\varphi_m + \gamma_m(C))] |m\rangle + a_n \exp[i(\varphi_n + \gamma_n(C))] |n\rangle . \tag{2.73}$$

Now, let A be a self-adjoint observable which does not commute with $H(0)$. Then one easily finds that

$$\begin{aligned} \langle \psi(T)|A|\psi(T)\rangle = &|a_m|^2 \langle m|A|m\rangle + |a_n|^2 \langle n|A|n\rangle \\ &+ \ 2\,\mathrm{Re}\left(a_m a_n^* \langle n|A|m\rangle \exp[i(\varphi_n - \varphi_m + \gamma_n(C) - \gamma_m(C))]\right) . \end{aligned} \tag{2.74}$$

Clearly, if A commutes with $H(0)$, then $\langle n|A|m\rangle \sim \delta_{nm}$ and hence the interference term vanishes for $m \neq n$. In this type of experiment we may detect the difference of geometric phases, $\gamma_n(C) - \gamma_m(C)$, provided we know the corresponding dynamical phases φ_n and φ_m.

2.2.4 Berry–Simon connection

Just after M. Berry derived his celebrated formula, B. Simon (Simon 1983) observed that the quantal adiabatic Berry phase has an elegant mathematical interpretation as the holonomy[2] of a certain connection in the appropriate fibre bundle. It is a special case of a general situation described in section 2.1.3, where we observed that adiabatic evolution may be given a purely geometric interpretation.

Let us construct the nth spectral bundle: the base space is a manifold of external parameters M and the fibre at $x \in M$ is a complex line $\mathcal{H}_n(x)$, i.e., the nth eigenspace of $H(x)$. We assume that the nth eigenvalue of the Hamiltonian is nondegenerate. The case of degeneracy will be analyzed in section 2.3. Restricting ourselves to normalized vectors in $\mathcal{H}_n$, we may equivalently consider a fibre at x defined by

$$F_x := \left\{ \, e^{i\alpha} |n(x)\rangle \, \Big| \, \alpha \in \mathbb{R} \, \right\} \cong U(1) \,. \tag{2.75}$$

This construction gives rise to the principal $U(1)$-bundle $(P, M, U(1))$ with the total space

$$P = \bigcup_{x \in M} F_x \,.$$

Having defined a bundle one then needs a connection. It turns out that there exists a *mathematically natural* connection on $(P, M, U(1))$. Note that it is *natural* to call a vector $|h\rangle$ a horizontal one if it is orthogonal (in the sense of the scalar product in $\mathcal{H}$) to the corresponding fibre F_x, i.e., if

$$\langle n(x)|h\rangle = 0 \,. \tag{2.76}$$

A connection defined in this way is usually called a *Berry–Simon connection*. Let us see what is the physical meaning of this connection. Consider a curve C in M and let $t \longrightarrow \psi(t)$ be a horizontal lift of C with respect to the Berry–Simon connection, i.e.,

$$\langle n|\dot{\psi}\rangle = 0 \,. \tag{2.77}$$

The above formula is a special case of (2.38). Indeed, it may be rewritten as

$$\langle n|d\psi\rangle = 0 \,, \tag{2.78}$$

which is equivalent to

$$P_n d\psi = 0 \,, \tag{2.79}$$

with $P_n = |n\rangle\langle n|$. In particular, let us note that if $|n\rangle$ is in the Born–Fock gauge, i.e., $\langle n|\dot{n}\rangle = 0$, then the curve $t \longrightarrow |n(x_t)\rangle$ defines a horizontal lift of the original

---

curve C in the parameter space M. Equivalently, a family $|n(x)\rangle$ is in the Born–Fock gauge if, for any curve $t \longrightarrow x_t$ in M, the corresponding eigenvector $|n(x_t)\rangle$ defines a parallel transport of $|n\rangle$.

In this way, we establish a striking correspondence between the physical notion of adiabatic evolution and the purely mathematical concept of parallel transport with respect to the natural Berry–Simon connection:

$$\text{adiabatic evolution} \quad \longleftrightarrow \quad \text{parallel transport} .$$

Now, let us turn to a $u(1)$-valued connection form. If

$$M \ni x \longrightarrow |n(x)\rangle \in \mathcal{H}_n(x)$$

is a local section of the nth spectral bundle, then a local connection form (in a gauge $|n(x)\rangle$) is given by a $u(1)$-valued one-form

$$i A^{(n)} = -\langle n|dn\rangle ,$$

that is, a *Berry–Simon* connection one-form is (up to a factor of i) equal to a Berry potential one-form. Hence, the Berry phase factor corresponding to the closed curve C corresponds to an element from the holonomy group of the Berry–Simon connection:

$$\text{Berry's phase factor } e^{\gamma_n(C)} = \text{holonomy of } C .$$

One obviously has

$$e^{i\gamma_n(C)} \;=\; \exp\left(i \oint_C A^{(n)} \right) \;=\; \exp\left(i \int_\Sigma F^{(n)} \right) , \tag{2.80}$$

where Σ is any two-dimensional region such that $\partial\Sigma = C$.

2.2.5 Examples

Now, it is time to illustrate the appearance of Berry's phase in physical systems. We shall present the spinning quantum particle interacting with a slowly-varying magnetic field, and the parameter-dependent harmonic oscillator.

Example 2.2.1 (Spin-half in a magnetic field) Consider an adiabatic evolution of the spin-half particle in a slowly-varying magnetic field $\mathbf{B}$. It turns out that this system naturally leads to the celebrated monopole bundle (see Section 1.4.2) or, equivalently, to the Hopf fibration $S^3 \longrightarrow S^2$. To show this, let us note that the corresponding Hamiltonian is given by

$$H(\mathbf{B}) = \frac{1}{2} \mu \boldsymbol{\sigma} \cdot \mathbf{B} , \tag{2.81}$$

where $\boldsymbol{\sigma} = (\sigma_1, \sigma_2, \sigma_3)$ is a three-vector of Pauli matrices. The magnetic field $\mathbf{B} \in \mathbb{R}^3$ plays the role of the external parameter. To find the eigenvalues and eigenvectors of the instantaneous Hamiltonian let us write the magnetic field in spherical coordinates, as

$$\mathbf{B} = B(\sin\theta\cos\varphi,\ \sin\theta\sin\varphi,\ \cos\theta)\ .$$

Clearly, if $\mathbf{B} = B\mathbf{e}_3$, then the corresponding eigenvalue problem is solved by

$$H(B\mathbf{e}_3)|\pm(\mathbf{e}_3)\rangle = E_\pm(B)|\pm(\mathbf{e}_3)\rangle\ , \tag{2.82}$$

with

$$E_\pm(B) = \pm\frac{1}{2}\mu B\ , \tag{2.83}$$

and

$$|+(\mathbf{e}_3)\rangle = \begin{pmatrix} 1 \\ 0 \end{pmatrix},\qquad |-(\mathbf{e}_3)\rangle = \begin{pmatrix} 0 \\ 1 \end{pmatrix}. \tag{2.84}$$

Now, if an arbitrary $\mathbf{B}$ is parametrized by spherical angles (θ,φ), then it may be obtained from $\mathbf{e}_3$ by the following $SO(3)$ rotation:

$$R(\theta,\varphi) = R_3(\varphi)\cdot R_2(\theta)\cdot R_3(-\varphi)\ , \tag{2.85}$$

where $R_k(\alpha)$ denotes the rotation about the k-axis by an angle α.[3] The corresponding unitary operator acting in $\mathbb{C}^2$ is given by

$$U(\theta,\varphi) = U_3(\varphi)\cdot U_2(\theta)\cdot U_3(-\varphi)\ , \tag{2.86}$$

where

$$U_k(\alpha) = e^{-i\alpha J_k}\ , \tag{2.87}$$

and $J_k = \frac{1}{2}\sigma_k$ are the $so(3)\cong su(2)$ generators. Let $|\pm(\mathbf{B})\rangle \equiv |\pm(\theta,\varphi)\rangle$ denote eigenvectors of the original Hamiltonian. Using a spherical parametrization,

$$|\pm\mathbf{B}\rangle = |\pm(\theta,\varphi)\rangle = U(\theta,\varphi)|\pm(\mathbf{e}_3)\rangle\ . \tag{2.88}$$

Clearly,

$$H(\mathbf{B})|\pm(\theta,\varphi)\rangle = E_\pm(B)|\pm(\theta,\varphi)\rangle\ , \tag{2.89}$$

---

[3]They are given by

$$R_3(\varphi) = \begin{pmatrix} \cos\varphi & -\sin\varphi & 0 \\ \sin\varphi & \cos\varphi & 0 \\ 0 & 0 & 1 \end{pmatrix},\qquad R_2(\theta) = \begin{pmatrix} \cos\theta & 0 & \sin\theta \\ 0 & 1 & 0 \\ -\sin\theta & 0 & \cos\theta \end{pmatrix}.$$

Actually, the rotation $R_3(-\varphi)$ acts trivially on $\mathbf{e}_3$. However, we included it in the definition of $R(\theta,\varphi)$ in order to have $R(0,\varphi)\mathbf{e}_3 = \mathbf{e}_3$.

with $E_\pm(B)$ given by (2.83). Now, using the well-known formula

$$e^{i\alpha\mathbf{n}\cdot\boldsymbol{\sigma}} = \cos\alpha + i\mathbf{n}\cdot\boldsymbol{\sigma}\sin\alpha \,, \tag{2.90}$$

with $\mathbf{n}$ being a unit vector in $\mathbb{R}^3$, one easily finds that

$$|+(\theta,\varphi)\rangle = \left|\begin{array}{c}\cos\frac{\theta}{2} \\ e^{i\varphi}\sin\frac{\theta}{2}\end{array}\right\rangle \,, \quad |-(\theta,\varphi)\rangle = \left|\begin{array}{c}-\sin\frac{\theta}{2} \\ e^{i\varphi}\cos\frac{\theta}{2}\end{array}\right\rangle \,. \tag{2.91}$$

Note that the eigenvectors $|\pm\rangle$ do not depend on $|\mathbf{B}|$, and, therefore, it is natural to take as the parameter space a two-dimensional sphere $M = S^2$. Obviously, the parametrization of $|\pm\rangle$ defined in (2.91) is not global. For $\theta = 0$ one finds

$$|+(0,\varphi)\rangle = |+(\mathbf{e}_3)\rangle \,,$$

whereas

$$|-(0,\varphi)\rangle = e^{i\varphi}|-(\mathbf{e}_3)\rangle \,,$$

which shows that at the north pole, $(0, 0, B)$, the eigenvector $|-(\mathbf{e}_3)\rangle$ evaluates via $U(\theta,\varphi)$ to the whole family of vectors $e^{i\varphi}|-(\mathbf{e}_3)\rangle$. Similarly, at $\theta = \pi$,

$$|-(\pi,\varphi)\rangle = |-(\mathbf{e}_3)\rangle \,,$$

while

$$|+(\pi,\varphi)\rangle = e^{i\varphi}|-(\mathbf{e}_3)\rangle \,,$$

which shows that $|+(\pi,\varphi)\rangle$ is ill-defined at the south pole, $(0, 0, -B)$. Hence we have two well-defined maps:

$$|+\rangle \,:\, S^2 - \{(0, 0, -B)\} \longrightarrow S^3 \subset \mathbb{C}^2 \,, \tag{2.92}$$

and

$$|-\rangle \,:\, S^2 - \{(0, 0, B)\} \longrightarrow S^3 \subset \mathbb{C}^2 \,. \tag{2.93}$$

Note that the point $\mathbf{B} = 0$ corresponds to a degeneracy of the spectrum, i.e., $E_+(0) = E_-(0) = 0$, but it does not belong to our parameter space $M = S^2$.

Now, inserting formulae (2.91) for $|\pm(\theta,\varphi)\rangle$ into (2.55) one obtains the following formulas for the Berry–Simon connection:

$$\begin{aligned}
A^{(+)} &= i\langle +(\theta,\varphi)|d|+(\theta,\varphi)\rangle \\
&= i\langle +(\theta,\varphi)|\partial_\theta|+(\theta,\varphi)\rangle d\theta + i\langle +(\theta,\varphi)|\partial_\varphi|+(\theta,\varphi)\rangle d\varphi \\
&= -\frac{1}{2}(1-\cos\theta)d\varphi \,,
\end{aligned} \tag{2.94}$$

a)

b)

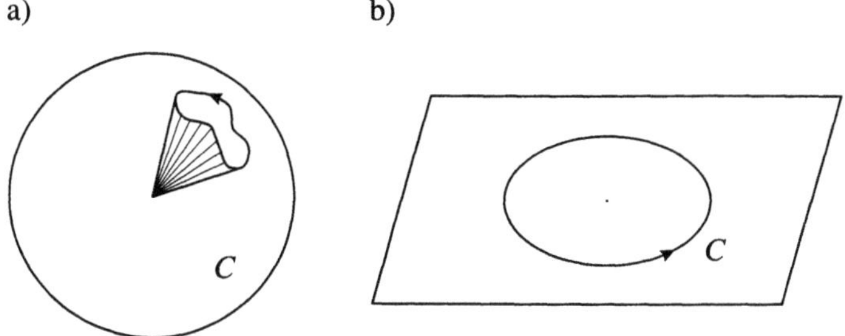

Figure 2.1: a) Solid angle $\Omega(C)$; b) a circuit around degeneracy $\mathbf{B} = 0$

and

$$A^{(-)} = i\langle -(\theta, \varphi)|d| - (\theta, \varphi)\rangle = \frac{1}{2}(1 + \cos\theta)d\varphi . \tag{2.95}$$

Obviously, $A^{(+)}$ and $A^{(-)}$ correspond to the potential one-forms of the magnetic monopole of strength $g = \mp 1/2$ placed at the origin $\mathbf{B} = 0$, cf. formulae (1.243) and (1.244). The corresponding Berry curvature reads:

$$F^{(+)} = -\frac{1}{2}\sin\theta\, d\theta \wedge d\varphi \tag{2.96}$$

and

$$F^{(-)} = \frac{1}{2}\sin\theta\, d\theta \wedge d\varphi , \tag{2.97}$$

on the appropriate patches of M. Clearly

$$F^{(+)} + F^{(-)} = 0 , \tag{2.98}$$

in agreement with (2.69). Finally, one obtains the following formula for the Berry phase for a spin-half particle:

$$\gamma_\pm(C) = \int_\Sigma F^{(\pm)} = \mp\frac{1}{2}\Omega(C) , \tag{2.99}$$

where $\Omega(C)$ is the solid angle subtended by C on the sphere S^2 — see Fig. 2.1. In particular, for $\theta = \frac{\pi}{2}$, i.e., if the circuit C stays in the xy-plane and encloses the degeneracy point $\mathbf{B} = 0$ (see Fig. 2.1), we have

$$\gamma_\pm(C) = \mp\pi , \tag{2.100}$$

and, therefore,

$$e^{i\gamma_\pm(C)} = -1 , \tag{2.101}$$

i.e., the wave function of a spin-half particle changes sign after coming back to the initial point of the parameter manifold.

In this way we have constructed two monopole bundles characterized by magnetic number or, equivalently, Chern number:[4]

$$\frac{1}{2\pi} \int_{S^2} F^{(\pm)} = \mp 1 \, . \tag{2.102}$$

The same result follows directly from the formula (2.68):

$$
\begin{aligned}
F^{(+)} &= \frac{-1}{(\mu B)^2} \, \mathrm{Im} \, \langle +|dH|-\rangle \wedge \langle -|dH|+\rangle \\
&= \frac{-1}{B^2} \sum_{k,l=1}^{3} \mathrm{Im} \left[\langle +|\sigma_k|-\rangle \langle -|\sigma_l|+\rangle - (k \rightleftharpoons l) \right] dB^k \wedge dB^l \, , \tag{2.103}
\end{aligned}
$$

where for simplicity we have omitted the argument of $|\pm\,\rangle$, and used

$$dH = \frac{\mu}{2} \sum_{k=1}^{3} \sigma_k \, dB^k \, .$$

Let us compute the above expression at the point $\mathbf{B} = (0, 0, B) = B\mathbf{e}_3$. Taking into account the following:

$$\mathrm{Im} \left(\langle +(\mathbf{e}_3)|\sigma_2| - (\mathbf{e}_3)\rangle \langle -(\mathbf{e}_3)|\sigma_3| + (\mathbf{e}_3)\rangle \right) = 0 \, ,$$

$$\mathrm{Im} \left(\langle +(\mathbf{e}_3)|\sigma_3| - (\mathbf{e}_3)\rangle \langle -(\mathbf{e}_3)|\sigma_1| + (\mathbf{e}_3)\rangle \right) = 0 \, ,$$

$$\mathrm{Im} \left(\langle +(\mathbf{e}_1)|\sigma_1| - (\mathbf{e}_3)\rangle \langle -(\mathbf{e}_3)|\sigma_2| + (\mathbf{e}_3)\rangle \right) = 1 \, ,$$

one easily finds that

$$F_{12}^{(+)} = -\frac{1}{2B^2} \, , \qquad F_{13}^{(+)} = F_{23}^{(+)} = 0 \, . \tag{2.104}$$

Hence, for an arbitrary point $\mathbf{B}$ one has

$$F_{kl}^{(+)} = \frac{-1}{2|\mathbf{B}|^3} \, \epsilon_{klm} B^m \, , \tag{2.105}$$

which reproduces the field of a magnetic pole with $g = -1/2$. $\qquad\qquad \diamond$

Example 2.2.2 (Arbitrary spin J) Let us study the generalization of the previous example to a particle having an arbitrary spin J. The Hamiltonian is then given by

$$H(\mathbf{B}) = \mu \mathbf{J} \cdot \mathbf{B} \, , \tag{2.106}$$

---

[4]Note that F is a real quantity and hence, contrary to (1.222), there is no 'i' in this formula.

where $\mathbf{J} = (J_1, J_2, J_3)$ defines a representation of $su(2)$ in $\mathbb{C}^{2J+1}$; for $J = 1/2$, one has $J_k = \sigma_k/2$ and one arrives at (2.81). To evaluate the matrix elements in (2.68) let us temporarily rotate the axes in $\mathbb{R}^3$ so that the 3rd axis points along $\mathbf{B}$, i.e., $\mathbf{B} = (0, 0, B)$. The standard theory of angular momentum (cf., e.g., Schiff 1968) implies

$$J_3|J, m\,\rangle = m|J, m\,\rangle\,, \tag{2.107}$$

with $m = -J, -J+1, \ldots, J-1, J$. In the rotated frame $H = \mu B J_3$ and, therefore, the corresponding eigenvalues E_m, such that

$$H|J, m\,\rangle = E_m|J, m\,\rangle\,, \tag{2.108}$$

are given by

$$E_m = m\mu B\,. \tag{2.109}$$

Moreover, the only nonvanishing matrix elements of J_1 and J_2 in the basis $|J, m\,\rangle$ read

$$\langle\, J, m \pm 1|J_1|J, m\,\rangle = \frac{1}{2}\sqrt{J(J+1) - m(m \pm 1)}\,, \tag{2.110}$$

$$\langle\, J, m \pm 1|J_2|J, m\,\rangle = \mp\frac{i}{2}\sqrt{J(J+1) - m(m \pm 1)}\,. \tag{2.111}$$

Now, formula (2.68) leads to

$$
\begin{aligned}
F_{kl}^{(m)} \;=\; & -\frac{1}{|\mathbf{B}|^2}\,\mathrm{Im}\,\Big(\langle\, J, m|J_k|J, m-1\,\rangle\langle\, J, m-1|J_l|J, m\,\rangle \\
& + \;\langle\, J, m|J_k|J, m+1\,\rangle\langle\, J, m+1|J_l|J, m\,\rangle - (k \rightleftharpoons l)\Big)\,.
\end{aligned}
\tag{2.112}
$$

Clearly, this implies that

$$F_{13}^{(m)} = F_{23}^{(m)} = 0\,, \tag{2.113}$$

since $\langle\, J, m|J_3|J, m \pm 1\,\rangle = 0$. The only nonvanishing component of $F^{(m)}$ reads

$$F_{12}^{(m)} = -\frac{m}{|\mathbf{B}|^2}\,. \tag{2.114}$$

Reverting now to the unrotated axes one obtains

$$F_{kl}^{(m)} = -\frac{m}{|\mathbf{B}|^3}\,\epsilon_{kli}\,B^i\,, \tag{2.115}$$

i.e., the field of a magnetic pole '$-m$' placed at $\mathbf{B} = 0$. Note that

$$\sum_{m=-J}^{J} F_{kl}^{(m)} = -F_{kl}^{(1)}\sum_{m=-J}^{J} m = 0\,, \tag{2.116}$$

in agreement with (2.69). The geometric Berry phase is given therefore by

$$\gamma_m(C) = -m\Omega(C) , \tag{2.117}$$

where $\Omega(C)$ is the solid angle that the circuit C subtends at $\mathbf{B} = 0$. Note, that $\gamma_m(C)$ depends only on the eigenvalue, m, of the spin component along $\mathbf{B}$ and not of the spin, J of the particle, so that the geometric phase is insensitive to the strength '$2J + 1$' of the degeneracy at $\mathbf{B} = 0$. Again, each eigenvalue of J_3 defines a monopole bundle characterized by the magnetic number (= Chern number)

$$2m = \frac{1}{2\pi} \int_{S^2} F^{(m)} \in \mathbb{Z} , \tag{2.118}$$

with $m = -J, -J + 1, \ldots , J$. $\qquad\qquad\qquad\qquad\qquad\qquad\qquad\qquad \diamondsuit$

Example 2.2.3 (Harmonic oscillator) Consider the quantum system defined by the following Hamiltonian:

$$\hat{H}(\mathbf{R}) = \frac{1}{2}\left[X\,\hat{q}^2 + Y(\hat{q}\hat{p} + \hat{p}\hat{q}) + Z\,\hat{p}^2\right] , \tag{2.119}$$

usually called a *generalized harmonic oscillator*. The Hamiltonian depends on the set of external parameters $\mathbf{R} := (X, Y, Z) \in \mathbb{R}^3$. For fixed values of the parameters $\mathbf{R}$, the eigenvalue equation

$$\hat{H}(\mathbf{R})\psi_n(\mathbf{R}) = E_n(\mathbf{R})\psi_n(\mathbf{R}) , \tag{2.120}$$

takes the following form:

$$-\frac{Z\hbar^2}{2}\frac{d^2\psi_n}{dq^2} - i\hbar Yq\frac{d\psi_n}{dq} + \left(\frac{Xq^2}{2} - i\hbar\frac{Y}{2}\right)\psi_n = E_n\,\psi_n . \tag{2.121}$$

The normalized solution of the above equation reads

$$\psi_n(q; \mathbf{R}) = \left(\frac{\omega}{Z\hbar}\right)^{1/4} \chi_n\left(q\sqrt{\frac{\omega}{Z\hbar}}\right) \exp\left(\frac{-iYq^2}{2Z\hbar}\right) , \tag{2.122}$$

where the parameter-dependent frequency

$$\omega := (XZ - Y^2)^{1/2} , \tag{2.123}$$

and the so-called nth Hermite function χ_n is defined as follows:

$$\chi_n(x) = (n!\,2^n\,\sqrt{\pi})^{-1/2}e^{-x^2/2}H_n(x) , \tag{2.124}$$

where H_n denotes nth Hermite polynomial satisfying the following equation:

$$\frac{d^2H_n(x)}{dx^2} + (2n + 1 - x^2)H_n(x) = 0 . \tag{2.125}$$

Note that formula (2.123) implies

$$XZ > Y^2 \,. \tag{2.126}$$

This means that the corresponding parameter manifold M is the following subset of $\mathbb{R}^3$:

$$M := \{ (X, Y, Z) \in \mathbb{R}^3 \mid XZ > Y^2 \} \,.$$

The oscillator energy eigenvalues are given by the usual oscillator formula:

$$E_n := \hbar\omega \left(n + \frac{1}{2} \right) \,. \tag{2.127}$$

Inserting the wave function (2.122) into the formula (2.61) defining the two-form $F^{(n)}$, one finds[5]

$$
\begin{aligned}
F^{(n)} &= -\operatorname{Im} d_{\mathbf{R}} \int dq \, \psi_n^*(q; \mathbf{R}) \, d_{\mathbf{R}} \psi_n(q; \mathbf{R}) \\
&= -\frac{1}{2\hbar} d_{\mathbf{R}} \left\{ \sqrt{\frac{\omega}{Z\hbar}} \int_{-\infty}^{\infty} dq \, \chi_n^2 \left(q \sqrt{\frac{\omega}{Z\hbar}} \right) q^2 d_{\mathbf{R}} \left(\frac{Y}{Z} \right) \right\} \,.
\end{aligned}
\tag{2.128}
$$

Introducing

$$\xi := q \sqrt{\frac{\omega}{Z\hbar}} \,, \tag{2.129}$$

and using the following property of the Hermite functions (see, e.g., Schiff 1968):

$$\int_{-\infty}^{\infty} d\xi \, \xi^2 \chi_n^2(\xi) = n + \frac{1}{2} \,, \tag{2.130}$$

one obtains the following formula for $F^{(n)}$:

$$F^{(n)}(\mathbf{R}) = -\frac{n + \frac{1}{2}}{2} d_{\mathbf{R}} \left(\frac{Z}{\omega} \right) \wedge d_{\mathbf{R}} \left(\frac{Y}{Z} \right) \,. \tag{2.131}$$

A straightforward calculation leads to the result

$$F^{(n)}(\mathbf{R}) = \frac{n + \frac{1}{2}}{2} \frac{X d_{\mathbf{R}} Y \wedge d_{\mathbf{R}} Z + Y d_{\mathbf{R}} Z \wedge d_{\mathbf{R}} X + Z d_{\mathbf{R}} X \wedge d_{\mathbf{R}} Y}{4(XZ - Y^2)^{3/2}} \,, \tag{2.132}$$

and the Berry phase for the adiabatic cyclic evolution reads

$$\gamma_n(C) = \int_\Sigma F^{(n)} \,, \tag{2.133}$$

where Σ is an arbitrary two-dimensional surface in M with C as its boundary. $\diamond$

---

[5] To distinguish between dq and, e.g., dX, we denote by $d_{\mathbf{R}}$ the exterior derivative in the parameter space.

2.2.6 Quantum geometric tensor

As we have seen, the adiabatic evolution gives rise to a two-form $F^{(n)} \in \Lambda^2(M)$ on the manifold of external parameters M. It is defined by

$$F_{ij}^{(n)} = -\mathrm{Im}\left(\langle \partial_i n | \partial_j n \rangle - \langle \partial_j n | \partial_i n \rangle \right). \tag{2.134}$$

This antisymmetric tensor is invariant under the gauge transformation (2.49) and defines the Berry curvature of the corresponding nth spectral bundle. Berry shows (Berry 1989a, 1989b) that there is another natural gauge-invariant tensor on M, the so-called *quantum geometric tensor*, defined by

$$T_{ij}^{(n)} := \langle \partial_i n | (\mathbb{1} - P_n) | \partial_j n \rangle, \tag{2.135}$$

with $P_n = |n\rangle\langle n|$. It is easy to see that $T_{ij}^{(n)}$ is gauge-invariant and hermitian, i.e.,

$$T_{ij}^{(n)*} = T_{ji}^{(n)}, \tag{2.136}$$

and it is clear that the imaginary part of $T^{(n)}$ reproduces $F^{(n)}$:

$$\mathrm{Im}\, T_{ij}^{(n)} = -\frac{1}{2} F_{ij}^{(n)}. \tag{2.137}$$

Let us define

$$g_{ij}^{(n)} := \mathrm{Re}\, T_{ij}^{(n)}. \tag{2.138}$$

One can easily show that $g_{ij}^{(n)}$ defines a symmetric tensor on M. It turns out that $g_{ij}^{(n)}$ enables one to measure distances along paths in the parameter space. For that reason it is called the *quantum metric tensor*. To see why this is so let us consider two nearby states, $|n(x)\rangle$ and $|n(x+dx)\rangle$. Define a distance $\Delta(x, dx)$ between point x and $x+dx$ in M by

$$\Delta^2(x, x + dx) = 1 - |\langle n(x)|n(x + dx)\rangle|^2. \tag{2.139}$$

Having a distance function lets us define the corresponding metric tensor $G_{ij}(x)$ according to:

$$\Delta^2(x, x + dx) =: G_{ij}(x)dx^i dx^j. \tag{2.140}$$

To show that $G_{ij} = g_{ij}^{(n)}$, consider the Taylor expansion of $|n(x + dx)\rangle$:

$$|n(x + dx)\rangle = |n(x)\rangle + |\partial_i n(x)\rangle dx^i + \frac{1}{2}|\partial_i \partial_j n(x)\rangle dx^i dx^j + \dots. \tag{2.141}$$

It follows that

$$\langle n(x)|n(x + dx)\rangle = 1 + \langle n(x)|\partial_i n(x)\rangle dx^i + \frac{1}{2}\langle n(x)|\partial_i \partial_j n(x)\rangle dx^i dx^j + \dots. \tag{2.142}$$

and hence, up to second order terms,

$$|\langle n(x)|n(x+dx)\rangle|$$
$$= 1 + \frac{1}{2}\mathrm{Re}\left(\langle n(x)|\partial_i\partial_j n(x)\rangle + \langle \partial_i n(x)|n(x)\rangle\langle n(x)|\partial_j n(x)\rangle\right)dx^i dx^j .$$

Using

$$\mathrm{Re}\langle n|\partial_i\partial_j n\rangle = -\mathrm{Re}\langle \partial_i n|\partial_j n\rangle , \tag{2.143}$$

we finally obtain

$$G_{ij} = \mathrm{Re}\left(\langle \partial_i n|\partial_j n\rangle - \langle \partial_i n|n\rangle\langle n|\partial_j n\rangle\right) = g_{ij}^{(n)} . \tag{2.144}$$

Actually, $g_{ij}^{(n)}$ is only a positive semidefinite, and hence, strictly speaking, does not define a metric tensor on M. The length of any curve C on M joining $x(0)$ and $x(T)$ is given by

$$\mathrm{Length}(C) = \int_C (g_{ij}^{(n)}dx^i dx^j)^{1/2} = \int_0^T (g_{ij}^{(n)}\dot{x}^i\dot{x}^j)^{1/2}\,dt \ \geq 0 . \tag{2.145}$$

To illustrate the concept of the quantum metric tensor let us consider once more a spin-half in a magnetic field (cf. Example 2.2.1).

Example 2.2.4 The Hamiltonian of the system is given by (2.81) and the parameter space is

$$M = \mathbb{R}^3 - \{0\} ,$$

where we have removed the degeneracy point $\mathbf{B} = 0$. The corresponding eigenvectors $|\psi^\pm(\mathbf{B})\rangle$ are defined in (2.91):

$$|\psi^+(\mathbf{B})\rangle = \left|\begin{array}{c} \cos\frac{\theta}{2} \\ e^{i\varphi}\sin\frac{\theta}{2} \end{array}\right\rangle , \quad |\psi^-(\mathbf{B})\rangle = \left|\begin{array}{c} -\sin\frac{\theta}{2} \\ e^{i\varphi}\cos\frac{\theta}{2} \end{array}\right\rangle . \tag{2.146}$$

One can easily compute that:

$$\langle \psi^+|\partial_\varphi \psi^+\rangle = i\langle \partial_\varphi\psi^+|\partial_\varphi\psi^+\rangle = i\sin^2\frac{\theta}{2} ,$$
$$\langle \psi^-|\partial_\varphi \psi^-\rangle = i\langle \partial_\varphi\psi^-|\partial_\varphi\psi^-\rangle = i\cos^2\frac{\theta}{2} ,$$
$$\langle \psi^+|\partial_\theta \psi^+\rangle = \langle \psi^-|\partial_\theta\psi^-\rangle = 0 ,$$
$$\langle \partial_\varphi\psi^+|\partial_\theta \psi^+\rangle = \overline{\langle \partial_\varphi\psi^-|\partial_\theta\psi^-\rangle} = -\frac{i}{4}\sin\theta ,$$
$$\langle \partial_\theta\psi^+|\partial_\theta \psi^+\rangle = \langle \partial_\theta\psi^-|\partial_\theta\psi^-\rangle = \frac{1}{4} ,$$

and, clearly, all derivatives with respect to the radial parameter B vanish. Hence, the corresponding components of the quantum metric tensors $g_{ij}^{(\pm)}$ are given by

$$g_{ij}^{(+)} = g_{ij}^{(-)} = \frac{1}{4} \begin{pmatrix} 0 & 0 & 0 \\ 0 & 1 & 0 \\ 0 & 0 & \sin^2\theta \end{pmatrix} , \tag{2.147}$$

where $i,\, j = B, \theta, \varphi$. Note that $g_{BB}^{(\pm)} = 0$, which means that the distance $\Delta(A, B) = 0$ for any two points A and B lying on the same straight line passing through $\mathbf{B} = 0$. $\diamond$

Remark 2.2.1 The tensor g_{ij} was studied in another context by Provost and Vallee (1980). They considered a family of quantum states parametrized by points from a parameter manifold M. This manifold is often called a *manifold of collective states*. Consider, for example, a manifold of coherent states for a harmonic oscillator. Each coherent state is uniquely defined by a complex number $\alpha \in \mathbb{C}$, as follows:

$$|\alpha\rangle = \exp\left(-\frac{1}{2}|\alpha|^2\right) \sum_{n=0}^{\infty} \frac{\alpha^n}{\sqrt{n!}} |n\rangle ,$$

where $|n\rangle$ denote the eigenvectors of the harmonic oscillator Hamiltonian. It is easy to check that formula (2.138) implies, in this case, that the quantum metric tensor is the euclidean metric on $\mathbb{R}^2 \cong \mathbb{C}$. For more examples of interesting quantum metrics see Provost and Vallee 1980. $\diamond$

2.2.7   Quantal phase and geometry — a simple illustration

In this section we present a simple illustration of the geometric origin of the quantal Berry phase. Consider a two-dimensional sphere S^2 together with its tangent bundle TS^2. Suppose that we are going to perform a parallel transport of a vector $\mathbf{e}_0 \in T_x S^2$ along a curve C in S^2 from x to y, that is, we look for an operation

$$\mathbf{T}_C : T_x S^2 \longrightarrow T_y S^2 .$$

Let C be described by

$$[0, 1] \ni t \longrightarrow \mathbf{r}(t) \in S^2 \subset \mathbb{R}^3 ,$$

with $\mathbf{r}(0) = x$ and $\mathbf{r}(1) = y$. Denote by $\mathbf{e}(t)$ the vector $\mathbf{e}_0$ after parallel transport to the point $\mathbf{r}(t)$. How to determine $\mathbf{e}(t)$? Clearly, $\mathbf{e}(t)$ remains tangent to S^2 and hence

$$\mathbf{e}(t) \cdot \mathbf{r}(t) = 0 . \tag{2.148}$$

Moreover, we would like as much as possible to imitate an ordinary parallel transport on a plane. In this case parallel transport is simply a rigid translation without any rotation. We demand that during a parallel transport, $\mathbf{e}(t)$ does not rotate on S^2 — that is, it does not rotate about its instantaneous position $\mathbf{r}(t)$, and its length remains constant. These properties uniquely determine the law of transport: we have $\dot{\mathbf{e}} \cdot \mathbf{e} = 0$, and therefore

$$\dot{\mathbf{e}} = \mathbf{\Omega} \times \mathbf{e} \,, \tag{2.149}$$

for some vector $\mathbf{\Omega} = \mathbf{\Omega}(t)$. Now, since $(\mathbf{r}, \dot{\mathbf{r}}, \mathbf{r} \times \dot{\mathbf{r}})$ defines an orthogonal basis in $\mathbb{R}^3$, the general form of $\mathbf{\Omega}$ reads

$$\mathbf{\Omega} = a\,\mathbf{r} + b\,\dot{\mathbf{r}} + c\,\dot{\mathbf{r}} \times \dot{\mathbf{r}} \,. \tag{2.150}$$

The definition of parallel transport implies that $a = 0$ (i.e., $\mathbf{e}$ does not rotate around $\mathbf{r}$). Moreover, $\mathbf{e}$, being tangent to S^2, remains orthogonal to $\mathbf{r}$. Hence

$$\begin{aligned} 0 &= \frac{d}{dt}\left(\mathbf{e} \cdot \mathbf{r}\right) = \dot{\mathbf{e}} \cdot \mathbf{r} + \mathbf{e} \cdot \dot{\mathbf{r}} \\ &= (\mathbf{\Omega} \times \mathbf{e}) \cdot \mathbf{r} + \mathbf{e} \cdot \dot{\mathbf{r}} = (-\mathbf{\Omega} \times \mathbf{r}) \cdot \mathbf{e} + \mathbf{e} \cdot \mathbf{r} \\ &= b\,(\dot{\mathbf{r}} \times \mathbf{r}) \cdot \mathbf{e} - c\,\dot{\mathbf{r}} \cdot \mathbf{e} + \mathbf{e} \cdot \dot{\mathbf{r}} \,. \end{aligned} \tag{2.151}$$

This implies that $b = 0$ and $c = 1$, and hence

$$\mathbf{\Omega} = \mathbf{r} \times \dot{\mathbf{r}} \,, \tag{2.152}$$

which shows that the law of parallel transport reads

$$\dot{\mathbf{e}} = -(\mathbf{e} \cdot \dot{\mathbf{r}})\mathbf{r} \,. \tag{2.153}$$

Thus, the unique solution of (2.153) with $\mathbf{e}(0) = \mathbf{e}_0$ defines a parallel transport of $\mathbf{e}_0$ along a curve $\mathbf{r}(t) \in S^2$.

Clearly, we may use the above law to perform a parallel transport of an orthonormal frame $(\mathbf{e}_1, \mathbf{e}_2)$ attached at $\mathbf{r}(0)$. Note that if $(\mathbf{e}_1, \mathbf{e}_2)$ is an orthonormal frame on S^2, then a parallel transport implies

$$\mathbf{e}_i \cdot \dot{\mathbf{e}}_j = 0 \,, \qquad i, j = 1, 2 \,. \tag{2.154}$$

In this way we define a connection in a frame bundle FS^2. This is an $SO(2)$-principal fibre bundle over S^2, cf. Example 1.3.3. Equivalently, since $SO(2) \cong U(1)$, it is a natural connection in the Hopf bundle $S^3 \longrightarrow S^2$. To see the relation to the Hopf $U(1)$-bundle more clearly, let us take an arbitrary orthonormal frame $(\mathbf{e}_1, \mathbf{e}_2)$ attached at a point $\mathbf{r}_0 \in S^2$, and define a unit complex vector

$$\boldsymbol{\phi} := \frac{\mathbf{e}_1 + i\mathbf{e}_2}{\sqrt{2}} \,. \tag{2.155}$$

Now, if $\mathbf{e}_1$ and $\mathbf{e}_2$ are parallel transported along a curve $\mathbf{r} = \mathbf{r}(t)$, then, using (2.153), one finds

$$\mathrm{Im}\left(\boldsymbol{\phi}^* \cdot \dot{\boldsymbol{\phi}}\right) = 0 \,. \tag{2.156}$$

The above equation is an analog of the Born–Fock gauge condition (2.16), which is the defining equation of the Berry–Simon connection.

Now, consider a parallel transport of a frame $(\mathbf{e}_1, \mathbf{e}_2)$ (or equivalently, of a complex vector $\boldsymbol{\phi}$) along a closed curve C on S^2. In general, a parallel transport leads to an orthonormal frame $(\mathbf{e}_1', \mathbf{e}_2')$ which is rotated with respect to the initial one. This rotation (an element of $SO(2)$) is actually the holonomy of the closed curve C. Thus a basis $(\mathbf{e}_1(t), \mathbf{e}_2(t))$ is not single-valued along C. To obtain a single-valued basis let us perform a gauge transformation

$$(\mathbf{e}_1, \mathbf{e}_2) \;\longrightarrow\; (\widetilde{\mathbf{e}}_1, \widetilde{\mathbf{e}}_2) \,,$$

defined by

$$\begin{pmatrix} \widetilde{\mathbf{e}}_1 \\ \widetilde{\mathbf{e}}_2 \end{pmatrix} = \begin{pmatrix} \cos\alpha & \sin\alpha \\ -\sin\alpha & \cos\alpha \end{pmatrix} \begin{pmatrix} \mathbf{e}_1 \\ \mathbf{e}_2 \end{pmatrix} , \tag{2.157}$$

where $\alpha = \alpha(t)$ is the t-dependent angle of the $SO(2)$ rotation. Clearly, it is always possible to choose $\alpha(t)$ such that

$$(\widetilde{\mathbf{e}}_1(0), \widetilde{\mathbf{e}}_2(0)) = (\widetilde{\mathbf{e}}_1(1), \widetilde{\mathbf{e}}_2(1)) \,.$$

Let $(\widetilde{\mathbf{e}}_1, \widetilde{\mathbf{e}}_2)$ be a single-valued frame along C and define

$$\mathbf{n} := \frac{\widetilde{\mathbf{e}}_1 + i\widetilde{\mathbf{e}}_2}{\sqrt{2}} \,. \tag{2.158}$$

One immediately finds that

$$\boldsymbol{\phi}(\mathbf{r}(t)) = e^{i\alpha(t)} \, \mathbf{n}(\mathbf{r}(t)) \,. \tag{2.159}$$

Equation (2.156) implies that

$$\mathrm{Im}\left[i\dot{\alpha}\,(\mathbf{n}^* \cdot \mathbf{n}) + \mathbf{n}^* \cdot \dot{\mathbf{n}} \right] = 0 \,, \tag{2.160}$$

and, since $\mathbf{n}^* \cdot \mathbf{n} = 1$, the term $\mathbf{n}^* \cdot \dot{\mathbf{n}}$ must be purely imaginary, which leads to

$$\dot{\alpha} = -\mathrm{Im}\,(\mathbf{n}^* \cdot \dot{\mathbf{n}}) \,. \tag{2.161}$$

This construction enables us to introduce a one-form on S^2,

$$A := -\mathrm{Im}\,(\mathbf{n}^* \cdot d\mathbf{n}) \,, \tag{2.162}$$

which is an analog of the Berry connection form. Finally, the $SO(2)$ rotation — the holonomy $\Delta(C)$ of C — is given by

$$\Delta(C) := \alpha(1) = -\mathrm{Im} \int_0^1 \mathbf{n}^* \cdot \dot{\mathbf{n}}\, dt = \oint_C A \,. \tag{2.163}$$

Due to the Stokes theorem one has

$$\Delta(C) = \int_{\Sigma} F \, , \tag{2.164}$$

where $F = dA$, and Σ is any two-dimensional region in S^2 with boundary $C = \partial\Sigma$. The reader can easily show that if $(\widetilde{\mathbf{e}}_1', \widetilde{\mathbf{e}}_2')$ is another single-valued frame along C, i.e., there exists $\lambda(t)$ such that

$$\mathbf{n}'(t) = e^{i\lambda(t)} \, \mathbf{n}(t) \, , \tag{2.165}$$

then

$$A' = A - d\lambda \quad \text{and} \quad F' = F \, . \tag{2.166}$$

The final step is to calculate $\Delta(C)$. To do this we have to specify a single-valued frame $(\widetilde{\mathbf{e}}_1(\mathbf{r}), \widetilde{\mathbf{e}}_2(\mathbf{r}))$. Let $\mathbf{e}_3$ be a fixed unit vector along the z-axis in $\mathbb{R}^3$. Define

$$\widetilde{\mathbf{e}}_1(\mathbf{r}) := \frac{\mathbf{r} \times \mathbf{e}_3}{|\mathbf{r} \times \mathbf{e}_3|} \quad \text{and} \quad \widetilde{\mathbf{e}}_2(\mathbf{r}) := \frac{\mathbf{r} \times \widetilde{\mathbf{e}}_1(\mathbf{r})}{r} \, , \tag{2.167}$$

with $r = |\mathbf{r}|$. The above formulae give rise to a field of frames in $\mathbb{R}^3$ if and only if $\mathbf{r}$ is not parallel to $\mathbf{e}_3$. In particular, if $r = 1$ they define a field of frames on S^2, minus the north and south poles. A simple calculation leads to

$$A = \frac{z}{r(x^2 + y^2)} \, (ydx - xdy) \, , \tag{2.168}$$

where (x, y, z) are cartesian coordinates in $\mathbb{R}^3$. Finally, one finds the following expression for F:

$$F = \frac{1}{r^3} \, (xdy \wedge dz + zdx \wedge dy + ydz \wedge dx) \, , \tag{2.169}$$

or, using a vector notation,

$$\mathbf{F} := (F_{yz}, F_{zx}, F_{xy}) = \frac{\mathbf{r}}{r^3} \, , \tag{2.170}$$

which reproduces the field of a magnetic pole. In this way we have shown that a parallel transport of orthonormal frames in FS^2 is equivalent to a parallel transport with respect to a natural connection in a monopole bundle. Therefore, the holonomy $\Delta(C)$ reads

$$\Delta(C) = \int_{\partial\Sigma=C} F = \Omega(C) \, , \tag{2.171}$$

where $\Omega(C)$ denotes the solid angle subtended by the closed curve C.

2.3 Non-abelian Wilczek–Zee phase

2.3.1 Standard derivation

A non-abelian Wilczek–Zee phase factor (Wilczek and Zee 1984) is a natural generalization of the Berry phase for systems with degenerate spectra. Suppose that the nth eigenvalue of the Hamiltonian is N-times degenerate, i.e.,

$$H(x)\psi_{na}(x) = E_n(x)\psi_{na}(x)\,, \qquad a = 1, 2, \ldots, N\,, \tag{2.172}$$

i.e., the nth eigenspace,

$$\mathcal{H}_n(x) = \left\{ \sum_{a=1}^{N} c_a \psi_{na}(x) \,\Big|\, c_a \in \mathbb{C} \right\}\,,$$

has dimension N. We may always choose the eigenvectors $\psi_{na}(x)$ such that

$$\langle \psi_{na}(x)|\psi_{nb}(x)\rangle = \delta_{ab}\,. \tag{2.173}$$

Obviously, this choice is not unique; one may perform a unitary, x-dependent transformation

$$\psi_{na}(x) \longrightarrow \psi'_{na}(x) = \sum_{b=1}^{N} U_{ab}(x)\psi_{nb}(x)\,, \tag{2.174}$$

to another orthonormal basis $\psi'_{na}(x)$ in $\mathcal{H}_n(x)$. Consider an adiabatic evolution of the state vector $\psi(t)$, corresponding to an adiabatic change of the external parameters:

$$[0, T] \ni t \longrightarrow x_t \in M\,.$$

Suppose that $\psi(0) \in \mathcal{H}_n(x_0)$. If the adiabatic evolution is cyclic, i.e., $x_0 = x_T$, then the adiabatic theorem (Theorem 2.1.2) implies that $\psi(T) \in \mathcal{H}_n(x_T) = \mathcal{H}_n(x_0)$, which means that $\psi(0)$ and $\psi(T)$ are unitary related, i.e.,

$$\psi(T) = V\psi(0)\,, \tag{2.175}$$

for some unitary operator $V \in U(N)$. To find the unitary matrix V note that during an adiabatic evolution, the state vector $\psi(t)$ stays in $\mathcal{H}_n(x_t)$. Suppose, that $\psi(0) = \psi_{na}$ for some $1 \le a \le N$. Therefore, for $t > 0$, a solution to the Schrödinger equation in the adiabatic approximation has the following form:

$$\psi(t) = \exp\left(-\frac{i}{\hbar}\int_0^t E_n(\tau)d\tau\right) \sum_{b=1}^{N} U_{ab}^{(n)}(t)\psi_{nb}(x_t), \tag{2.176}$$

where $U^{(n)}$ is an $N \times N$ unitary matrix. As usual, we have separated out a dynamical phase factor. Inserting (2.176) into the Schrödinger equation, one obtains the following equation for the time dependent matrix $U^{(n)}$:

$$(U^{(n)-1}\dot{U}^{(n)})_{ab} = -\langle \psi_{na}|\dot{\psi}_{nb}\rangle\,. \tag{2.177}$$

Defining the following one-form, called the *Wilczek–Zee potential*:

$$A_{ab}^{(n)} := i \langle \psi_{nb} | d | \psi_{na} \rangle \,, \tag{2.178}$$

one finds the following formula for V:

$$V = V_{\text{dyn}} \cdot V_{\text{geo}} \,, \tag{2.179}$$

where the dynamical factor reads

$$V_{\text{dyn}} = \exp\left(-\frac{i}{\hbar} \int_0^T E_n(\tau) d\tau \right) \mathbb{1}_N \,, \tag{2.180}$$

and the *geometric Wilczek–Zee factor* is given by the following path-ordered integral:

$$V_{\text{geo}} = U^{(n)}(T) = \text{P} \exp\left(i \oint_C A^{(n)} \right) \,. \tag{2.181}$$

Note that $A^{(n)}$ is hermitian, i.e., $A^{(n)*} = A^{(n)}$. Obviously, when there is no degeneracy, i.e., $N = 1$, the non-abelian Wilczek–Zee factor (2.181) reproduces an abelian Berry's phase factor (2.58).

2.3.2   Fibre bundle approach

It is evident that the Wilczek–Zee factor may be reformulated as a holonomy element in an appropriate fibre bundle. Each point of a parameter space, $x \in M$, gives rise to an N-dimensional spectral space $\mathcal{H}_n(x)$, and hence we may define the following spectral bundle:

$$E^{(n)} = \bigcup_{x \in M} \mathcal{H}_n(x) \tag{2.182}$$

over M, with a typical fibre $F = \mathbb{C}^N$. Note that by fixing N unit vectors $\varphi_1, \ldots, \varphi_N \in \mathcal{H}_n(x)$ and defining a fibre as follows:

$$F_x^{(n)} := \left\{ \sum_b U_{ab}\varphi_b \;\middle|\; U \in U(N) \right\} \cong U(N) \,, \tag{2.183}$$

one may equivalently consider a $U(N)$-principle bundle over M:

$$P^{(n)} := \bigcup_{x \in M} F_x^{(n)} \,.$$

Clearly, the nth spectral bundle $E^{(n)}$ is an associated vector bundle to the $U(N)$-principal bundle $P^{(n)}$. Consider now a curve $t \longrightarrow C(t)$ in M, and let

$$t \longrightarrow \{ \varphi_1(t), \ldots, \varphi_N(t) \} \,,$$

be a lift of C to $E^{(n)}$. We call $\varphi_a(t)$ a horizontal lift with respect to a *Wilczek–Zee connection* if

$$\langle \psi_{nb}|\dot\varphi_a \rangle = 0 , \tag{2.184}$$

for any $a, b = 1, \dots , N$. This formula may be equivalently written as

$$\langle \psi_{nb}|d\varphi_a \rangle = 0 , \tag{2.185}$$

which is an analog of (2.78). Let P_n be a projection on $\mathcal{H}_n$. Then the above formula may be rewritten as follows:

$$P_n\, d\varphi_a = 0 , \quad a = 1, \dots , N , \tag{2.186}$$

where d stands for exterior differentiation on M, that is, the last system of equations define another example of our basic geometric formula (2.38). Clearly, we may generalize the Born–Fock gauge condition (2.16) to the non-abelian case. One says that the family ψ_{na} is in the Born–Fock gauge, iff

$$\langle \psi_{na}|\dot\psi_{nb} \rangle = 0 , \quad \text{for } a, b = 1, 2, \dots , N . \tag{2.187}$$

As in the abelian case, to be in the Born–Fock gauge is equivalent to being parallel transported with respect to the natural Wilczek–Zee connection.

To find the geometric meaning of the Wilczek–Zee potential let us perform a unitary transformation (2.174) and calculate Wilczek–Zee potential using a transformed basis, ψ'_{na}. One easily finds that

$$\begin{aligned}
A'^{(n)}_{ab} = i\langle \psi'_{nb}|d|\psi'_{na} \rangle\ &= i \sum_{c,d=1}^{N} \overline{U_{bd}}\Big(dU_{ac}\langle \psi_{nd}|\psi_{nc} \rangle + U_{ac}\langle \psi_{nd}|d|\psi_{nc} \rangle\Big) \\
&= \sum_{c,d=1}^{N} \Big(U_{ac}A^{(n)}_{cd}\overline{U_{bd}} + i(dU_{ac})\delta_{cd}\overline{U_{bd}}\Big) \\
&= (U \cdot A \cdot U^* + i(dU) \cdot U^*)_{ab} .
\end{aligned} \tag{2.188}$$

Clearly, $A^{(n)}$ transforms exactly as a gauge potential in the non-abelian gauge theory. Hence, a degenerate spectrum leads in a natural way to a non-abelian $U(N)$ gauge theory, with N being the degree of degeneracy. We may, therefore, define the corresponding gauge field,

$$F^{(n)} = dA^{(n)} - iA^{(n)} \wedge A^{(n)} . \tag{2.189}$$

Let us observe that $iF^{(n)}$ is a $u(N)$-valued two-form on a parameter manifold M. Using local coordinates $(x^1, \dots , x^n)$ on M, one obtains

$$(F^{(n)}_{kl})_{ab} = \partial_k(A^{(n)}_l)_{ab} - \partial_l(A^{(n)}_k)_{ab} - i[A^{(n)}_k, A^{(n)}_l]_{ab} , \tag{2.190}$$

with

$$(A_k^{(n)})_{ab} = i \langle \psi_{nb} | \partial_k | \psi_{na} \rangle \,. \tag{2.191}$$

Clearly, the non-abelian Wilczek–Zee factor V_{geo} is a holonomy of the Wilczek–Zee connection.

Example 2.3.1 To illustrate the structure of the non-abelian geometric phase let us consider the following generalization of the model Hamiltonian (2.81) corresponding to the spin-half particle in a magnetic field. The generalization (Biswas 1989, see also Arodź and Babiuch 1989) consists in replacing Pauli matrices σ_k by the Dirac matrices γ_k:

$$\gamma_k := \sigma_1 \otimes \sigma_k = \begin{pmatrix} 0 & \sigma_k \\ \sigma_k & 0 \end{pmatrix} \,, \tag{2.192}$$

that is, instead of (2.81) we take

$$H(\mathbf{B}) = \frac{1}{2} \mu \mathbf{B} \cdot \boldsymbol{\gamma} \,. \tag{2.193}$$

The above Hamiltonian has two doubly-degenerate eigenvalues:

$$E_\pm = \pm \frac{1}{2} \mu B \,.$$

The reader can easily check that the corresponding eigenvectors,

$$H(\mathbf{B}) \, \psi_{\pm a} = E_\pm \, \psi_{\pm a} \,, \qquad a = 1, 2 \,,$$

are given by

$$\psi_{+1} = \frac{1}{\sqrt{2}} \begin{pmatrix} 1 \\ 0 \\ \cos\theta \\ \sin\theta e^{i\varphi} \end{pmatrix} \,, \qquad \psi_{+2} = \frac{1}{\sqrt{2}} \begin{pmatrix} 0 \\ 1 \\ \sin\theta e^{-i\varphi} \\ -\cos\theta \end{pmatrix} \,,$$

and

$$\psi_{-1} = \frac{1}{\sqrt{2}} \begin{pmatrix} \cos\theta \\ \sin\theta e^{i\varphi} \\ 1 \\ 0 \end{pmatrix} \,, \qquad \psi_{-2} = \frac{1}{\sqrt{2}} \begin{pmatrix} \sin\theta e^{-i\varphi} \\ -\cos\theta \\ 0 \\ 1 \end{pmatrix} \,,$$

where as usual θ and φ are the standard spherical angles in $\mathbb{R}^3$. Now let us compute the Wilczek–Zee gauge potential corresponding to the E_+ energy level, i.e.,

$$A_{ab}^{(+)} = i \langle \psi_{+b} | d | \psi_{+a} \rangle \,.$$

One easily finds that

$$(A_\theta^{(+)})_{11} = (A_\theta^{(+)})_{22} = 0\,, \qquad (A_\theta^{(+)})_{12} \;=\; \frac{i}{2}e^{i\varphi}\,,$$

and

$$(A_\varphi^{(+)})_{11} = -(A_\varphi^{(+)})_{22} = -\frac{1}{2}\sin^2\theta\,, \qquad (A_\varphi^{(+)})_{12} \;=\; \frac{1}{2}\sin\theta\cos\theta\, e^{-i\varphi}\,.$$

Therefore, the non-abelian one-form potential $A^{(+)}$ reads

$$A^{(+)} = A_\theta^{(+)}\, d\theta + A_\varphi^{(+)}\, d\varphi\,,$$

with

$$A_\theta^{(+)} = \frac{i}{2}\begin{pmatrix} 0 & e^{-i\varphi} \\ -e^{i\varphi} & 0 \end{pmatrix}\,, \tag{2.194}$$

and

$$A_\varphi^{(+)} = \frac{\sin\theta}{2}\begin{pmatrix} -\sin\theta & \cos\theta\, e^{-i\varphi} \\ \cos\theta\, e^{i\varphi} & \sin\theta \end{pmatrix}\,. \tag{2.195}$$

Having found the gauge potential $A^{(+)}$ one easily finds the corresponding gauge field $F^{(+)}$. The only nonvanishing component of $F^{(n)}$ reads

$$F_{\theta\varphi}^{(+)} = \partial_\theta A_\varphi^{(+)} - \partial_\varphi A_\theta^{(+)} - i[A_\theta^{(+)}, A_\varphi^{(+)}]\,. \tag{2.196}$$

Inserting the formulae for $A_\theta^{(+)}$ and $A_\varphi^{(+)}$, one gets

$$F_{\theta\varphi}^{(+)} = -\frac{1}{2}\frac{\mathbf{B}\cdot\boldsymbol{\sigma}}{B}\sin\theta\, d\theta\wedge d\varphi\,, \tag{2.197}$$

or, using cartesian coordinate,

$$F^{(+)} = -\frac{1}{2}\frac{\mathbf{B}\cdot\boldsymbol{\sigma}}{B^4}\epsilon_{ijk}\, B^i dB^j\wedge dB^k\,. \tag{2.198}$$

Note that F still has a form of the field strength of a magnetic pole, where now the magnetic charge g is replaced by a matrix $g\mathbf{B}\cdot\boldsymbol{\sigma}/B$. The corresponding spectral bundle over $M = S^2$ is trivial, and hence its Chern number vanishes:

$$\frac{1}{2\pi}\int_{S^2}\mathrm{Tr}\, F^{(+)} = 0\,.$$

Nevertheless, the geometric phase factor (or, equivalently, the holonomy group) is nontrivial. $\diamond$

2.3.3   Non-abelian phase in quadrupole resonance

An elegant example of a physical system displaying the non-abelian Wilczek–Zee geometric phase is the nuclear quadrupole resonance (Tycko 1987, Zee 1988). Consider the spin quadrupole Hamiltonian describing the interaction of the nuclear quadrupole with a magnetic field $\mathbf{B}$, i.e.,

$$H = \mu(\mathbf{J} \cdot \mathbf{B})^2 \, , \tag{2.199}$$

where μ plays a role of the coupling constant. More precisely, a quadrupole system is described by

$$H(Q) := \sum_{k,l=1}^{3} Q_{kl} J_k J_l \, , \tag{2.200}$$

where Q_{kl} are the components of a real 3×3, symmetric and traceless matrix — a quadrupole matrix. Our Hamiltonian (2.199) is related to $H(Q)$ as follows

$$H = H(Q) + \frac{1}{3}\mu B^2 J^2 \, , \tag{2.201}$$

with

$$Q_{kl} = \mu \left(B_k B_l - \frac{1}{3} B^2 \delta_{kl} \right) \, , \tag{2.202}$$

that is, H differs from $H(Q)$ by a term of the form 'const. $\cdot J^2$,' which is a constant of motion, since $[H, J^2] = 0$.[6] The Hamiltonian (2.199) is reminiscent of the spin dipole Hamiltonian '$\mu \mathbf{J} \cdot \mathbf{B}$' considered in Example 2.2.2. The quadrupole Hamiltonian possesses an additional symmetry $\mathbf{J} \longrightarrow -\mathbf{J}$ which implies that the energy eigenvalues are doubly degenerate. Indeed, if $|J, m\rangle$ are eigenvectors of J_3, i.e.,

$$J_3|J, m\rangle = m|J, m\rangle \, , \qquad m = -J, -J + 1, \ldots , J - 1, J \, ,$$

then the corresponding energy eigenvalues E_m, defined by

$$H|J, \pm m\rangle = E_m|J, \pm m\rangle \, ,$$

are given by $E_m = \mu(mB)^2$. Note that for $J = \frac{1}{2}$ one has $J_k = \sigma_k/2$ and hence the Hamiltonian is trivial: $H = (\mu B^2/2)\,\mathbb{1}_2$. Therefore, to obtain a nontrivial quadrupole system one needs $J \geq 1$. In the experiment performed by Tycko, a spin $J = 3/2$ Cl atom in an NaCl crystal was used. The case $J = \frac{3}{2}$ gives rise to two doublets corresponding to $m = \pm\frac{3}{2}$ and $m = \pm\frac{1}{2}$.

---

[6]For a more detailed discussion of quadrupole systems we refer the reader to section 6.3.2.

Let us assume that the magnetic-field varies slowly in time, such that $B = |\mathbf{B}| = $ const. Hence, the parameter space is $M = S^2$. Parametrizing S^2 by the standard spherical angles θ and φ, we shall denote the corresponding eigenvectors of (2.199) by $|m(\theta, \varphi)\rangle$, with $m = \pm\frac{1}{2}, \pm\frac{3}{2}$. Clearly,

$$|m(\theta, \varphi)\rangle = U(\theta, \varphi)|m(\mathbf{e}_3)\rangle , \qquad (2.203)$$

where $U(\theta, \varphi)$ is the unitary operation defined in (2.86), and $|m(\mathbf{e}_3)\rangle := |\frac{3}{2}, m\rangle$.

The non-abelian Wilczek–Zee connection $A^{(m)}$ on the parameter manifold $M = S^2$ reads

$$A^{(m)}(\theta, \varphi) = A_\theta^{(m)}(\theta, \varphi)\, d\theta + A_\varphi^{(m)}(\theta, \varphi)\, d\varphi , \qquad (2.204)$$

where the 2×2 matrices $A_\theta^{(m)}$ and $A_\varphi^{(m)}$ are given by

$$\left(A_\theta^{(m)}\right)_{ba} = i\left\langle a(\theta, \varphi)\left|\frac{\partial}{\partial\theta}\right|b(\theta, \varphi)\right\rangle = \left\langle a(\mathbf{e}_3)\left|iU^*(\theta, \varphi)\cdot\frac{\partial}{\partial\theta}U(\theta, \varphi)\right|b(\mathbf{e}_3)\right\rangle ,$$
$$(2.205)$$

and

$$\left(A_\varphi^{(m)}\right)_{ba} = i\left\langle a(\theta, \varphi)\left|\frac{\partial}{\partial\varphi}\right|b(\theta, \varphi)\right\rangle = \left\langle a(\mathbf{e}_3)\left|iU^*(\theta, \varphi)\cdot\frac{\partial}{\partial\varphi}U(\theta, \varphi)\right|b(\mathbf{e}_3)\right\rangle ,$$
$$(2.206)$$

with $a, b = \pm|m|$. Using the well-known Baker–Campbell–Hausdorff formula

$$e^B \cdot A \cdot e^{-B} = A + [B, A] + \frac{1}{2}[B, [B, A]] + \frac{1}{3!}[B, [B, [B, A]]] + \dots , \qquad (2.207)$$

one easily finds that

$$\begin{aligned}
U_2(\theta) \cdot J_3 \cdot U_2^*(\theta) &= J_3\cos\theta + J_1\sin\theta , \\
U_3(\varphi) \cdot J_1 \cdot U_3^*(\varphi) &= J_1\cos\varphi + J_2\sin\varphi , \\
U_3(\varphi) \cdot J_2 \cdot U_3^*(\varphi) &= J_2\cos\varphi - J_1\sin\varphi .
\end{aligned} \qquad (2.208)$$

Hence

$$\begin{aligned}
iU^*(\theta, \varphi) \cdot \frac{\partial}{\partial\theta}U(\theta, \varphi) &= U_3(\varphi) \cdot J_2 \cdot U_3^*(\varphi) \\
&= J_2\cos\varphi - J_1\sin\varphi ,
\end{aligned} \qquad (2.209)$$

and

$$iU^*(\theta, \varphi) \cdot \frac{\partial}{\partial\varphi}U(\theta, \varphi) = U_3(\varphi) \cdot U_2^*(\theta) \cdot J_3 \cdot U_2(\theta) \cdot U_3^*(\varphi) - J_3$$
$$= -(J_1\cos\varphi + J_2\sin\varphi)\sin\theta + J_3(\cos\theta - 1) . \qquad (2.210)$$

This leads to the following formulae for $A^{(m)}$:

$$\left(A_\theta^{(m)}\right)_{ba} = \langle\, a(\mathbf{e}_3)|J_2\cos\varphi - J_1\sin\varphi|b(\mathbf{e}_3)\,\rangle\,, \tag{2.211}$$

and

$$\left(A_\varphi^{(m)}\right)_{ba} = \langle\, a(\mathbf{e}_3)| - (J_1\cos\varphi + J_2\sin\varphi)\sin\theta + J_3(\cos\theta - 1)|b(\mathbf{e}_3)\,\rangle\,. \tag{2.212}$$

Taking into account (2.110) and (2.111), one finds that

$$\left\langle\tfrac{3}{2},-\tfrac{1}{2}\Big|J_2\Big|\tfrac{3}{2},\tfrac{1}{2}\right\rangle = -i\left\langle\tfrac{3}{2},-\tfrac{1}{2}\Big|J_1\Big|\tfrac{3}{2},\tfrac{1}{2}\right\rangle = -\frac{i}{4}\,,$$

$$\left\langle\tfrac{3}{2},-\tfrac{3}{2}\Big|J_2\Big|\tfrac{3}{2},\tfrac{3}{2}\right\rangle = \left\langle\tfrac{3}{2},-\tfrac{3}{2}\Big|J_1\Big|\tfrac{3}{2},\tfrac{3}{2}\right\rangle = 0\,,$$

and $\langle\, a(\mathbf{e}_3)|J_3|b(\mathbf{e}_3)\,\rangle = a\delta_{ab}$. Thus, for $m = \frac{3}{2}$ we obtain[7]

$$A_\theta^{\left(\frac{3}{2}\right)} = 0\,, \tag{2.213}$$

$$A_\varphi^{\left(\frac{3}{2}\right)} = \frac{3}{2}(1 - \cos\theta)\,\sigma_3\,, \tag{2.214}$$

and, for $m = \frac{1}{2}$,

$$A_\theta^{\left(\frac{1}{2}\right)} = \frac{1}{4}(-\cos\varphi\,\sigma_2 - \sin\varphi\,\sigma_1) = -\frac{1}{4}\begin{pmatrix} 0 & ie^{i\varphi} \\ -ie^{-i\varphi} & 0 \end{pmatrix}\,, \tag{2.215}$$

$$A_\varphi^{\left(\frac{1}{2}\right)} = \frac{1}{2}\left[(1 - \cos\theta)\sigma_3 + \frac{1}{2}\sin\theta(-\cos\varphi\,\sigma_1 + \sin\varphi\,\sigma_2)\right]$$

$$= \frac{1}{2}\begin{pmatrix} (1 - \cos\theta) & -\frac{1}{2}\sin\theta\,e^{i\varphi} \\ -\frac{1}{2}\sin\theta\,e^{-i\varphi} & -(1 - \cos\theta) \end{pmatrix}\,. \tag{2.216}$$

Hence for $m = \frac{1}{2}$ one obtains a truly non-abelian structure. We shall meet a similar structure in studying a diatomic particle in section 6.3.4.

Further reading

Section 2.1. A detailed discussion of the adiabatic approximation may be found, e.g., in Messiah 1961. Kato's results were extended in Nenciu 1980 and Avron et al. 1987. See also the review articles by Avron et al. (1988) and Richter and Seiler (2000).

---

[7]We use the following convention to enumerate the matrix elements of 2×2 matrices:

$$X_{ab} = \begin{pmatrix} x & y \\ u & v \end{pmatrix}\,,$$

with $x := X_{-|m|,-|m|}$, $y := X_{-|m|,|m|}$, $u := X_{|m|,-|m|}$, $v := X_{|m|,|m|}$.

Section 2.2. Quantal Berry's phase is discussed in standard quantum mechanics courses (see, e.g., the monograph by Bohm (1993a)). We also recommend useful reviews by Berry (1988a), (1989a), (1989b); Zwanziger et al. (1990); Bohm et al. (1991), and Bohm (1993b).

Other examples of Berry's phase may be found in: Bouchiat 1987, 1989; Chaturvedi, Sriram and V. Srinivasan 1987; Cervero and Lejarreta 1989; and Cheng and Fung 1989. Garrison and Wright (1988) generalized the geometric Berry's phase factor to dissipative evolution equations phenomenologically described by nonhermitian Hamiltonians. In this case, (real) Berry's phase is replaced by complex geometrical multipliers. De Polaviejaa and Sjöqvist (1998) extended the quantal adiabatic phase to noncyclic motions — see also Pati 1998; Mostafazadeh 1999; and Zhu, Wang, and Zhang 2000.

For the classification of bundles arising in quantum mechanical problems we refer the reader to Kiritsis 1987; Bohm et al. 1993; Mostafazadeh and Bohm 1993; and Mostafazadeh 1996.

Problems

2.1. Show that $|\tilde{n}\rangle$, defined in (2.15), satisfies

$$\langle \tilde{n}|\dot{\tilde{n}}\rangle = 0 \,.$$

2.2. Find the Kato Hamiltonian corresponding to the one-dimensional projector $P_n(s) = |n(s)\rangle\langle n(s)|$, where $H(s)|n(s)\rangle = E(s)|n(s)\rangle$. Show that H_{Kato} is gauge invariant and that $\psi(s) = c(s)|n(s)\rangle$ solves the Kato equation with $\psi(0) = |n(0)\rangle$.

2.3. Show that Berry's phase is gauge invariant, i.e.,

$$\oint \langle n|dn\rangle = \oint \langle n'|dn'\rangle \,.$$

2.4. Find the formulae for the matrix elements of $\langle +(\theta,\varphi)|\sigma_k| - (\theta,\varphi)\rangle$, where $|\pm(\theta,\varphi)\rangle$ are defined in (2.88), and compute $F^{(+)}$ using formula (2.103).

2.5. Show that the quantum geometric tensor $T_{ij}^{(n)}$ is hermitian and gauge invariant.

2.6. Prove that the quantum metric tensor $g_{ij}^{(n)}$ defined in (2.138) is symmetric.

2.7. Show that the quantity $\Delta(x, dx)$ introduced in (2.139) does satisfy all requirements of the distance function.

2.8. Let P_n be the N-dimensional projector onto the nth eigenspace $\mathcal{H}_n$. Find the corresponding Kato Hamiltonian and show that under the gauge transformation

$$\psi_{na}(x) \longrightarrow \psi'_{na}(x) = \sum_{b=1}^{N} U_{ab}(x)\psi_{nb}(x) \,,$$

the Kato Hamiltonian transforms according to

$$H_{\text{Kato}} \longrightarrow U \cdot H_{\text{Kato}} \cdot U^* .$$

2.9. Try to generalize Proposition 2.2.1 to the degenerate case.

2.10. Prove that the Wilczek–Zee curvature $F^{(n)}$ transforms under the gauge transformation (2.174) in a tensorial way, that is,

$$F^{(n)} \longrightarrow F'^{(n)} = U \cdot F^{(n)} \cdot U^* .$$

2.11. Following Example 2.3.1, compute $A^{(-)}$ and $F^{(-)}$. Show that

$$F^{(+)} + F^{(-)} = 0 .$$

2.12. Verify formula (2.168) for the connection form A.

2.13. Show that dA, with A given by (2.168), reproduces the field of a magnetic pole.

2.14. Find the adiabatic Wilczek–Zee connection for the systems governed by (Arodż and Babiuch 1989)

(1) $H(\mathbf{n}) = \sum_{i,j,k=1}^{3} \epsilon_{ijk} n^i \sigma^j \otimes \sigma^k$,

(2) $H(\mathbf{n}) = \sum_{k=1}^{3} n^k \sigma^k \otimes \sigma^k$,

(3) $H^{(\pm)}(\mathbf{n}) = \sum_{i=1}^{3} n^i (\mathbb{1}_2 \otimes \sigma^i \pm \sigma^i \otimes \mathbb{1}_2)$,

where σ^i are Pauli matrices and $\mathbf{n} = (n^1, n^2, n^3) \in S^2$ are the adiabatic parameters.

2.15. Consider the spin quadrupole system discussed in section 2.3.3. Define

$$|m(\theta, \varphi) \rangle' = U'(\theta, \varphi)|m(\mathbf{e}_3) \rangle ,$$

where $U'(\theta, \varphi) = U(\theta, \varphi) \cdot U_3(\varphi) = U_3(\varphi) \cdot U_2(\theta)$. Find the corresponding non-abelian connection form $A'^{(m)}$. Derive the gauge transformation relating $A^{(m)}$ and $A'^{(m)}$.

2.16. Compute the Wilczek–Zee curvature for the spin quadrupole system $H = \mu (\mathbf{J} \cdot \mathbf{B})^2$.

2.17. Use the results of section 2.3.3 to derive the solid angle formula $\gamma_m(C) = -m\Omega(C)$ for the Berry phase of the spin system.

2.18. Using the Baker–Campbell–Hausdorff formula (2.207), prove (2.208).

3
Adiabatic Phases in Classical Mechanics

3.1 Hamiltonian systems

3.1.1 What we mean by a phase in classical mechanics

What could be a classical analog of the quantum geometric phase? An obvious candidate, which is even called a phase, is the phase of harmonic motion:

$$x(t) = A \cos(\omega t + \varphi_0) .$$

One may define the phase difference between two positions $x(t = 0)$ and $x(t = T)$ by

$$\Delta\varphi = \omega T .$$

Clearly, if ω depends on time, then $\Delta\varphi$ is replaced by

$$\Delta\varphi_{\text{dyn}} = \int_0^T \omega(t)dt ,$$

and, in analogy to quantum mechanics, it may be called a dynamical phase (or a dynamical angle). However, if the oscillator is coupled to a time-dependent environment, then one may expect that if the state of the environment is slowly (adiabatically) cycled, then, as in the quantum case, there is an additional phase (angle) which depends on a closed trajectory C in the environment phase space, i.e.,

$$\Delta\varphi_{\text{total}} = \Delta\varphi_{\text{dyn}} + \Delta\varphi(C) .$$

This additional angle $\Delta\varphi(C)$ is a classical analog of the quantum geometric phase and for this reason is called the *classical geometric phase*.

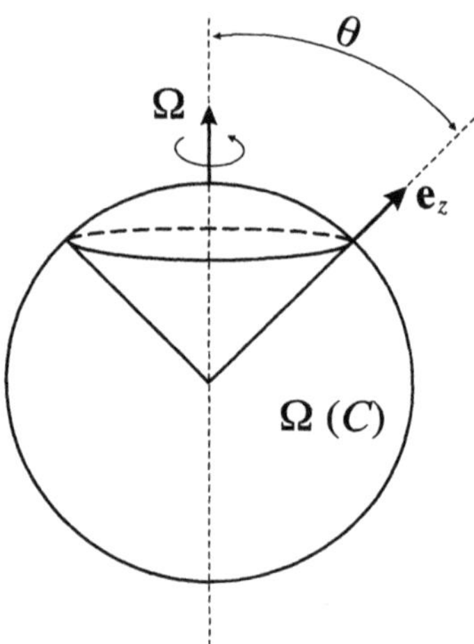

Figure 3.1: Foucault pendulum

Example 3.1.1 (Foucault pendulum) The precession of a Foucault pendulum is perhaps the most transparent illustration of the appearance of the classical geometric phase. The standard treatment describes the pendulum motion with respect to a noninertial frame rotating with the surface of the Earth (see any book on classical mechanics, e.g., Goldstein 1950, Landau and Lifshitz 1976), with the following equation of motion:

$$m\ddot{\mathbf{q}} = \mathbf{F} - m\left[2\boldsymbol{\Omega} \times \dot{\mathbf{q}} + \boldsymbol{\Omega} \times (\boldsymbol{\Omega} \times \mathbf{q})\right] , \tag{3.1}$$

where $\boldsymbol{\Omega}$ stands for the constant angular velocity of the Earth. In the above formula, $\mathbf{F}$ is simply the gravitational force and the next two terms on the r.h.s. are the Coriolis and centrifugal forces, in that order. Let ω denote the frequency of the small oscillations of the pendulum. It is evident that

$$\omega \gg \Omega = |\boldsymbol{\Omega}| . \tag{3.2}$$

Take the coordinate axes $\mathbf{e}_x$, $\mathbf{e}_y$, $\mathbf{e}_z$ of the noninertial frame such that $\mathbf{e}_x$ and $\mathbf{e}_y$ are tangent to the surface of the Earth, and $\mathbf{e}_z = \mathbf{e}_x \times \mathbf{e}_y$, i.e., $\mathbf{e}_z$ points along the Earth's radius (see Fig. 3.1).

In the regime of small oscillations one has $\dot{z} \approx 0$. Moreover, one may neglect the centrifugal force, which behaves like Ω^2. Therefore, in such an approximation, the dynamics of the pendulum is described by the following set of equations:

$$\begin{aligned} \ddot{x} &= -\omega^2 x + 2\dot{y}\Omega_z , \\ \ddot{y} &= -\omega^2 y - 2\dot{x}\Omega_z , \end{aligned} \tag{3.3}$$

with $\Omega_z = \Omega \cos\theta$, and θ is a constant latitude measured relative to the north pole — see Fig. 3.1. Defining a complex variable $w := x + iy$, one may rewrite the above system of equations as follows:

$$\ddot{w} + 2i\Omega_z \dot{w} + \omega^2 w = 0 \ . \tag{3.4}$$

Looking for a solution of the form $w(t) = e^{i\lambda t}$, one easily finds

$$\lambda = -\Omega_z \pm \sqrt{\Omega_z^2 + \omega^2} \approx -\Omega_z \pm \omega \ , \tag{3.5}$$

where, once again, we have used (3.2). Hence, the approximate solution to equation (3.4) reads

$$w(t) = e^{-i\Omega_z t} \left(c_1 e^{i\omega t} + c_2 e^{-i\omega t} \right) \ . \tag{3.6}$$

Note that the role of the Coriolis force is to rotate the invariant plane of oscillation of the pendulum with angular velocity Ω_z. Therefore, after one revolution of the Earth, i.e., after $T = 2\pi/\Omega = 24\mathrm{h}$, the net rotation of the plane of oscillation will be

$$\Delta\varphi = \Omega_z T = 2\pi \cos\theta \ . \tag{3.7}$$

A remarkable feature of this result is that it is independent of Ω (provided formula (3.2) holds), i.e., the net rotation does not depend on the rate at which the closed curve $C = \{\theta = \mathrm{const.}\}$ is traversed by the pendulum. This is a typical feature of the geometric phase. Note that modulo 2π, $\Delta\varphi$ equals to the solid angle enclosed by C:

$$\Delta\varphi = \Omega(C) \quad \mathrm{modulo} \ 2\pi \ . \tag{3.8}$$

This shows that the net rotation $\Delta\varphi$ of the invariant plane of the Foucault pendulum is a purely geometric effect.
$\diamond$

It is clear that the simple picture above holds only for periodic systems. However, there is an important class of classical hamiltonian systems that share this property — they are periodic (or multiply periodic) in the classical phase space $\mathcal{P}$. The crucial feature of these systems, which are called *integrable systems*, is that the evolution takes place on an n-dimensional torus T^n (with n being the number of degrees of freedom), in which case one has n frequencies $(\omega_1, \dots, \omega_n)$ and n classical phases (angles) $\Delta\varphi_1, \dots, \Delta\varphi_n$.

This fact was first observed by John Hannay at the university of Bristol in 1985 and, hence, the additional geometric phases (angles) $\Delta\varphi_i(C)$ are called *Hannay's angles*. The present chapter derives the formula for these angles together with a necessary introduction to integrable systems and the adiabatic theorem in classical mechanics.

3.1.2 Symplectic geometry and Hamiltonian dynamics

Any dynamical system contains two ingredients: a phase space $\mathcal{P}$, i.e., a space of physical states, and a vector field X on $\mathcal{P}$ that defines the dynamics. Then the evolution of the state, $t \longrightarrow x(t)$, satisfies the following dynamical equation on $\mathcal{P}$:

$$\dot{x}(t) = X(x(t)) \ , \qquad x(0) = x_0 \ . \tag{3.9}$$

Hamiltonian systems define a particular class of dynamical systems. A hamiltonian system is usually defined by a set of the Hamilton equations:

$$\dot{q}^i = \frac{\partial H}{\partial p_i}\,, \qquad \dot{p}_i = -\frac{\partial H}{\partial q^i}\,, \tag{3.10}$$

where $(q^1, \ldots, q^n)$ are generalized coordinates and $(p_1, \ldots, p_n)$ are the corresponding conjugate momenta. It turns out that the above canonical equations follow from a beautiful geometric structure encoded into the space of states $\mathcal{P}$.

Definition 3.1.1 *A symplectic manifold is a pair $(\mathcal{P}, \Omega)$ where $\mathcal{P}$ is a manifold and Ω is a closed, nondegenerate two-form on $\mathcal{P}$, that is,*

- *$d\Omega = 0$, and*

- *if $\Omega_p(u, v) = 0$ for any vector $u \in T_p\mathcal{P}$, then $v = 0$.*

One calls Ω a symplectic form.

The nondegeneracy of Ω implies that $\mathcal{P}$ is of even dimension, say $2n$, and

$$\overbrace{\Omega \wedge \ldots \wedge \Omega}^{n \text{ times}}$$

defines a volume element on $\mathcal{P}$. Let $(x^1, \ldots, x^{2n})$ be a local coordinate system on $\mathcal{P}$. The condition $d\Omega = 0$ is equivalent to the following set of equations for the components Ω_{ij}:

$$\partial_i \Omega_{jk} + \partial_k \Omega_{ij} + \partial_j \Omega_{ki} = 0\,, \tag{3.11}$$

where as usual

$$\Omega = \frac{1}{2} \sum_{i,j=1}^{2n} \Omega_{ij}\, dx^i \wedge dx^j\,.$$

Let us consider some examples of symplectic manifolds.

Example 3.1.2 (Symplectic vector space) Let $\mathcal{P} = \mathbb{R}^{2n}$ and define Ω as follows:

$$\Omega = \begin{pmatrix} 0 & \mathbb{1}_n \\ -\mathbb{1}_n & 0 \end{pmatrix}\,. \tag{3.12}$$

Since Ω is constant it is evidently closed. $\diamond$

Our next example plays a prominent role in classical mechanics.

Example 3.1.3 (Cotangent bundle) Let Q be an n-dimensional manifold and $\mathcal{P} = T^*Q$ be the corresponding cotangent bundle (cf. Example 1.3.2). Introducing the canonical bundle projection

$$\pi_Q : T^*Q \longrightarrow Q\,,$$

one defines the following one-form on $\mathcal{P}$:

$$\Theta_p(v) := p(T\pi_Q(v)) , \qquad (3.13)$$

where $p \in \mathcal{P}$, $v \in T_p\mathcal{P}$ and

$$T\pi_Q : T\mathcal{P} \longrightarrow TQ ,$$

is a tangent map of π_Q. Now, the point $p \in \mathcal{P}$ defines a one-form on Q and the r.h.s. of (3.13) denotes the evaluation of p on a vector $T\pi_Q(v)$ tangent to Q. One defines

$$\Omega = -d\Theta , \qquad (3.14)$$

from which it follows that Ω is a symplectic form on T^*Q. Choosing a local coordinate system on T^*Q: $(x^i) = (q^1, \ldots, q^n, p_1, \ldots, p_n)$ it is easy to see that

$$\Theta = \sum_{i=1}^{n} p_i dq^i ,$$

and hence one obtains the following well-known formula for Ω:

$$\Omega = \sum_{i=1}^{n} dq^i \wedge dp_i , \qquad (3.15)$$

which proves that (T^*Q, Ω) is a symplectic manifold. $\qquad\qquad\diamond$

Most of the hamiltonian systems from classical textbooks have cotangent bundles as their phase spaces. The manifold Q is called a configuration space and the fibre T_q^*Q contains all possible momenta. Note that Ω is constant in local coordinates (q^i, p_i) and has the same form as (3.12). This is not a coincidence; due to the Darboux theorem any symplectic form is locally constant, i.e., it is given by the formula (3.15), which is called the *canonical form* of Ω. The corresponding coordinates (q^i, p_i) are called *canonical coordinates*. Clearly, the symplectic vector space $\mathbb{R}^{2n} = T\mathbb{R}^n$ also defines a cotangent bundle, and cartesian coordinates on $\mathbb{R}^{2n}$ are canonical ones. However, there are also important symplectic manifolds which are not cotangent bundles.

Example 3.1.4 (Sphere S^2) A unit two-dimensional sphere S^2 is a symplectic manifold with the symplectic form being proportional to the standard area element:

$$\Omega = a \sin\theta \, d\theta \wedge d\varphi , \quad a \in \mathbb{R} . \qquad (3.16)$$

Note, that the spherical angles θ and φ are not canonical. Introducing a new variable $I = -a\cos\theta$, one finds

$$\Omega = dI \wedge d\varphi ,$$

which shows that (φ, I) in fact defines a canonical pair. $\qquad\qquad\diamond$

Consider now a smooth function f on $\mathcal{P}$; one calls f a classical observable.

Definition 3.1.2 *The vector field X_f on $\mathcal{P}$ defined by:*

$$i_{X_f}\Omega \equiv \Omega(X_f,\,\cdot\,) = df \,, \tag{3.17}$$

is called the hamiltonian vector field corresponding to f.

In local coordinates (x^i) on $\mathcal{P}$, the components of X_f with respect to the coordinate basis $\partial/\partial x^i$, are given by

$$(X_f)^i = \sum_{j=1}^{2n} \Omega^{ji}\,\partial_j f \,, \tag{3.18}$$

where Ω^{ij} is the inverse of Ω_{ij} (its existence is guaranteed by the nondegeneracy of Ω), i.e.,

$$\sum_{i=1}^{2n} \Omega_{ki}\,\Omega^{im} = \delta^m_{\,k} \,. \tag{3.19}$$

The above construction enables one to introduce the Poisson bracket in the space of classical observables, as follows:

$$\{\,,\,\} : C^\infty(\mathcal{P}) \times C^\infty(\mathcal{P}) \longrightarrow C^\infty(\mathcal{P}) \,,$$

defined by

$$\{F, G\} := \Omega(X_F, X_G) = \sum_{i,j=1}^{2n} \Omega^{ij}\,\frac{\partial F}{\partial x^i}\frac{\partial G}{\partial x^j} \,, \tag{3.20}$$

where X_F and X_G are hamiltonian vector fields corresponding to F and G, respectively. In local canonical coordinates (q^i, p_i), one recovers the standard formula

$$\{F, G\} = \sum_{i=1}^{n} \left(\frac{\partial F}{\partial q^i}\frac{\partial G}{\partial p_i} - \frac{\partial G}{\partial q^i}\frac{\partial F}{\partial p_i} \right) \,.$$

Now we are ready to present the general definition of a hamiltonian system: it is a triple $(\mathcal{P}, \Omega, H)$, with $(\mathcal{P}, \Omega)$ being a symplectic manifold, and H a smooth function on $\mathcal{P}$. The Hamiltonian dynamics on $(\mathcal{P}, \Omega)$ is defined by

$$\dot{x} = X_H(x) \,, \tag{3.21}$$

where X_H is the hamiltonian vector field corresponding to H. In local canonical coordinates (q^i, p_i) formula (3.21) reduces to the standard canonical equations (3.10):

$$\dot{q}^i = \frac{\partial H}{\partial p_i} = \{q^i, H\} \,, \qquad \dot{p}_i = -\frac{\partial H}{\partial q^i} = \{p_i, H\} \,, \tag{3.22}$$

for $i = 1, \ldots, n$.

Let us consider two symplectic manifolds, (P_1, Ω_1) and (P_2, Ω_2). A mapping

$$\varphi : P_1 \longrightarrow P_2$$

is called *symplectic* or *canonical* if

$$\varphi^* \Omega_2 = \Omega_1 . \tag{3.23}$$

In particular, if $P_1 = P_2 = P$ one speaks of canonical (or symplectic) transformations of P. These are transformations preserving the symplectic form Ω on P. Obviously they form a group that is a subgroup of $\mathrm{Diff}(P)$.

Example 3.1.5 Consider that symplectic vector space $\mathbb{R}^{2n}$ from Example 3.1.2. Linear canonical transformations of $\mathbb{R}^{2n}$ form a group called a *symplectic group*:

$$Sp(n) := \{ A \in GL(2n, \mathbb{R}) \mid A \cdot \Omega \cdot A^T = \Omega \} , \tag{3.24}$$

with Ω given by (3.12). $\diamond$

3.1.3 Integrable systems

Let (P, Ω, H) be a hamiltonian system. Recall that a function $F : P \to \mathbb{R}$ is a first integral, or a constant of the motion of the hamiltonian system, if

$$\{F, H\} = 0 . \tag{3.25}$$

The existence of first integrals plays a crucial role in solving the corresponding evolution equation, since it enables one to reduce the number of degrees of freedom. A hamiltonian system always has at least one constant of motion — a Hamiltonian. A system with n degrees of freedom has maximally $2n$ independent constants of motion. If we know them then the trajectory is completely determined. However, it turns out that frequently one needs only n first integrals to solve the system completely.

Definition 3.1.3 *A hamiltonian system is said to be integrable iff there exists* $n = (\dim P)/2$ *constants of motion* F_i, *such that*

- $\{F_i, F_j\} = 0 , \quad i, j = 1, 2, \ldots, n$;

- *the functions* F_i *are independent on a level set of* $\mathbf{F} = (F_1, \ldots, F_n)$ *at a point* $\mathbf{f} = (f_1, \ldots, f_n) \in \mathbb{R}^n$. *That is, on*

$$\mathbf{F}^{-1}(\mathbf{f}) := \{x \in P \mid F_i(x) = f_i ; i = 1, \ldots, n \} ,$$

we have

$$dF_1 \wedge dF_2 \wedge \ldots \wedge dF_n \neq 0 .$$

Most books on classical mechanics emphasize the treatment of integrable systems, which form an exceptional class of hamiltonian systems, in that they are soluble. If a system is integrable then the n functions F_i can be used as new coordinates.

Theorem 3.1.1 (Liouville) *For any integrable system there exist n functions φ_i on $\mathcal{P}$, such that*

$$\{\varphi_i, \varphi_j\} = 0 \quad \text{and} \quad \{\varphi_i, F_j\} = \delta_{ij} \, .$$

The functions φ_i are determined up to the following transformation:

$$\varphi_i \longrightarrow \varphi_i + \frac{\partial K}{\partial F_i} \, ,$$

with an arbitrary function $K : \mathbb{R}^n \longrightarrow \mathbb{R}$.

Theorem 3.1.2 (Arnold) *Suppose that a level set $\mathbf{F}^{-1}(\mathbf{f})$ is compact and connected. Then it is diffeomorphic to an n-dimensional torus:*

$$\mathbf{F}^{-1}(\mathbf{f}) \cong T^n = \{ \, (\varphi_1, \ldots, \varphi_n) \mod 2\pi \, \} \, .$$

Moreover, the hamiltonian flow on $\mathbf{F}^{-1}(\mathbf{f})$ is quasi-periodic, i.e.,

$$\frac{d\varphi_i}{dt} = \omega_i(\mathbf{f}) \, , \quad i = 1, \ldots, n \, . \tag{3.26}$$

The quantities $\omega = (\omega_1, \ldots, \omega_n)$ are called the frequencies of the quasi-periodic motion. They are independent if

$$k_1\omega_1 + \ldots + k_n\omega_n = 0 \, , \quad \text{with } k_i \in \mathbb{Z} \, , \tag{3.27}$$

implies $k_1 = \ldots = k_n = 0$. If the above condition does not hold, the frequency vector Ω is called *resonant*. One can show that for independent frequencies the trajectory is dense on T^n, whereas in the resonant case the motion is strictly periodic, i.e., the trajectory is closed.

Let γ_i be a basic one-dimensional cycle of $\mathbf{F}^{-1}(\mathbf{f}) \cong T^n$, i.e.,

$$\oint_{\gamma_i} d\,\varphi_j = 2\pi\delta_{ij} \, , \tag{3.28}$$

for $i, j = 1, \ldots, n$. Two cycles γ_1 and γ_2 for a two-dimensional torus T^2 are shown in Fig. 3.2. Define the following quantities:

$$I_k(\mathbf{f}) := \frac{1}{2\pi} \oint_{\gamma_k} \sum_{i=1}^{n} p_i dq^i \, , \quad k = 1, \ldots, n \, . \tag{3.29}$$

They are called the *standard action* variables. If $\mathrm{Det}(\partial I_k/\partial f_l) \neq 0$, then the f's can be expressed locally in terms of I's. The canonical set $(I_1, \ldots, I_n, \varphi_1, \ldots, \varphi_n)$ is known as the (set of) action-angle variables. There exists a canonical transformation,

$$(q^1, \ldots, q^n, p_1, \ldots, p_n) \longrightarrow (I_1, \ldots, I_n, \varphi_1, \ldots, \varphi_n) \, ,$$

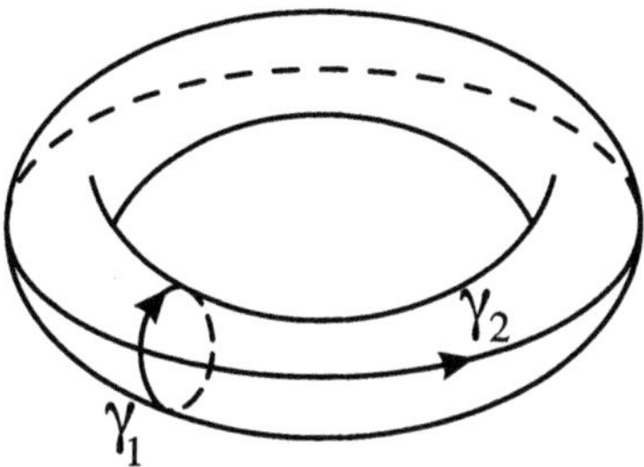

Figure 3.2: Cycles of T^2

with a generating function $S = S(\mathbf{q}, \mathbf{I})$, such that

$$\sum_{i=1}^{n} p_i dq^i + \sum_{i=1}^{n} \varphi_i dI_i = dS(\mathbf{q}, \mathbf{I}) \ . \tag{3.30}$$

The function $S(\mathbf{q}, \mathbf{I})$ may be constructed as follows: Note that the symplectic form Ω vanishes on the level set $\mathbf{F}^{-1}(\mathbf{f}) \cong T^n$. Therefore, the function $\tilde{S} : \mathbf{F}^{-1}(\mathbf{f}) \longrightarrow \mathbb{R}$ defined by

$$\tilde{S}(x) = \int_{x_0}^{x} \sum_{i=1}^{n} p_i dq^i \ ,$$

where x_0 is some point in $\mathbf{F}^{-1}(\mathbf{f})$, does not depend on the path connecting x_0 and x. Hence, the function $\tilde{S}$ is multi-valued on T^n. In particular, for a basic cycle γ_i, one obtains the following periods:

$$\Delta \tilde{S}_i = \oint_{\gamma_i} d\tilde{S} = 2\pi I_i \ .$$

Now, let us assume that in a vicinity of x_0 one may choose n coordinates $(q^1, \ldots, q^n)$ such that the level set T^n is defined by

$$p_i = p_i(\mathbf{q}, \mathbf{I}) \ , \quad i = 1, \ldots, n \ .$$

Hence, in a simply-connected neighborhood of $\mathbf{q}_0 = \mathbf{q}(x_0) \subset \mathbb{R}^n$ one may define a single-valued function $S : \mathbb{R}^n \times \mathbb{R}^n \longrightarrow \mathbb{R}$, given by

$$S(\mathbf{q}, \mathbf{I}) = \int_{\mathbf{q}_0}^{\mathbf{q}} \sum_{i=1}^{n} p_i(\mathbf{q}, \mathbf{I}) \, dq^i \ , \tag{3.31}$$

which is a generating function of the canonical transformation $(q, p) \longrightarrow (\varphi, I)$, that is,

$$p_i = \frac{\partial S}{\partial q^i} \ , \qquad \varphi_i = \frac{\partial S}{\partial I_i} \ , \tag{3.32}$$

for $i = 1, 2, \ldots, n$. The Hamilton equations have, in action-angle variables, the especially simple form

$$\frac{dI_i}{dt} = 0, \qquad \frac{d\varphi_i}{dt} = \omega_i, \quad i = 1, \ldots, n, \tag{3.33}$$

with an obvious solution:

$$\mathbf{I}(t) = \mathbf{I}(0), \qquad \boldsymbol{\varphi}(t) = \boldsymbol{\varphi}(0) + \boldsymbol{\omega}t. \tag{3.34}$$

The term "angle variables" makes sense in that φ_j changes by 2π if the jth cycle of T^n is circled. Note the striking similarity of angle variables and quantum phases.

Example 3.1.6 Consider a harmonic oscillator with one degree of freedom. Introducing cartesian coordinates $q := x^1$ and $p := x^2$ in $\mathcal{P} = \mathbb{R}^2$, one has $\Omega = dp \wedge dq$ and the Hamiltonian of the oscillator reads

$$H(q, p) = \frac{p^2}{2m} + \frac{m\omega^2 q^2}{2}.$$

Defining a new canonical set

$$Q = \sqrt{m\omega}\, q, \qquad P = \frac{p}{\sqrt{m\omega}},$$

one finds that

$$H(Q, P) = \frac{\omega}{2}(P^2 + Q^2),$$

and the solution of the Hamilton's equations reads

$$Q(t) = A \cos \varphi(t), \qquad P(t) = A \sin \varphi(t), \tag{3.35}$$

where $\varphi(t) = \omega t + \varphi_0$. In the above formulae, the amplitude $A = \sqrt{Q_0^2 + P_0^2}$, where (Q_0, P_0) stands for an initial state. Integral curves of the hamiltonian flow define circles in $\mathbb{R}^2$: $Q^2 + P^2 = A^2$. Each circle is nothing but a one-dimensional torus from the Arnold theorem.

Let us construct the action-angle variables. Note that in polar coordinates (r, φ) on $\mathbb{R}^2$, the symplectic form reads:

$$\Omega = dP \wedge dQ = r dr \wedge d\varphi = d\left(\frac{1}{2}r^2\right) \wedge d\varphi, \tag{3.36}$$

that is, $(r^2/2, \varphi)$ define the canonical coordinates. Also note that the Hamiltonian is

$$H = \omega \frac{r^2}{2}, \tag{3.37}$$

and $r^2/2$ defines the standard action:

$$\frac{1}{2}r^2 = \frac{1}{2\pi}\pi r^2 = \frac{1}{2\pi}\int_{Q^2+P^2\leq r^2} r\,dr \wedge d\varphi$$

$$= \frac{1}{2\pi}\int_{Q^2+P^2\leq r^2} dP \wedge dQ = \frac{1}{2\pi}\oint_{Q^2+P^2=r^2} P\,dQ =: I \ , \quad (3.38)$$

where we have used the Stokes theorem on $\mathbb{R}^2$. Hence, the standard action has the following simple physical interpretation:

$$I = \frac{1}{2}(P^2 + Q^2) = \frac{\text{Energy}}{\omega} \ , \qquad (3.39)$$

and the Hamiltonian

$$H(I, \varphi) = \omega I \ ,$$

gives

$$\dot{I} = 0 \ , \qquad \dot{\varphi} = \omega \ .$$

Hence the oscillator frequency ω coincides with the frequency of the trajectory on the Arnold torus. $\diamond$

3.2   Adiabatic phase of Hannay

3.2.1   Averaging principle

Consider an integrable hamiltonian system defined on a $2n$-dimensional phase space $\mathcal{P}$. Introducing local action-angle variables $(\mathbf{I}, \boldsymbol{\varphi})$, the dynamics of the system is described by the following set of equations:

$$\dot{\mathbf{I}} = 0 \ , \quad \dot{\boldsymbol{\varphi}} = \boldsymbol{\omega} = \frac{\partial H_0}{\partial \mathbf{I}} \ , \qquad (3.40)$$

where H_0 denotes the Hamiltonian. The solution to (3.40) defines a quasi-periodic motion on an n-dimensional torus, i.e.,

$$\boldsymbol{\varphi}(t) = \boldsymbol{\varphi}_0 + \boldsymbol{\omega}t \ .$$

Consider now a "slightly" perturbed system, in which

$$\dot{\boldsymbol{\varphi}} = \boldsymbol{\omega}(\mathbf{I}) + \epsilon\,\mathbf{f}(\mathbf{I}, \boldsymbol{\varphi}) \ ,$$
$$\dot{\mathbf{I}} = \epsilon\,\mathbf{g}(\mathbf{I}, \boldsymbol{\varphi}) \ , \qquad (3.41)$$

where $\mathbf{f} = (f_1, \ldots, f_n)$ and $\mathbf{g} = (g_1, \ldots, g_n)$. In the unperturbed system the action variables $\mathbf{I}$ are constants of motion. This is no longer true in the perturbed system.

However, if the perturbation is "small" (i.e., $\epsilon \ll 1$) one expects that $\mathbf{I}$ will evolve "slowly" in time, i.e., much slower than, e.g., the angle variables $\boldsymbol{\varphi}$. Let us replace the real system described by (3.41) by the averaged one:

$$\dot{\mathbf{J}} = \epsilon \, \langle \mathbf{g} \rangle (\mathbf{J}) \,, \tag{3.42}$$

where we denote by $\mathbf{J}$ the averaged form of I, and where $\langle \mathbf{g} \rangle$ denotes the torus average of $\mathbf{g}$:

$$\langle \mathbf{g} \rangle := \frac{1}{(2\pi)^n} \int_0^{2\pi} \cdots \int_0^{2\pi} \mathbf{g}(\varphi_1, \ldots \varphi_n) d\varphi_1 \ldots d\varphi_n \,. \tag{3.43}$$

Perceived wisdom says that the averaged system (3.42) defines a *good approximation* to the real one. It should be stressed that this is not a mathematical theorem but rather a statement based on physical intuition. Moreover, in general this statement is not true! However, for a system with one degree of freedom, the averaging principle becomes a theorem.

Theorem 3.2.1 *Consider the following system on* $\mathbb{R}^2$:

$$\begin{aligned}
\dot{\varphi} &= \omega(I) + \epsilon \, f(I, \varphi) \,, \\
\dot{I} &= \epsilon \, g(I, \varphi) \,,
\end{aligned}$$

where both f and g are periodic, i.e., $f(I, \varphi) = f(I, \varphi + 2\pi)$ and the same holds for g. Moreover, one assumes some regularity conditions upon the functions ω, f and g (see Arnold 1989 for details). If $\omega \neq 0$, then the difference between the real, $I(t)$, and averaged, $J(t)$, motions satisfies

$$|I(t) - J(t)| < C \epsilon \,, \tag{3.44}$$

for all $0 \leq t \leq 1/\epsilon$, where the constant C does not depend on ϵ.

To illustrate the above theorem, called also the *averaging principle*, in accordance with the discussion above, consider the following

Example 3.2.1 Define a perturbed system on $\mathbb{R}^2$ by:

$$\dot{\varphi} = \omega \,, \quad \dot{I} = \epsilon \, (a + b \cos \varphi) \,,$$

i.e., $g(I, \varphi) = a + b \cos \varphi$. One easily finds the following solution:

$$I(t) = I_0 + \epsilon \, a t + \epsilon \, b \frac{\sin(\omega t + \varphi_0)}{\omega} \,, \tag{3.45}$$

where (I_0, φ_0) denotes the initial condition. The corresponding averaged system is given by

$$\dot{J} = \epsilon \, a \,,$$

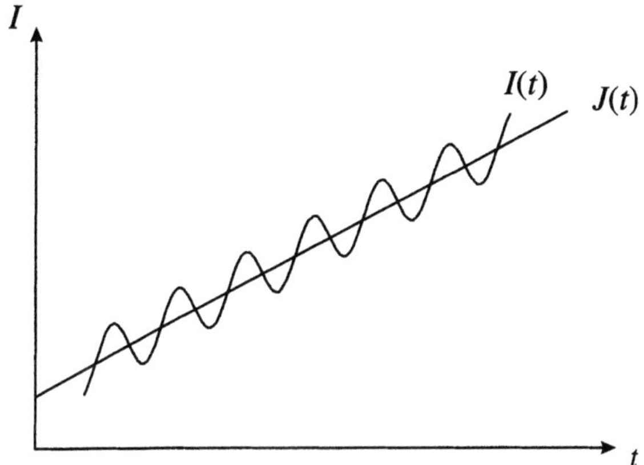

Figure 3.3: Real and averaged motions

and its solution reads

$$J(t) = I_0 + \epsilon\, a t .$$

Hence, the difference between the real and averaged motions,

$$J(t) - I(t) = \epsilon\, b \frac{\sin(\omega t + \varphi_0)}{\omega} , \qquad (3.46)$$

is a "small" oscillatory term (see Fig. 3.3). $\diamond$

3.2.2 Adiabatic invariants

To formulate the adiabatic limit of classical Hamiltonian dynamics it is convenient to replace physical time t by a rescaled time $\tau := \epsilon t$.[1] Using τ, the Hamilton equations take the following form:

$$\mathbf{q}'(\tau) = \frac{1}{\epsilon}\frac{\partial H}{\partial \mathbf{p}} , \quad \mathbf{p}'(\tau) = -\frac{1}{\epsilon}\frac{\partial H}{\partial \mathbf{q}} , \qquad (3.47)$$

where $\mathbf{q}' = d\mathbf{q}/d\tau$. The classical adiabatic limit, or the limit of "infinitely slow" changes of the Hamiltonian, corresponds to the limit $\epsilon \to 0$.

Definition 3.2.1 *A quantity $F(\mathbf{q}(t), \mathbf{p}(t); \epsilon t)$ is called an adiabatic invariant of the system (3.47), if for any $\kappa > 0$ there exists $\epsilon_0 > 0$ such that, for any $\epsilon < \epsilon_0$ and $0 < t < 1/\epsilon$, the following inequality holds:*

$$|F(\mathbf{q}(t), \mathbf{p}(t); \epsilon t) - F(\mathbf{q}(0), \mathbf{p}(0); 0)| < \kappa . \qquad (3.48)$$

---

[1] In Chapter 3 we used t/T. However, most authors prefer to use ϵt when dealing with the classical adiabatic theorem.

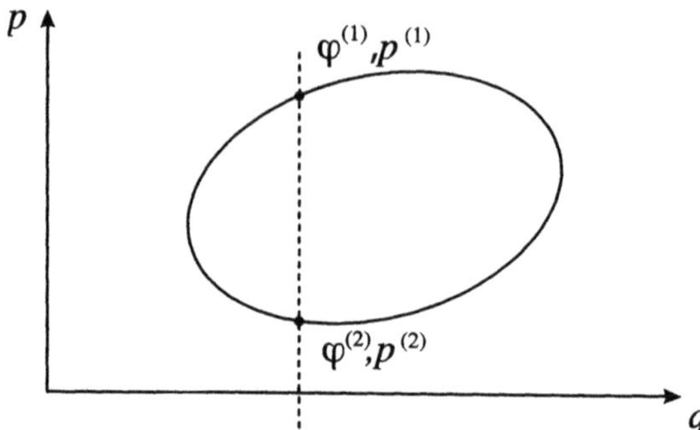

Figure 3.4: Multivaluedness of the map $q \longrightarrow \varphi$

Consider a time-dependent hamiltonian system with one degree of freedom, and suppose that the time dependence of the Hamiltonian $H = H(t)$ enters via the time dependence of some external parameters. Let M be the parameter manifold and consider

$$H : \mathcal{P} \times M \longrightarrow \mathbb{R} , \tag{3.49}$$

where $\mathcal{P}$ denotes the system's phase space. We shall write $H = H(q, p; x)$ with $x \in M$. For a fixed x our system is completely integrable (as a system with one degree of freedom). Therefore it admits local action-angle variables with a standard action

$$I(q, p; x) = \frac{1}{2\pi} \oint p dq , \tag{3.50}$$

where the integral is performed along a closed integral curve (Arnold one-dimensional torus) in $\mathcal{P}$.

Theorem 3.2.2 (Classical adiabatic theorem) *If the frequency of the quasi-periodic motion $\omega(I; x) = \partial H(I; x)/\partial I$ does not vanish, the standard action $I(q, p, x)$ defines an adiabatic invariant.*

Proof. For any $x \in M$ we may perform a canonical transformation to action-angle variables:

$$(q, p) \longrightarrow (I, \varphi) . \tag{3.51}$$

Let $S = S(q, I; x)$ be the x-dependent generating function of this transformation. This function is multi-valued since on a torus T^1, defined by the action variable $I(x)$, the corresponding angle variable φ is a multi-valued function of q (see Fig. 3.4). To label different branches of S we shall write $S^{(\alpha)}$. From the very definition of the generating function one has

$$p^{(\alpha)}dq + \varphi^{(\alpha)}dI = dS^{(\alpha)}(q, I; x) , \tag{3.52}$$

which implies that

$$p^{(\alpha)} = \frac{\partial S^{(\alpha)}}{\partial q} , \qquad \varphi^{(\alpha)} = \frac{\partial S^{(\alpha)}}{\partial I} . \tag{3.53}$$

The new, transformed Hamiltonian $K(I, \varphi; x)$ reads

$$K(I, \varphi; x) = H_0(I; x) + \frac{\partial}{\partial t} S^{(\alpha)}(q(I, \varphi; x), I; x) , \tag{3.54}$$

where

$$H_0(I; x) \equiv H(q(I, \varphi; x), p(I, \varphi; x); x) \tag{3.55}$$

is just the original Hamiltonian expressed in terms of the action variable I. To make K single-valued we have to make the $S^{(\alpha)}$ single-valued by specifying the values of φ to which q in $S^{(\alpha)}$ refers. Note that $\mathfrak{S}$, defined by

$$\mathfrak{S}(\varphi, I; x) := S^{(\alpha)}(q(\varphi, I; x), I; x) , \qquad 0 \le \varphi < 2\pi , \tag{3.56}$$

is single-valued; indeed, x, φ and I uniquely define q and p. Now take a local co-ordinate system $(x^1, \ldots, x^n)$ on M and let $x^i = x^i(t)$ describe the evolution of the external parameters. To compute $\partial_t S^{(\alpha)}$, note that

$$\frac{\partial S^{(\alpha)}}{\partial t} = \sum_{i=1}^{n} \frac{\partial S^{(\alpha)}}{\partial x^i} \dot{x}^i .$$

Moreover, using the definition of $\mathfrak{S}$ we have:

$$\frac{\partial \mathfrak{S}}{\partial x^i} = \frac{\partial S^{(\alpha)}}{\partial q} \frac{\partial q}{\partial x^i} + \frac{\partial S^{(\alpha)}}{\partial x^i} = p^{(\alpha)} \frac{\partial q}{\partial x^i} + \frac{\partial S^{(\alpha)}}{\partial x^i} , \tag{3.57}$$

and hence

$$\frac{\partial S^{(\alpha)}}{\partial t} = \sum_{i=1}^{n} \left[\frac{\partial \mathfrak{S}}{\partial x^i} - p^{(\alpha)} \frac{\partial q}{\partial x^i} \right] \dot{x}^i . \tag{3.58}$$

Note, however, that $p^{(\alpha)}$ is uniquely defined by (x, φ, I), and hence we may omit the index α. Finally, we obtain the following formula for the new Hamiltonian K:

$$K(\varphi, I; x) = H_0(I; x) + \sum_{i=1}^{n} \left[\frac{\partial \mathfrak{S}}{\partial x^i} - p \frac{\partial q}{\partial x^i} \right] \dot{x}^i , \tag{3.59}$$

where position q and momentum p are uniquely defined by

$$p = p(I, \varphi; x) , \qquad q = q(I, \varphi; x) . \tag{3.60}$$

Therefore, the Hamilton equations have the following form:

$$\dot{\varphi} = \frac{\partial K}{\partial I} = \omega(I; x) + \frac{\partial}{\partial I} \sum_{i=1}^{n} \left[\frac{\partial \mathfrak{S}}{\partial x^i} - p \frac{\partial q}{\partial x^i} \right] \dot{x}^i , \qquad (3.61)$$

$$\dot{I} = -\frac{\partial K}{\partial \varphi} = -\frac{\partial}{\partial \varphi} \sum_{i=1}^{n} \left[\frac{\partial \mathfrak{S}}{\partial x^i} - p \frac{\partial q}{\partial x^i} \right] \dot{x}^i , \qquad (3.62)$$

with $\omega(I; x) := \partial H_0 / \partial I$. Note that using the rescaled time $\tau = \epsilon t$, the Hamilton equations (3.61) and (3.62) become:

$$\dot{\varphi} = \omega(I; x) + \epsilon f(I, \varphi; x) , \qquad (3.63)$$
$$\dot{I} = \epsilon g(I, \varphi; x) , \qquad (3.64)$$

with

$$f = \frac{\partial}{\partial I} \sum_{i=1}^{n} \left[\frac{\partial \mathfrak{S}}{\partial x^i} - p \frac{\partial q}{\partial x^i} \right] \frac{dx^i}{d\tau} , \qquad (3.65)$$

and

$$g = -\frac{\partial}{\partial \varphi} \sum_{i=1}^{n} \left[\frac{\partial \mathfrak{S}}{\partial x^i} - p \frac{\partial q}{\partial x^i} \right] \frac{dx^i}{d\tau} . \qquad (3.66)$$

Consider now an averaged system, with

$$\dot{J} = \epsilon \langle g \rangle . \qquad (3.67)$$

Clearly, due to (3.66), we have

$$\langle g \rangle \equiv 0 ,$$

and hence J defines a constant of motion, i.e.,

$$J(t) = J(0) \equiv I(0) .$$

Applying the *averaging principle*, that is Theorem 3.2.1, one obtains

$$|I(t) - I(0)| = |I(t) - J(t)| < C\epsilon \qquad (3.68)$$

for all $0 \leq t \leq 1/\epsilon$. Therefore, the action variable is an adiabatic invariant. $\qquad \square$

Example 3.2.2 Consider once more the harmonic oscillator from Example 3.1.6. Clearly, if the oscillator frequency ω depends on time, then the energy of the oscillator is not conserved. However, if ω varies in time sufficiently slowly then the adiabatic theorem says that there is another quantity, the action,

$$J = \frac{\mathrm{Energy}(t)}{\omega(t)} ,$$

which is conserved. Hence, in the adiabatic limit the above combination of two time-dependent quantities (energy and frequency) is time independent. Geometrically, this means that the area of "the phase space" ellipse defined by

$$\frac{p^2}{2m} + \frac{m\omega^2(t)q^2}{2} = E(t) ,$$

is an adiabatic invariant.

Example 3.2.3 Consider a charged (nonrelativistic) particle interacting with a homogeneous magnetic field $\mathbf{B}$ pointing in the z-direction, i.e., $\mathbf{B}(t) = B(t)\mathbf{e}_z$. The system is defined by the following Lagrangian function:

$$L(\mathbf{x}, \mathbf{v}) = \frac{m\mathbf{v}^2}{2} + \frac{e}{c}\mathbf{v} \cdot \mathbf{A} , \tag{3.69}$$

where $\mathbf{A}$ is a vector potential, and e, m, c denote the electric charge and mass and the velocity of light, respectively. The momentum $\mathbf{P}$ canonically conjugate to the particle position $\mathbf{x}$ reads

$$\mathbf{P} := \frac{\partial L}{\partial \mathbf{v}} = \mathbf{p} + \frac{e}{c}\mathbf{A} , \tag{3.70}$$

where $\mathbf{p} = m\mathbf{v}$ denotes the standard kinetic momentum. The particle motion is described by the Lorentz equation, i.e.,

$$\frac{d\mathbf{p}}{dt} = \frac{e}{c}\mathbf{v} \times \mathbf{B} . \tag{3.71}$$

It is well known (see, e.g., Jackson 1999) that in the case of a static magnetic field, the projection of the particle motion onto the plane perpendicular to $\mathbf{B}$ is circular, with a frequency

$$\omega_B = \frac{eB}{mc} . \tag{3.72}$$

Now, for a uniform field $\mathbf{B}$ the vector potential is (up to a gauge transformation) given by $\mathbf{A} = \frac{1}{2}\mathbf{x} \times \mathbf{B}$ and, therefore, using polar coordinates (r, φ) in the xy-plane, one easily finds from (3.70) that

$$P_\varphi = mr^2\dot\varphi + \frac{e}{2c}Br^2 = -\frac{e}{2c}Br^2 , \tag{3.73}$$

where we have used $\dot\varphi = -\omega_B$. Hence, the J_φ action variable reads

$$J_\varphi = \frac{1}{2\pi} \oint P_\varphi \, d\varphi = -\frac{e}{2c}r^2B = -\frac{e}{2\pi c}\Phi_B , \tag{3.74}$$

where Φ_B denotes the magnetic flux through the circle of radius r. Therefore, the magnetic flux Φ_B defines an adiabatic invariant. This means that if the external magnetic field changes sufficiently slowly, then Φ_B is constant.

3.2.3 Hannay's angles

Consider, now, the time evolution of the angle variable φ described by (3.61). Simple integration leads to the following formula for $\varphi(t)$:

$$\varphi(t) - \varphi_0 = \int_0^t \omega(I; x)\, dt' + \frac{\partial}{\partial I} \int_0^t \sum_{i=1}^n [\partial_i \mathfrak{S} - p\, \partial_i q]\, \dot{x}^i \, dt' . \tag{3.75}$$

The first term in (3.75) is the total "dynamical angle." For an integrable time-independent system it would read

$$\varphi(t) - \varphi_0 = \omega(I) \cdot t .$$

There is, however, an additional shift defined by

$$\Delta \varphi(I, t) = \frac{\partial}{\partial I} \int_0^t \sum_{i=1}^n [\partial_i \mathfrak{S} - p\partial_i q]\, \dot{x}^i \, dt'. \tag{3.76}$$

We have already shown, in the previous section, that this extra shift comes from performing the canonical transformation $(q, p) \longrightarrow (I, \varphi)$ in the correct manner, i.e., taking into account the fact that the generating function $S^{(\alpha)}$ depends explicitly on time through the time dependence of the external parameters $x^i = x^i(t)$, so that we have to add its derivative $\partial_t S^{(\alpha)}$ to H_0 in formula (3.54) to get the correctly transformed Hamiltonian K.

Let us study the formula for $\Delta\varphi$ more carefully. Note that the integrand in (3.76) depends on time implicitly through the phase space variables q and p, and explicitly through the external parameters x^i. As we are interested in the adiabatic limit, the more natural quantity is the averaged one, so we consider letting

$$\partial_i \mathfrak{S} - p\partial_i q \longrightarrow \langle \partial_i \mathfrak{S} - p\partial_i q \rangle . \tag{3.77}$$

Let d_{M} denote the external derivative on M, i.e.,

$$d_{\mathrm{M}} f := \sum_{i=1}^n \partial_i f \, dx^i ,$$

for any function $f \in C^\infty(M)$. In this way, we may define the following one-form on M:

$$\langle d_{\mathrm{M}}\mathfrak{S} - p d_{\mathrm{M}}q \rangle := \sum_{i=1}^n \langle \partial_i \mathfrak{S} - p\partial_i q \rangle\, dx^i . \tag{3.78}$$

Now, as in the quantum case, let us consider an adiabatic, cyclic change of the external parameters, i.e., let $t \longrightarrow x_t$ be a closed curve C in M. Equation (3.76) implies that

$$\Delta \varphi(I; C) := \frac{\partial}{\partial I} \oint_C \langle d_{\mathrm{M}}\mathfrak{S} - p d_{\mathrm{M}}q \rangle = -\frac{\partial}{\partial I} \oint \langle p\, d_{\mathrm{M}}q \rangle , \tag{3.79}$$

since the integral along a closed curve of the exact one-form $\langle d_M \mathfrak{S} \rangle$ vanishes. Therefore, we may rewrite the above equation in the more transparent form

$$\Delta \varphi(I; C) = -\frac{\partial}{\partial I} \oint_C \mathcal{A}(I) , \qquad (3.80)$$

with

$$\mathcal{A}(I) := \langle p \, d_M q \rangle . \qquad (3.81)$$

The quantity $\Delta \varphi(I; C)$ is called the *Hannay angle* (Hannay 1985) corresponding to a closed curve C in the parameter space M. Therefore, the total change of the angle variable during the cyclic adiabatic evolution splits into two parts, as follows:

$$\Delta\varphi(T) = \underbrace{\int_0^T \omega(I; x_t) \, dt}_{\text{dynamical angle}} + \underbrace{\Delta\varphi(I; C)}_{\text{geometric angle}} , \qquad (3.82)$$

in perfect analogy with formula (2.59). By applying the Stokes theorem to (3.80) we can transform it into a surface integral over a two-dimensional region Σ such that $\partial \Sigma = C$, i.e.,

$$\Delta \varphi(I; C) = -\frac{\partial}{\partial I} \int_\Sigma \mathcal{W}(I) , \qquad (3.83)$$

with

$$\mathcal{W}(I) := d_M \mathcal{A}(I) = \langle d_M p \wedge d_M q \rangle . \qquad (3.84)$$

Note that on each one-dimensional torus (parametrized by the values of the external parameters), the action-angle variables (I, φ) are not uniquely defined; one may always perform a parameter-dependent angle transformation:

$$\varphi \longrightarrow \varphi + \lambda(x) . \qquad (3.85)$$

Proposition 3.2.3 *Under the angle transformation the quantity $\mathcal{A}(I)$ transforms in the following way:*

$$\frac{\partial}{\partial I} \mathcal{A}(I) \longrightarrow \frac{\partial}{\partial I} \mathcal{A}(I) + d_M \lambda , \qquad (3.86)$$

and, therefore $\partial_I \mathcal{W}(I) = d_M(\partial_I \mathcal{A}(I))$ is, in this sense, invariant.

Note, that (3.86) describes the well-known formula for the transformation of a gauge potential in an abelian gauge theory. It is an analog of the phase transformation (2.49) in quantum theory. Moreover, we have a striking correspondence between classical and quantum adiabatic objects, namely, the one-forms $A^{(n)}$ and $\mathcal{A}(I)$:

$$A^{(n)} = -\text{Im} \langle n|d_M n \rangle \longleftrightarrow \mathcal{A}(I) = \langle p \, d_M q \rangle ,$$

and their curvature two-forms $F^{(n)}$ and $\mathcal{W}(I)$:

$$F^{(n)} = -\mathrm{Im}\,\langle d_\mathrm{M}n| \wedge |d_\mathrm{M}n\rangle \longleftrightarrow \mathcal{W}(I) = \langle d_\mathrm{M}p \wedge d_\mathrm{M}q\rangle \,.$$

What about systems with many degrees of freedom? Physicists usually say that the generalization to such systems is straightforward. Note, however, that this is not so. As we have already stressed, the averaging principle is in general not true for such systems, and so neither is Hannay's formula (3.80). Nevertheless, in many important physical examples the above procedure works perfectly well, and the "straightforward" generalization is

$$\mathcal{A}(I) = \sum_{i=1}^{n} \langle p_i\, d_\mathrm{M}\,q^i \rangle \,, \tag{3.87}$$

with the jth Hannay angle defined by

$$\Delta\,\varphi_j(I;C) := -\frac{\partial}{\partial I_j}\oint_C \mathcal{A}(I) = -\frac{\partial}{\partial I_j}\int_\Sigma \mathcal{W}(I)\,, \tag{3.88}$$

where $\mathcal{W}(I) = d_\mathrm{M}\,\mathcal{A}(I)$.

3.2.4 Berry's phase versus Hannay's angle

At this stage it is interesting to study the connection between the quantal Berry phase $\gamma_n(C)$ and the classical Hannay angle $\Delta\varphi(I;C)$. In particular we want to answer the following two questions:

1. If the classical system develops a Hannay angle, will it also possesses a Berry phase when quantized?

2. If the quantum system has a Berry phase, will its classical version have a Hannay angle?

To answer these questions let us expand the two-form $F^{(n)}$ in powers of $\hbar$. Recall that

$$
\begin{aligned}
F^{(n)} &= -\mathrm{Im}\,[\,d_\mathrm{M}\,\langle n(x)|d_\mathrm{M}\,n(x)\rangle\,] \\
&= -\mathrm{Im}\left[\,d_\mathrm{M}\int d^N q\,\psi_n^*(q;x)d_\mathrm{M}\psi_n(q;x)\right],
\end{aligned}
\tag{3.89}
$$

where N denotes the number of degrees of freedom, and the wave function $\psi_n(q;x)$ corresponds to the state vector $|n(x)\rangle$ in the position representation, i.e.,

$$\psi_n(q;x) := \langle q|n(x)\rangle \,.$$

Let us first expand the wave function in powers of $\hbar$ keeping only the leading order term. It turns out (see, e.g., Berry 1983) that in the semiclassical approximation the wave function ψ_n may be expressed in the following form:

$$\psi_n(q;x) = \sum_{\alpha} a_{(\alpha)}(q,I;x)\exp\left[\frac{i}{\hbar}S^{(\alpha)}(q,I;x)\right], \tag{3.90}$$

where

$$\left(a_{(\alpha)}(q, I; x)\right)^2 := \frac{1}{(2\pi)^N} \det\left[\frac{\partial \varphi_i^{(\alpha)}}{\partial q^j}\right],$$

(3.91)

$i, j = 1, \ldots, N$. Inserting the wave formulae (3.90) and (3.91) into (3.89), and noting that the terms corresponding to different α's are strongly oscillating, one obtains the following formula for $F^{(n)}$:

$$F^{(n)} = -\frac{1}{\hbar} d_M \left\{ \int \frac{d^N q}{(2\pi)^N} \sum_\alpha \det\left[\frac{\partial \varphi_i^{(\alpha)}}{\partial q^j}\right] d_M S^{(\alpha)}(q, I; x) + O(\hbar) \right\},$$

(3.92)

where $O(\hbar)$ denotes terms at least of order $\hbar$. To calculate the integral over q's let us make the change of variables $q \longrightarrow \varphi$. Using (3.57), one has

$$
\begin{aligned}
F^{(n)} &= -\frac{1}{\hbar} d_M \left[\oint \frac{d^N \varphi}{(2\pi)^N} \left(d_M \mathfrak{S} - \sum_{k=1}^N p_k \, d_M \, q^k \right) \right] \\
&= \frac{1}{\hbar} \sum_{k=1}^N \langle d_M \, p_k \wedge d_M \, q^k \rangle = \frac{1}{\hbar} W(I) .
\end{aligned}
$$

(3.93)

Now, let C be a closed curve in the parameter manifold M. One defines the quantal Berry phase (cf. (2.60)) by

$$\gamma_n(C) = \int_\Sigma F^{(n)} ,$$

(3.94)

and the classical Hannay angles (cf. (3.88)) by

$$\Delta \varphi_j(I; C) = -\frac{\partial}{\partial I_j} \int_\Sigma W(I) , \quad j = 1, 2, \ldots, N,$$

(3.95)

where Σ is any two-dimensional region with boundary C. It follows from (3.93), and the formulae for $\gamma_n(C)$ and $\Delta \varphi_j(I; C)$, that

$$\frac{\partial \gamma_n(C)}{\partial I_j} = -\frac{1}{\hbar} \left(\Delta \varphi_j(I; C) + O(\hbar) \right) , \quad j = 1, 2, \ldots, N .$$

(3.96)

Recall, now, that in the semiclassical approximation the classical action variables I_j are quantized according to the celebrated Bohr–Sommerfeld rule, i.e.,

$$I_j = \hbar(n_j + \mu_j/4) ,$$

(3.97)

where μ_j are purely topological quantities called *Maslov indices*, and $n_j \in \mathbb{Z}$.[2] Using the Bohr–Sommerfeld rule, one obtains, to leading order in $\hbar$,

$$\frac{\partial \gamma_n(C)}{\partial n_j} = -\Delta \varphi_j(I; C) + O(\hbar) , \quad j = 1, 2, \ldots , N . \tag{3.98}$$

Clearly, in the above formula the quantum numbers n_j are treated as real parameters. Now we are ready to answer the two questions posed at the beginning of this section. The answer to the first question is in the affirmative; formula (3.98) implies that $\gamma_n(C)$ will differ from 0 by a term of order $\hbar$ whenever $\Delta \varphi_j(I; C) \neq 0$. Note, however, that it is possible to have a nonzero Berry phase yet vanishing Hannay angles for the corresponding classical system, since the higher order terms in the expansion defined by (3.98) may be different from zero. The only systems for which the second question also has an affirmative answer are quadratic ones. For such system all higher order terms necessarily vanish.

3.3 Classical geometric phases – examples

3.3.1 "Classical spin"

Let us consider a "classical spin," i.e., a magnetic moment $\mathbf{S}$ precessing about the direction of a magnetic field $\mathbf{B}$ (see Berry 1986). This system is described by the following equation:

$$\frac{d}{dt} \mathbf{S} = \mathbf{B} \times \mathbf{S} . \tag{3.99}$$

It follows from the above formula that $|\mathbf{S}|$ is constant in time, and hence the vector $\mathbf{S}$ moves on the surface of a two-dimensional sphere. This sphere, S^2, defines the phase space of our system. Let $\mathbf{b}$ be a unit vector in the direction of $\mathbf{B}$, i.e., $\mathbf{B} = B\mathbf{b}$, and let $(\mathbf{e}_1, \mathbf{e}_2)$ be an orthonormal basis in the plane orthogonal to $\mathbf{b}$. Clearly, the triple $(\mathbf{b}, \mathbf{e}_1, \mathbf{e}_2)$ defines an orthonormal basis in $\mathbb{R}^3$. Note that the basis $(\mathbf{e}_1, \mathbf{e}_2)$ is defined only up to an arbitrary $SO(2)$ rotation around the vector $\mathbf{b}$. Let us observe that the following quantities:

$$I = \mathbf{S} \cdot \mathbf{b} , \tag{3.100}$$

---

[2] As an example of Maslov indices consider the harmonic oscillator. The energy spectrum is given by

$$E_n = \hbar \omega \left(n + \frac{1}{2} \right) .$$

Now, for the oscillator the action I is simply the ratio (cf. Example 3.1.6)

$$I = \frac{E_n}{\omega} = \hbar \left(n + \frac{1}{2} \right) ,$$

and, therefore, the Maslov index is $\mu = 2$.

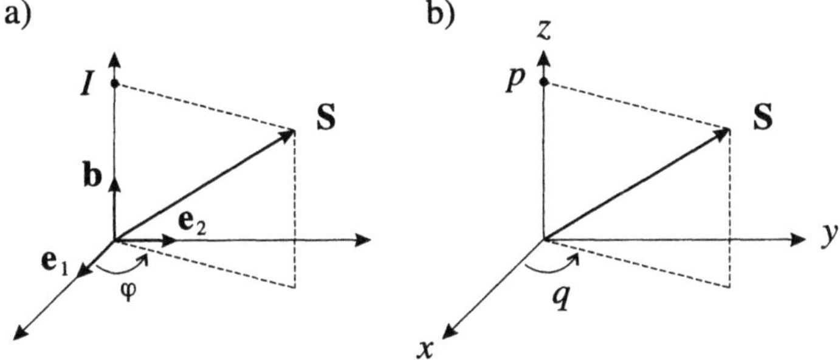

Figure 3.5: Canonical variables for spin: a) variables (φ, I), b) variables (q, p)

and

$$\varphi := \text{ azimuthal angle in the plane } (\mathbf{e}_1, \mathbf{e}_2) \tag{3.101}$$

define canonical action-angle variables on $\mathcal{P} = S^2$ (see Fig. 3.5). Indeed, one has

$$dI \wedge d\varphi = Sd\cos\theta \wedge d\varphi = -S\sin\theta\, d\theta \wedge d\varphi\,, \tag{3.102}$$

which is proportional to the area element on S^2, and, therefore, defines a symplectic form on $\mathcal{P}$ (cf. Example 3.1.4). Now, since the energy of a magnetic moment $\mathbf{S}$ interacting with a magnetic field $\mathbf{B}$ equals $\mathbf{S} \cdot \mathbf{B}$, the Hamiltonian, rewritten in the action-angle variables, has the following form:

$$H(I; \mathbf{B}) = BI\,. \tag{3.103}$$

The magnetic field $\mathbf{B}$ plays the role of the external parameter. Suppose now, that $\mathbf{B}$ is slowly cycled. At the end of the cycle C the magnetic moment $\mathbf{S}$ is back on its circle (one-dimensional torus in S^2), its position shifted by

$$\varphi(T) - \varphi(0) = \int_0^T B(t)dt + \Delta\,\varphi(I; C)\,, \tag{3.104}$$

where $\Delta\,\varphi(I; C)$ stands for the corresponding Hannay angle. However, it seems that the specification of $\Delta\varphi(I; C)$ by (3.104) is incomplete, because the basis $(\mathbf{e}_1, \mathbf{e}_2)$, relative to which the angle φ is defined, has not been specified (one has still a freedom to perform any $SO(2)$ rotation in the $(\mathbf{e}_1, \mathbf{e}_2)$–plane). The remarkable fact is that such a specification is not necessary. To find $\Delta\,\varphi(I; C)$ let us compute $\mathcal{W}(I) = \langle\, d_\mathbf{B}\, p \wedge d_\mathbf{B}\, q\,\rangle$, or, more explicitly,

$$\mathcal{W}(I) = \frac{1}{2\pi} \int_0^{2\pi} d\varphi\, d_\mathbf{B}\, p(\varphi, I; \mathbf{B}) \wedge d_\mathbf{B}\, q(\varphi, I; \mathbf{B})\,, \tag{3.105}$$

where for an obvious reason we have changed $d_\mathrm{M} \longrightarrow d_\mathbf{B}$. Let us assume, as in the quantum case (see Example 2.2.1), that $\mathbf{B}$ evolves on a two-sphere $|\mathbf{B}| = \text{const.}$, i.e., the parameter space $M = S^2$. Therefore, to compute $\mathcal{W}(I)$ one needs to define a **B**-dependent canonical pair (q, p) on the phase space $\mathcal{P} = S^2$. To do so, take any cartesian coordinate system (x, y, z) in $\mathbb{R}^3$ and define

$$
\begin{aligned}
p &:= S_z, \\[2mm]
q &:= \arctan\left[S_y/S_x\right] = \text{azimuthal angle in the } xy\text{-plane.}
\end{aligned}
\tag{3.106}
$$

It is easy to show that $dp \wedge dq$ is proportional to the area element on S^2, and hence that the pair (q, p) defines a canonical coordinate system. Using these coordinates one easily finds

$$
d_{\mathrm{B}}\, p \wedge d_{\mathrm{B}}\, q = \frac{d_{\mathrm{B}}\, S_z \wedge (S_x d_{\mathrm{B}}\, S_y - S_y d_{\mathrm{B}}\, S_x)}{S_x^2 + S_y^2} .
\tag{3.107}
$$

The spin vector **S**, expressed in terms of q and p, takes the following form:

$$
\begin{aligned}
S_x &= \sqrt{S^2 - p^2} \cos q , \\[1mm]
S_y &= \sqrt{S^2 - p^2} \sin q , \\[1mm]
S_z &= p .
\end{aligned}
\tag{3.108}
$$

Now, we shall compute our basic quantity $\mathcal{W}(I)$, defined in (3.105). To compute the exterior d_{B}-derivatives $(d_{\mathrm{B}}\, S_x, d_{\mathrm{B}}\, S_y, d_{\mathrm{B}}\, S_z)$ let us represent **S** in the basis $(\mathbf{b}, \mathbf{e}_1, \mathbf{e}_2)$:

$$
\mathbf{S} = I\mathbf{b} + \sqrt{S^2 - I^2} \cos \varphi\, \mathbf{e}_1 + \sqrt{S^2 - I^2} \sin \varphi\, \mathbf{e}_2 .
\tag{3.109}
$$

Clearly, the basis $(\mathbf{b}, \mathbf{e}_1, \mathbf{e}_2)$ does depend on **B**. To simplify the calculation let us choose the following instantaneous axes x, y, z along $(\mathbf{e}_1, \mathbf{e}_2, \mathbf{b})$:

$$
\mathbf{e}_1 = (1, 0, 0) , \quad \mathbf{e}_2 = (0, 1, 0) , \quad \mathbf{b} = \mathbf{e}_1 \times \mathbf{e}_2 = (0, 0, 1) .
$$

It is easy to compute the corresponding differentials, i.e., $d_{\mathrm{B}}\mathbf{e}_1$, $d_{\mathrm{B}}\mathbf{e}_1$ and $d_{\mathrm{B}}\mathbf{b}$. Note that $d_{\mathrm{B}}\mathbf{e}_1 \perp \mathbf{e}_1$, and hence, in an obvious notation,

$$
d_{\mathrm{B}}\mathbf{e}_1 = (0, d_{\mathrm{B}}e_{1y}, d_{\mathrm{B}}e_{1z}) .
$$

Similarly,

$$
d_{\mathrm{B}}\mathbf{e}_2 = (d_{\mathrm{B}}e_{2x}, 0, d_{\mathrm{B}}e_{2z}) ,
$$

and finally,

$$
d_{\mathrm{B}}\mathbf{b} = (d_{\mathrm{B}}\mathbf{e}_1) \times \mathbf{e}_2 + \mathbf{e}_1 \times d_{\mathrm{B}}\mathbf{e}_2 = (-d_{\mathrm{B}}e_{1z}, -d_{\mathrm{B}}e_{2z}, 0) .
$$

Therefore, formula (3.109) implies that

$$
\begin{aligned}
d_{\mathrm{B}}\, S_x &= -I d_{\mathrm{B}}\, e_{1z} + \sqrt{S^2 - I^2} \sin \varphi\, d_{\mathrm{B}}\, e_{2x} , \\[1mm]
d_{\mathrm{B}}\, S_y &= -I d_{\mathrm{B}}\, e_{2z} + \sqrt{S^2 - I^2} \cos \varphi\, d_{\mathrm{B}}\, e_{1y} , \\[1mm]
d_{\mathrm{B}}\, S_x &= \sqrt{S^2 - I^2}\, (\cos \varphi\, d_{\mathrm{B}}\, e_{1z} + \sin \varphi\, d_{\mathrm{B}}\, e_{2z}) .
\end{aligned}
\tag{3.110}
$$

By inserting these formulae into (3.107) and averaging over φ, one obtains

$$\mathcal{W}(I) = \langle\, d_B\, p \wedge d_B\, q \,\rangle = -I\, d_B\, e_{1z} \wedge d_B\, e_{2z}\ . \tag{3.111}$$

For a general set of axes, this becomes

$$\mathcal{W}(I) = -I \sum_{k=1}^{3} d_B\, e_{1k} \wedge d_B\, e_{2k} = -I\, d_B\, \mathbf{e}_1 \cdot \wedge d_B\, \mathbf{e}_2\ , \tag{3.112}$$

and since $\mathcal{W}(I)$ is linear in the action variable I, we have

$$-\frac{\partial}{\partial I}\, \mathcal{W}(I) = d_B\, \mathbf{e}_1 \cdot \wedge d_B\, \mathbf{e}_2 = d_B\, (\mathbf{e}_1 \cdot d_B\, \mathbf{e}_2)\ . \tag{3.113}$$

To compute the r.h.s. of the above formula one needs to know the explicit $\mathbf{B}$-dependence of the $\mathbf{e}$'s . One possible choice is the following:

$$\mathbf{e}_1 := \frac{\mathbf{b} \times \hat{\mathbf{z}}}{|\mathbf{b} \times \hat{\mathbf{z}}|}\ , \qquad \mathbf{e}_2 := \frac{\mathbf{b} \times \mathbf{e}_1}{|\mathbf{b} \times \mathbf{e}_1|}\ , \tag{3.114}$$

where $\hat{\mathbf{z}}$ is a unit vector along the z-axis. With this choice, one easily obtains the following explicit formulae for $\mathbf{e}_1$ and $\mathbf{e}_2$:

$$\mathbf{e}_1 = \frac{(B_y, -B_x, 0)}{(B_x^2 + B_y^2)^{1/2}}\ ,$$

$$\mathbf{e}_2 = \frac{(B_x B_z,\, B_y B_z,\, -B_x^2 - B_y^2)}{B(B_x^2 + B_y^2)^{1/2}}\ ,$$

and a direct calculation gives

$$-\frac{\partial}{\partial I}\, \mathcal{W}(I) = \frac{1}{|\mathbf{B}|^3} \sum_{k,l,m=1}^{3} \epsilon_{klm}\, B^k\, dB^l \wedge dB^m = \sin\theta\, d\theta \wedge d\varphi\ , \tag{3.115}$$

which is the familiar formula for the field of a unit magnetic pole placed at $\mathbf{B} = 0$. The Hannay angle is therefore given by a solid angle formula:

$$\Delta\varphi(I; C) = \Omega(C)\ , \tag{3.116}$$

where, as usual, $\Omega(C)$ is the solid angle subtended by the closed curve C on the parameter manifold $\mathbf{B} = \text{const}$.

Finally, let us investigate the relation between the Berry phase for a quantum spin (2.117) and the corresponding Hannay angle for a classical spin (3.116). We have a simple relation:

$$\Delta\varphi(I; C) = -\frac{\partial \gamma_n(C)}{\partial n}\ , \tag{3.117}$$

which is in perfect agreement with the general formula (3.98).

3.3.2 Classical oscillator

Consider a classical generalized harmonic oscillator defined by the following Hamiltonian (see Example 2.2.3):

$$H = \frac{1}{2}(X\,q^2 + 2Y\,qp + Z\,p^2)\,. \tag{3.118}$$

The Hamiltonian depends on the set of external parameters $\mathbf{R} := (X, Y, Z) \in \mathbb{R}^3$. Hamilton's equations of motion are given by

$$\frac{d}{dt}\begin{pmatrix} q \\ p \end{pmatrix} = \begin{pmatrix} Y & Z \\ -X & -Y \end{pmatrix}\begin{pmatrix} q \\ p \end{pmatrix}\,, \tag{3.119}$$

and they lead to the following 2nd order equation for q:

$$\ddot{q} = -(XZ - Y^2)\,q\,. \tag{3.120}$$

Equation (3.120) implies that the integral curves of our system are concentric, oblique ellipses for $XZ > Y^2$, or hyperbolae for $XZ < Y^2$. Since we are interested in closed contours, we shall consider the case

$$XZ > Y^2\,. \tag{3.121}$$

Then the corresponding parameter manifold M is the following subset of $\mathbb{R}^3$:

$$M := \{\,(X, Y, Z) \in \mathbb{R}^3 \mid XZ > Y^2\,\}\,,$$

i.e., the parameter space is the same as in the quantum case (see Example 2.2.3). Note, that due to (3.120), the quantity

$$\omega := (XZ - Y^2)^{1/2} \tag{3.122}$$

defines the frequency of the oscillatory motion, and reproduces formula (2.123) in the quantum case. Let us recall that for the harmonic oscillator the action variable I is defined by the ratio

$$I = \frac{E}{\omega} = \frac{E}{(XZ - Y^2)^{1/2}}\,, \tag{3.123}$$

with E being the energy (for fixed parameters the energy is constant along the integral curves). We shall find the corresponding Hannay angle using the formula

$$\Delta\varphi(I; C) = -\frac{\partial}{\partial I}\int_{\partial\Sigma=C} \mathcal{W}(I)\,, \tag{3.124}$$

with $\mathcal{W}(I) = \langle\, d_{\mathbf{R}}p \wedge d_{\mathbf{R}}q\,\rangle$. To find the two-form $\mathcal{W}(I)$ one needs explicit formulae for $q = q(\mathbf{R})$ and $p = p(\mathbf{R})$. This $\mathbf{R}$-dependence comes via the canonical transformation between (q, p) and (I, φ), which is itself $\mathbf{R}$-dependent. Solving oscillator equation (3.120), one obtains

$$q(t) = A\cos\varphi(t)\,, \tag{3.125}$$

where $\varphi(t)$ is an angle variable evolving according to

$$\varphi(t) = \varphi_0 + \omega t \ . \tag{3.126}$$

Therefore, using (3.119), one gets the following formula for $p(\mathbf{R})$:

$$p = -A \left(\frac{\omega}{Z} \sin \varphi + \frac{Y}{Z} \cos \varphi \right) \ . \tag{3.127}$$

To find the amplitude A of the oscillation one inserts the above formulae for q and p into (3.118) and finds the oscillator energy $E = \frac{1}{2} A^2 \omega^2 / Z$. Hence, due to (3.123) the amplitude A is given by

$$A = \left(\frac{2IZ}{\omega} \right)^{1/2} \ . \tag{3.128}$$

Now, using the following formulae for q and p:

$$q(I, \varphi; \mathbf{R}) = \left(\frac{2ZI}{\omega} \right)^{1/2} \cos \varphi \ , \tag{3.129}$$

$$p(I, \varphi; \mathbf{R}) = - \left(\frac{2ZI}{\omega} \right)^{1/2} \left(\frac{Y}{Z} \cos \varphi + \frac{\omega}{Z} \sin \varphi \right) \ , \tag{3.130}$$

we are ready to compute $\mathcal{W}(I)$, as follows:

$$
\begin{aligned}
\mathcal{W}(I) = \langle d_{\mathbf{R}} \, p \wedge d_{\mathbf{R}} \, q \rangle &= \frac{1}{2\pi} \int_0^{2\pi} d\varphi \, d_{\mathbf{R}} \, p(I, \varphi, \mathbf{R}) \wedge d_{\mathbf{R}} \, q(I, \varphi, \mathbf{R}) \\
&= -\frac{I}{\pi} \int_0^{2\pi} d\varphi \left\{ \cos^2 \varphi \, d_{\mathbf{R}} \left(\frac{Y}{Z} \sqrt{\frac{Z}{\omega}} \right) \wedge d_{\mathbf{R}} \sqrt{\frac{Z}{\omega}} + \sin \varphi \cos \varphi \, d_{\mathbf{R}} \sqrt{\frac{Z}{\omega}} \wedge d_{\mathbf{R}} \sqrt{\frac{\omega}{Z}} \right\} \\
&= \frac{I}{2} \, d_{\mathbf{R}} \left(\frac{Z}{\omega} \right) \wedge d_{\mathbf{R}} \left(\frac{Y}{Z} \right) \ .
\end{aligned}
\tag{3.131}
$$

A straightforward calculation leads to

$$\mathcal{W}(I) = -I \, \frac{X d_{\mathbf{R}} \, Y \wedge d_{\mathbf{R}} \, Z + Y d_{\mathbf{R}} \, Z \wedge d_{\mathbf{R}} \, X + Z d_{\mathbf{R}} \, X \wedge d_{\mathbf{R}} \, Y}{4(XZ - Y^2)^{3/2}} \ , \tag{3.132}$$

so the Hannay angle is given by

$$\Delta \varphi(I; C) = -\frac{1}{I} \int_{\partial \Sigma = C} \mathcal{W}(I) \ , \tag{3.133}$$

and, actually, does not depend on I.

Finally, let us compare formula for the quantum Berry phase (3.131) with the formula for the classical Hannay angle (2.131). One gets the following relation:

$$F^{(n)} = -\frac{n + \frac{1}{2}}{I} \mathcal{W}(I) \ , \tag{3.134}$$

which implies

$$\gamma_n(C) = -\left(n + \frac{1}{2}\right) \Delta \varphi(I; C), \qquad (3.135)$$

for any closed curve in the parameter space. Evidently

$$\Delta \varphi(I; C) = -\frac{\partial \gamma_n(C)}{\partial n}, \qquad (3.136)$$

which means that the formula (3.98) holds in this case. It is clear since we are dealing with a quadratic system.

3.3.3 Rotated rotator (Hannay's hoop)

Consider a particle of mass m that slides without friction around a planar hoop (Hannay 1985).[3] As the bead is sliding, the hoop is *slowly* rotated in the xy-plane, say, through an angle $\theta = \theta(t)$ with angular velocity $\boldsymbol{\omega} = \dot{\theta}\hat{\mathbf{z}}$ about a center O. One calls this system a *rotated rotator* or *Hannay's hoop*. Let s denote the arc length along the hoop measured from some reference point (on the hoop), and let $\mathbf{q} = \mathbf{q}(s)$ be the vector pointing from the origin O to the corresponding point on the hoop. The derivative

$$\mathbf{t}(s) := \mathbf{q}'(s) = \frac{d\mathbf{q}}{ds}(s) \qquad (3.137)$$

defines the unit vector tangent to the hoop at $\mathbf{q}(s)$ (see Fig. 3.6). Let R_θ denote the rotation, about the center O in the xy-plane, through an angle θ, such that the position of the particle relative to an inertial (nonrotating) frame is given by $\mathbf{Q}(t) := R_{\theta(t)}\mathbf{q}(s(t))$. The configuration space is a fixed closed curve (i.e., the hoop) and the corresponding Lagrangian is simply the kinetic energy of the particle:

$$L(s, \dot{s}, t) = \frac{m}{2}|\dot{\mathbf{Q}}|^2. \qquad (3.138)$$

One finds for the velocity vector,

$$\dot{\mathbf{Q}}(s(t)) = \frac{d}{dt}\left[R_{\theta(t)}\mathbf{q}(s(t))\right] = R_{\theta(t)}\left[\mathbf{t}(s(t))\dot{s} + \boldsymbol{\omega}(t) \times \mathbf{q}(s(t))\right], \qquad (3.139)$$

since

$$\dot{R}_\theta R_\theta^{-1}\mathbf{q} = R_\theta^{-1}\dot{R}_\theta\mathbf{q} = \boldsymbol{\omega} \times \mathbf{q}. \qquad (3.140)$$

Therefore, one obtains for the Lagrangian

$$L(s, \dot{s}, t) = \frac{m}{2}|\mathbf{t}\dot{s} + \boldsymbol{\omega} \times \mathbf{q}|^2, \qquad (3.141)$$

---

[3]In this section we follow the elegant exposition in Marsden and Ratiu 1994.

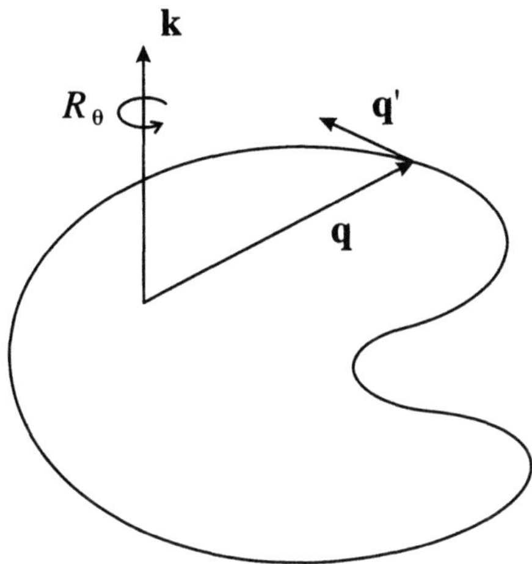

Figure 3.6: A bead sliding in the a rotating hoop

since R_θ represents an orthogonal matrix. Now, the Euler–Lagrange equation

$$\frac{d}{dt}\frac{\partial L}{\partial \dot{s}} - \frac{\partial L}{\partial s} = 0 \tag{3.142}$$

implies

$$\frac{d}{dt}\left[\mathbf{t}\cdot(\mathbf{t}\dot{s}+\boldsymbol{\omega}\times\mathbf{q})\right] - (\mathbf{t}\dot{s}+\boldsymbol{\omega}\times\mathbf{q})\cdot(\mathbf{t}'\dot{s}+\boldsymbol{\omega}\times\mathbf{t}) = 0 \,. \tag{3.143}$$

Taking into account that $\mathbf{t}\cdot\mathbf{t}' = 0$, one has

$$\ddot{s} = -\mathbf{t}\cdot\left[\dot{\boldsymbol{\omega}}\times\mathbf{q}+\boldsymbol{\omega}\times(\boldsymbol{\omega}\times\mathbf{q})\right] \,. \tag{3.144}$$

Moreover,

$$\mathbf{t}\cdot(\dot{\boldsymbol{\omega}}\times\mathbf{q}) = \ddot{\theta}q\sin\alpha \,, \tag{3.145}$$

with α being the angle between $\mathbf{q}$ and $\mathbf{t}$ (see Fig. 3.6), $q = |\mathbf{q}|$, and

$$\boldsymbol{\omega}\times(\boldsymbol{\omega}\times\mathbf{q}) = (\boldsymbol{\omega}\cdot\mathbf{q})\boldsymbol{\omega} - (\boldsymbol{\omega}\cdot\boldsymbol{\omega})\mathbf{q} = -\dot{\theta}^2\mathbf{q} \,, \tag{3.146}$$

since $\boldsymbol{\omega}\cdot\mathbf{q} = 0$. Therefore, equation (3.144) may be rewritten as follows:

$$\ddot{s} = \dot{\theta}^2\mathbf{q}\cdot\mathbf{t} - \ddot{\theta}q\sin\alpha \,. \tag{3.147}$$

A solution to the above equation is given by the following formula:

$$s(t) = s_0 + \dot{s}_0 t + \int_0^t (t-\tau)\left[\dot{\theta}(\tau)^2\mathbf{q}(s(\tau))\cdot\mathbf{t}(s(\tau)) - \ddot{\theta}(\tau)q(s(\tau))\sin\alpha(s(\tau))\right]d\tau \,, \tag{3.148}$$

where s_0 and $\dot{s}_0$ denote the initial position and velocity of the bead, respectively. Now, since the hoop rotates slowly, i.e., we assume that both $\dot{\theta}$ and $\ddot{\theta}$ are small with respect to particle velocity and acceleration, the particle makes many circuits while the hoop rotates a little. Therefore, the s-dependent quantities in the square brackets may be replaced by their averages around the hoop, giving the following formula for the position of the particle after the (long) time T, in which the hoop turns once:

$$s(T) = s_0 + \dot{s}_0 T$$
$$+ \int_0^T (T - \tau) \left[\dot{\theta}(\tau)^2 \frac{1}{\mathcal{L}} \int_0^{\mathcal{L}} \mathbf{q}(s) \cdot \mathbf{t}(s) ds - \ddot{\theta}(\tau) \frac{1}{\mathcal{L}} \int_0^{\mathcal{L}} q(s) \sin \alpha(s) ds \right] d\tau ,$$

where $\mathcal{L}$ denotes the length of the hoop. Note that the first integral over s vanishes, since $\mathbf{q} \cdot \mathbf{t} = \frac{d}{ds} |\mathbf{q}|^2$. The second integral

$$\int_0^{\mathcal{L}} q(s) \sin \alpha(s) ds = \int_0^{\mathcal{L}} |\mathbf{q}(s) \times \mathbf{t}(s)| ds = 2\mathcal{A} , \tag{3.149}$$

with $\mathcal{A}$ being the area enclosed by the hoop. Hence,

$$s(T) = s_0 + \dot{s}_0 T + \frac{2\mathcal{A}}{\mathcal{L}} \int_0^T (T - \tau) \ddot{\theta}(\tau) d\tau . \tag{3.150}$$

Integrating by parts and assuming, for simplicity, that $\dot{\theta}(0) = 0$, finally gives

$$s(T) = s_0 + \dot{s}_0 T - \frac{4\pi \mathcal{A}}{\mathcal{L}} . \tag{3.151}$$

The term $s_0 + \dot{s}_0 T$ is the standard formula for the arc length when the particle moves with constant velocity $\dot{s}_0$. However, there is another term, which is fully determined by the geometry of the hoop. Let us parameterize the position of the particle by an angle φ measured from the point s_0, that is, we perform the following change of variables:

$$\varphi(s) := 2\pi \frac{s - s_0}{\mathcal{L}} . \tag{3.152}$$

Therefore, the additional geometric factor in formula (3.151) may rewritten as

$$\Delta \varphi = -\frac{8\pi^2 \mathcal{A}}{\mathcal{L}^2} . \tag{3.153}$$

Writing this angle in the form

$$\Delta \varphi = -2\pi + 2\pi \left(1 - \frac{4\pi \mathcal{A}}{\mathcal{L}^2} \right) , \tag{3.154}$$

we see that the first term gives the expected phase shift resulting from the fact that the point s_0 has made a complete rotation. The nontrivial geometric aspect is embodied in the second term. The hoop returns to its original position but the particle does not. The additional *phase shift* in $\Delta\varphi$ is an example of the classical geometric phase. Note that this term vanishes for a circular hoop.

Now we show that the non trivial phase shift in the Hannay hoop may be interpreted as an adiabatic Hannay angle. The particle motion in the xy-plane can be described by the following Hamiltonian:

$$H(\mathbf{q}, \mathbf{p}; X) = \frac{\mathbf{p}^2}{2m} + V(\mathbf{q}; X) , \tag{3.155}$$

where V is a confining potential which is zero in a narrow strip centered on the hoop and very large elsewhere, and X is a parameter along the hoop.

For fixed X the particle moves with constant velocity v relative to the hoop. Denoting by $p = \mathbf{t} \cdot \mathbf{p} = mv$ the particle momentum, the action is given by

$$I = \frac{1}{2\pi} \oint_{\text{hoop}} p\, ds = \frac{1}{2\pi} p\mathcal{L} . \tag{3.156}$$

The corresponding angle variable is then given by (3.152). Of course, our problem has two degrees of freedom; however, the second one corresponds to transverse vibrations of the particle, and, therefore, does not play any role, due to the strong confining potential V. Let us compute the corresponding Hannay angle:

$$\begin{aligned}
\Delta\varphi(I; \text{hoop}) &= -\oint_{\text{hoop}} \frac{\partial}{\partial I} \langle p\, d_X q \rangle \\
&= -\frac{1}{2\pi} \frac{\partial}{\partial I} \int_0^{2\pi} dX \int_0^{2\pi} d\varphi\; p(I, \varphi; X) \frac{\partial}{\partial X} q(I, \varphi; X) .
\end{aligned} \tag{3.157}$$

Now,

$$p\frac{\partial q}{\partial X} = p\mathbf{t} \cdot \frac{\partial \mathbf{q}}{\partial X} = \frac{2\pi I}{\mathcal{L}} \mathbf{t} \cdot \frac{\partial \mathbf{q}}{\partial X} , \tag{3.158}$$

and, therefore,

$$\Delta\varphi(I; \text{hoop}) = -\frac{2\pi}{\mathcal{L}^2} \int_0^{2\pi} dX \int_0^{\mathcal{L}} ds\, \mathbf{t} \cdot \frac{\partial \mathbf{q}}{\partial X} , \tag{3.159}$$

where we have changed variables from φ to s according to (3.152). Observe that $d_X\mathbf{q} = \partial_X \mathbf{q}\, dX$ corresponds to an infinitesimal rotation of the hoop by an angle dX about the center O. Therefore,

$$\mathbf{t} \cdot \frac{\partial \mathbf{q}}{\partial X} dX = \mathbf{t} \cdot (\mathbf{q} \times \hat{\mathbf{z}})\, dX = \hat{\mathbf{z}} \cdot (\mathbf{t} \times \mathbf{q})\, dX = q \sin\alpha\, dX , \tag{3.160}$$

and, finally, using formula (3.149) one obtains

$$\Delta\varphi(I; \text{hoop}) = -\frac{8\pi^2 \mathcal{A}}{\mathcal{L}^2} , \tag{3.161}$$

in a perfect agreement with formula (3.153).

The quantized system corresponds to a quantum rotator parametrized by X — the orientation of the hoop. The Hamiltonian eigenfunctions for fixed X are given by (Berry 1985a)

$$\psi_n(\mathbf{q};\, X) = a(\mathbf{q};\, X)\exp\left[in\varphi(\mathbf{q};\, X)\right] = a(\mathbf{q};\, X)\exp\left[2\pi in\frac{s(\mathbf{q};\, X)}{\mathcal{L}}\right],\qquad (3.162)$$

where the amplitude $a(\mathbf{q};\, X)$ which confines the particle to the hoop, can be expressed in terms of a "coordinate perpendicular to the hoop," $\eta = \eta(\mathbf{q};\, X)$, as follows:

$$a^2 = \frac{\delta(\eta)}{\mathcal{L}}.\qquad (3.163)$$

Now, the adiabatic Berry phase $\gamma_n(\text{hoop})$ reads

$$\gamma_n(\text{hoop}) = -\mathrm{Im}\oint_{\text{hoop}}\int d\mathbf{q}\,\psi_n(\mathbf{q};\, X)d_X\,\psi_n^*(\mathbf{q};\, X).\qquad (3.164)$$

Inserting formula (3.162) and changing variables in the xy-plane from (x, y) to (s, η), one obtains

$$\gamma_n(\text{hoop}) = -\frac{2\pi n}{\mathcal{L}^2}\int_0^{2\pi} dX\int_0^{\mathcal{L}} ds\,\frac{\partial s}{\partial X}(\mathbf{q};\, X).\qquad (3.165)$$

Using equation (3.160), one has

$$\frac{\partial s}{\partial X} = \mathbf{t}\cdot\frac{\partial \mathbf{q}}{\partial X} = q\sin\alpha,\qquad (3.166)$$

and, therefore,

$$\gamma_n(\text{hoop}) = n\frac{8\pi^2\mathcal{A}}{\mathcal{L}^2} = -n\Delta\varphi(I;\,\text{hoop}).\qquad (3.167)$$

It then follows that

$$\frac{\partial \gamma_n(\text{hoop})}{\partial n} = -\Delta\varphi(I;\,\text{hoop}),\qquad (3.168)$$

in agreement with formula (3.98).

3.3.4   Motion in non-inertial frames

The Foucault pendulum considered in Example 3.1.1 is a special case of a more general situation displaying a geometric phase. Consider a particle constrained to move in a two-dimensional plane, which due the presence of some external forces, moves on another (curved) two-dimensional manifold $M \subset \mathbb{R}^3$, e.g., a sphere, as in the case of the Foucault pendulum. Suppose that the plane moves along a given curve C on M,

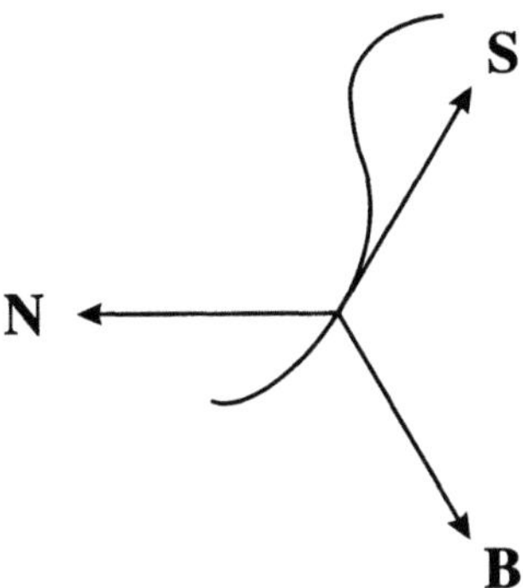

Figure 3.7: A moving orthogonal frame: **S** - tangent to the curve, **N**, **B** - normal and binormal, respectively.

such that it remains tangent to M at the same fixed point. Clearly, its orientation is uniquely determined by the unit vector $\mathbf{S} = \mathbf{S}(t)$ normal to M along C. Therefore, a curve C on M gives rise to a new curve

$$t \longrightarrow \mathbf{S}(t) \in \mathbb{R}^3 .$$

The elementary geometry of curves in $\mathbb{R}^3$ associates to any curve an orthogonal time-dependent triad $(\mathbf{S}, \mathbf{N}, \mathbf{B})$ such that

$$\mathbf{N} \| \dot{\mathbf{S}}, \quad \text{and} \quad \mathbf{B} = \mathbf{S} \times \mathbf{N} .$$

The time dependence of this triad is characterized by the Serret–Frenet formulae (cf. Fig 3.7):

$$\begin{aligned}
\frac{d\mathbf{S}}{dt} &= \kappa \mathbf{N} , \\
\frac{d\mathbf{N}}{dt} &= -\kappa \mathbf{S} + \tau \mathbf{B} , \\
\frac{d\mathbf{B}}{dt} &= -\tau \mathbf{N} .
\end{aligned} \qquad (3.169)$$

The parameters κ and τ denote the curvature and torsion of the curve $\mathbf{S}(t)$, respectively. Clearly, the plane on which the particle moves is spanned by $\mathbf{B}$ and $\mathbf{N}$. Parametrize this plane by a vector $\mathbf{r}$ such that $\mathbf{r} = 0$ denotes the tangency point of the plane with the manifold M. Moreover, let us introduce cartesian coordinates (x_0, y_0) as follows:

$$\mathbf{r} = x_0 \mathbf{N} + y_0 \mathbf{B} . \qquad (3.170)$$

Now, suppose that our particle moves under the influence of a potential $V = V(\mathbf{r})$. One finds the following equations of motion (in the plane:

$$m\ddot{x}_0 = -\frac{\partial V}{\partial x_0} + m\left[(\kappa^2 + \tau^2)x_0 + 2\tau\dot{y}_0 + \dot{\tau}y_0\right], \tag{3.171}$$

$$m\ddot{y}_0 = -\frac{\partial V}{\partial y_0} + m\left[\tau^2 y_0 - 2\tau\dot{x}_0 - \dot{\tau}x_0\right], \tag{3.172}$$

where m denotes the particle's mass. The above equations of motion are written in a co-moving noninertial frame. Note the presence of terms proportional to the velocity $\dot{\mathbf{r}} = (\dot{x}_0, \dot{y}_0)$ that are analogs of the Coriolis force. Now we are going to rewrite the equations of motion in a locally inertial frame. As in General Relativity, by a local inertial frame on M, we mean a frame spanned by two orthonormal vectors $\mathbf{U}_1$ and $\mathbf{U}_2$ that undergo parallel transport along the curve C on M. Note, that since $\mathbf{U}_i \cdot \mathbf{U}_j = \delta_{ij}$, one has that

$$\frac{d\mathbf{U}_i}{dt} \cdot \mathbf{U}_j + \mathbf{U}_i \cdot \frac{d\mathbf{U}_j}{dt} = 0, \qquad i, j = 1, 2. \tag{3.173}$$

The law of parallel transport imposes a stronger condition, however. Following our previous discussion of parallel transport on a two-dimensional sphere (cf. section 2.2.7), one can show that, in analogy with (2.154), one in fact has

$$\mathbf{U}_i \cdot \frac{d\mathbf{U}_j}{dt} = 0, \qquad i, j = 1, 2. \tag{3.174}$$

Clearly, the basis $(\mathbf{N}, \mathbf{B})$ does not satisfy this condition; instead one finds

$$\mathbf{B} \cdot \frac{d\mathbf{N}}{dt} = \tau, \quad \text{and} \quad \mathbf{N} \cdot \frac{d\mathbf{B}}{dt} = -\tau. \tag{3.175}$$

Let us construct a parallel transported frame by performing a time-dependent $SO(2)$ rotation, as follows:

$$\begin{pmatrix} \mathbf{U}_1 \\ \mathbf{U}_2 \end{pmatrix} = \begin{pmatrix} \cos\beta & -\sin\beta \\ \sin\beta & \cos\beta \end{pmatrix} \begin{pmatrix} \mathbf{N} \\ \mathbf{B} \end{pmatrix}. \tag{3.176}$$

One easily shows that

$$\mathbf{U}_2 \cdot \frac{d\mathbf{U}_1}{dt} = \tau - \dot{\beta}, \tag{3.177}$$

and hence, the condition for parallel transport is that

$$\dot{\beta} = \tau. \tag{3.178}$$

This means that $(\mathbf{U}_1, \mathbf{U}_2)$ rotates with respect to an instantaneous $(\mathbf{N}, \mathbf{B})$ frame with an angular velocity proportional to the torsion τ. Using the Serret–Frenet formulae one can easily show that if $(\mathbf{U}_1, \mathbf{U}_2)$ is parallel transported then

$$\frac{d\mathbf{U}_1}{dt} = -\kappa\cos\beta\,\mathbf{S}, \tag{3.179}$$

$$\frac{d\mathbf{U}_2}{dt} = -\kappa\sin\beta\,\mathbf{S}. \tag{3.180}$$

Let us introduce cartesian coordinates (x, y) in a locally inertial frame $(\mathbf{U}_1, \mathbf{U}_2)$, as follows:

$$\mathbf{r} = x\,\mathbf{U}_1 + y\,\mathbf{U}_2 .$$

The reader can easily derive the corresponding equations of motion:

$$m\ddot{x} \;=\; -\frac{\partial V}{\partial x} + m\kappa^2 \left[x\cos^2\beta + y\sin\beta\cos\beta \right] , \qquad (3.181)$$

$$m\ddot{y} \;=\; -\frac{\partial V}{\partial y} + m\kappa^2 \left[y\sin^2\beta + x\sin\beta\cos\beta \right] . \qquad (3.182)$$

Comparing these with the equations in the non-inertial frame (3.171)–(3.172), we see that the velocity-dependent terms are no longer present. The terms proportional to κ^2 correspond to the centrifugal force caused by the instantaneous rotation of the plane.

Now, let us assume that the plane changes its position adiabatically. Then $|\kappa| \ll 1$, and all terms proportional to κ^2 may be neglected in the adiabatic approximation. Hence

$$m\ddot{x} \;=\; -\frac{\partial V}{\partial x} , \qquad (3.183)$$

$$m\ddot{y} \;=\; -\frac{\partial V}{\partial y} . \qquad (3.184)$$

The above equations of motion have the same form as on the stationary plane. The only effect of rotation is encoded into the time dependence of the frame $(\mathbf{U}_1, \mathbf{U}_2)$. Note that if the curve C were closed on M, then the parallel transport of $(\mathbf{U}_1, \mathbf{U}_2)$ along C would lead to an $SO(2)$ rotation with respect to the $(\mathbf{N}, \mathbf{B})$ frame, with a total angle of rotation

$$\Delta\beta = \int_0^T \tau(t)\,dt , \qquad (3.185)$$

where T is the period of C.

Example 3.3.1 Let us see how this works in the case of the Foucault pendulum. Now M is a two-dimensional sphere — the surface of the Earth. Using standard spherical coordinates in $\mathbb{R}^3$, one has for the normal, $\mathbf{S}$,

$$\mathbf{S}(t) = \left(\sin\theta\cos\varphi(t),\, \sin\theta\sin\varphi(t),\, \cos\theta \right) ,$$

where θ is a constant latitude. Note that $\mathbf{N}||\dot{\mathbf{S}}$ and to satisfy $\mathbf{N}\cdot\mathbf{S} = 0$ one requires

$$\mathbf{N}(t) = \left(-\sin\varphi(t),\, \cos\varphi(t),\, 0 \right) .$$

Hence,

$$\mathbf{B}(t) = \mathbf{S}(t) \times \mathbf{N}(t) = \left(-\cos\theta\cos\varphi(t),\, -\cos\theta\sin\varphi(t),\, \sin\theta \right) .$$

The Serret–Frenet formulae give

$$\kappa = \Omega \sin\theta , \quad \text{and} \quad \tau = \Omega \cos\theta , \tag{3.186}$$

where

$$\Omega = |\mathbf{\Omega}| = \frac{d\varphi(t)}{dt} ,$$

denotes the angular velocity of Earth's rotation. Hence, the total angle of rotation after one turn of the Earth is, according to (3.185), given by

$$\Delta\beta = \int_0^T \tau(t)\, dt = 2\pi \cos\theta , \tag{3.187}$$

in perfect agreement with (3.7).[4]   ◇

3.3.5   Guiding center motion

Our next example concerns the motion of charged particles in strong magnetic fields (Littlejohn 1988). The corresponding (nonrelativistic) equations of motion are given by (cf. Example 3.2.3)

$$m\dot{\mathbf{v}} = \frac{q}{c}\, \mathbf{v} \times \mathbf{B} , \tag{3.188}$$

where q now denotes a particle's charge and c denotes the velocity of light. As is well known, in a constant magnetic field, say $\mathbf{B} = B_0 \hat{\mathbf{z}}$, a charged particle will move in a circular helix whose axis is parallel to the magnetic field direction. Let $\mathbf{x}$ be the position of the particle. The projection of the motion onto the plane perpendicular to $\mathbf{B}$, i.e., the xy-plane, is circular, with girofrequency $\Omega = qB_0/mc$. The projection of $\mathbf{x}$ onto the field line about which the particle spirals is denoted by $\mathbf{X}$ and is called the *guiding center* (see Fig. 3.8). In a sense, $\mathbf{X}$ corresponds to the average of $\mathbf{x}$ over a rapid oscillation with large girofrequency Ω. Of course, the guiding center $\mathbf{X}$ moves along the field line with constant velocity v_z. Now, the vector running from $\mathbf{X}$ to $\mathbf{x}$, i.e., $\mathbf{r} := \mathbf{x} - \mathbf{X}$, is called the *giroradius* vector. Note that its magnitude equals $r = v_\perp/\Omega$, where $v_\perp = \sqrt{v_x^2 + v_y^2}$ is the magnitude of the component of the particle's velocity perpendicular to $\mathbf{B}$. Finally, let us define the *girophase* θ as the angle in the perpendicular xy-plane between some reference direction, denoted by $\mathbf{e}$ in Fig. 3.8, and the giroradius vector $\mathbf{r}$. For the case of a uniform field $\mathbf{B}$, the vector $\mathbf{e}$ may conveniently be taken to be any constant unit vector in the xy-plane.

Now, let us turn to the case of a nonuniform (but time-independent) magnetic field. This case is important in plasma physics, and also has many applications in astrophysics. The fundamental question is then how the circular motion is perturbed by,

---

[4]Simple experiments demonstrating the appearance of geometric phases in noninertial frames were proposed by Kugler (1989) and Kugler and Shtrikman (1988).

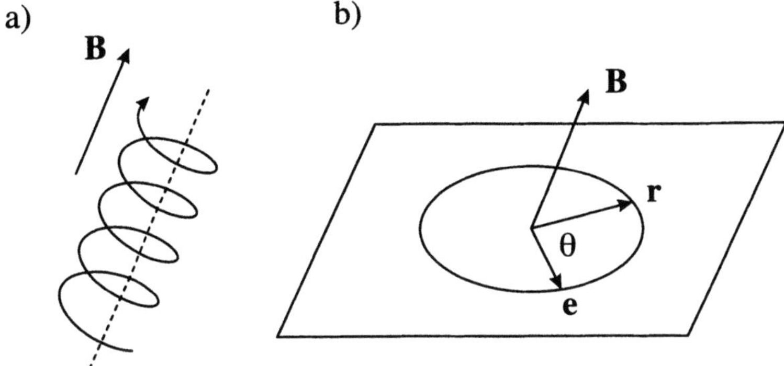

Figure 3.8: a) Particle motion in a uniform magnetic field; b) circular motion on a plane perpendicular to **B**

e.g., inhomogeneities of the magnetic field. Now the guiding center position $\mathbf{X}(t)$ no longer follows a field line. It turns out that if the inhomogeneity is "small," then $\mathbf{X}$ slowly drifts away from the field line and

$$\mathbf{\Omega}(\mathbf{x}) \approx \frac{q\mathbf{B}(\mathbf{x})}{mc} . \tag{3.189}$$

If the magnetic field is strong and sufficiently homogeneous (i.e., if it does not vary significantly over the radius of a cyclotron orbit), then to a first approximation the motion is still cyclotronic about the fields lines and the dominant contribution to the time evolution of the girophase is given by

$$\theta(t) = \int_0^t \Omega(\mathbf{X}(\tau))d\tau , \tag{3.190}$$

where, for simplicity, we take $\theta(0) = 0$, and $\Omega = |\mathbf{\Omega}|$. Note that in equation (3.190) we have evaluated the girofrequency at the guiding center position $\mathbf{X}$ rather than at the particle position $\mathbf{x}$. Note, however, that to uniquely define the gyrophase one needs to choose the reference direction $\mathbf{e}$. Unlike the case of uniform fields, it is no longer possible to choose a *constant* reference direction in the perpendicular plane; for non-uniform $\mathbf{B}(\mathbf{x})$ the reference direction is a function of position, i.e., $\mathbf{e} = \mathbf{e}(\mathbf{x})$. To see this more clearly, let us introduce an orthonormal triad (a *dreibein*) of vectors $(\mathbf{e}_1, \mathbf{e}_2, \mathbf{b})$, where $\mathbf{b} = \mathbf{B}/B$, and $\mathbf{e}_1$ and $\mathbf{e}_2$ are unit vectors spanning the plane perpendicular (to $\mathbf{B}$). We stress that all three vectors of the triad are space-dependent, i.e., they are functions of $\mathbf{x}$. Now, using these $\mathbf{x}$-dependent triads, we might, for example, define the gyrophase to be an angle between $\mathbf{e}_1$ and the gyroradius $\mathbf{r}$ — cf. Fig. 3.8. However, the vectors $\mathbf{e}_1$ and $\mathbf{e}_2$ are not uniquely defined; one can perform a rotation about $\mathbf{b}$ by an arbitrary angle. Therefore, the gyrophase defined by formula (3.190) cannot be the only contribution to the gyrophase, because the definition of the gyrophase depends upon the choice of $(\mathbf{e}_1, \mathbf{e}_2)$, and the formula (3.190) does not depend on any such particular choice.

Does there exist any privileged choice of vectors $(\mathbf{e}_1, \mathbf{e}_2)$? Often one chooses $(\mathbf{e}_1, \mathbf{e}_2)$ to be the principal normal and binormal vectors of the field line, i.e.,

$$\mathbf{e}_1 = \frac{(\mathbf{b} \cdot \nabla)\mathbf{b}}{|(\mathbf{b} \cdot \nabla)\mathbf{b}|} , \qquad \mathbf{e}_2 = \mathbf{b} \times \mathbf{e}_1 . \tag{3.191}$$

But this is only one particular choice out of many. Moreover, it has obvious disadvantages, e.g., it leaves $(\mathbf{e}_1, \mathbf{e}_2)$ undefined for straight field lines. Instead of concentrating on a particular choice of vectors $(\mathbf{e}_1, \mathbf{e}_2)$ let us consider the whole family of possible $\mathbf{e}$'s:

$$\begin{aligned} \mathbf{e}_1' &= \mathbf{e}_1 \cos \psi + \mathbf{e}_2 \sin \psi , \\ \mathbf{e}_2' &= -\mathbf{e}_1 \sin \psi + \mathbf{e}_2 \cos \psi , \end{aligned} \tag{3.192}$$

where ψ is an arbitrary $\mathbf{x}$-dependent function. One calls formula (3.192) the *gyrogauge transformation*. It is evident that all physical quantities should be gyrogauge invariant, i.e, invariant under gyrogauge transformations (3.192).

Obviously, the gyrophase is not gyrogauge invariant. One has, rather,

$$\theta'(\mathbf{x}) = \theta(\mathbf{x}) + \psi(\mathbf{x}) , \tag{3.193}$$

where θ' is defined relative to $\mathbf{e}_1'$. Now, since θ is not gyrogauge invariant, neither is $\dot{\theta}$. However, θ, defined in equation (3.190) is manifestly gyrogauge invariant. Therefore, there must be an additional term on the r.h.s. of formula (3.190) such that the corrected θ transforms according to (3.193). It turns out that the so-called guiding center expansion leads to the following formula (Littlejohn 1988):

$$\dot{\theta} = \Omega(\mathbf{X}) + \mathbf{R} \cdot \dot{\mathbf{X}} + \text{ gyrogauge invariant terms} , \tag{3.194}$$

where

$$\mathbf{R} := \nabla \mathbf{e}_1 \cdot \mathbf{e}_2 , \tag{3.195}$$

that is, $R_i = (\partial_i \mathbf{e}_1) \cdot \mathbf{e}_2$. Performing the gyrogauge transformation, one easily finds that

$$\mathbf{R}' = \nabla \mathbf{e}_1' \cdot \mathbf{e}_2' = \mathbf{R} + \nabla \psi , \tag{3.196}$$

i.e., we recover the analog of the gauge transformation for the vector potential. Therefore, due to (3.194), the gyrophase θ transforms according to (3.193):

$$\dot{\theta}' = \dot{\theta} + \nabla \psi \cdot \dot{\mathbf{X}} . \tag{3.197}$$

Therefore, the term $\mathbf{R} \cdot \dot{\mathbf{X}}$ in eq. (3.194) is necessary in order for both sides of the equation to have the same transformation properties under the gyrogauge transformations. It is clear that $\mathbf{R}$ is an analog of the connection and, therefore, that the analog of the curvature is nothing but the gyrogauge-invariant quantity $\nabla \times \mathbf{R}$. If the guiding center trajectory C is closed, i.e., $\mathbf{X}(0) = \mathbf{X}(T)$, then

$$\Delta\theta = \oint_C \mathbf{R} \cdot d\mathbf{X} = \int_{\partial\Sigma = C} (\nabla \times \mathbf{R}) \cdot d\mathbf{S} , \tag{3.198}$$

is a gyrogauge-invariant, purely geometric object. Therefore, it may be treated as a classical analog of the geometric phase. In such a case the total gyrophase is given by

$$\theta(T) = \int_0^T \Omega(\mathbf{X})(t)dt + \Delta\theta \, , \tag{3.199}$$

that is, $\theta(T)$ is a sum of the dynamical phase $\int \Omega dt$ and the geometric phase $\Delta\theta$.

3.3.6 Rigid bodies and the geometric phase

Consider the motion of a free rigid body in $\mathbb{R}^3$, governed by the Euler equations:

$$\dot{\mathbf{M}}_b = \mathbf{M}_b \times \boldsymbol{\omega}_b \, . \tag{3.200}$$

Here $\mathbf{M}_b$ and $\boldsymbol{\omega}_b$ denote the angular momentum and the angular velocity, respectively, in the body frame (cf. section 4.3.2 for more details of rigid body dynamics). Instead of solving the Euler equation, let us try to understand the problem from the geometric point of view. The angular momentum of the body, viewed from the inertial frame, $\mathbf{M}_s$, is constant in time, and that viewed from the body-fixed frame, $\mathbf{M}_b$, is periodic in time (for typical initial conditions). Therefore, after one period, say T, the body, as viewed from the inertial frame, must rotate about the direction of its angular momentum $\mathbf{M}_s$ by some angle $\Delta\theta$. As was shown by Montgomery (1991), this angle is given by the following formula:[5]

$$\Delta\theta = \frac{2ET}{J} - \Omega_b \, , \tag{3.201}$$

where E is the kinetic energy, J is the length of the angular momentum, i.e., $J = |\mathbf{M}_b| = |\mathbf{M}_s|$, and Ω_b is the solid angle swept out by the vector $\mathbf{M}_b$ on the two-dimensional sphere

$$S^2(J) := \left\{ \mathbf{M}_b \, \middle| \, |\mathbf{M}_b| = J \right\} \, .$$

Following the beautiful exposition of Montgomery (1991), we can prove this formula (3.201) using simple properties of the rotation group $SO(3)$.[6] The motion of the body is described by a trajectory on the rotation group,

$$t \longrightarrow R(t) \in SO(3) \, ,$$

such that for any t one has

$$\mathbf{M}_s = R(t)\mathbf{M}_b(t) \, , \tag{3.202}$$

---

[5] For an interesting history of formula (3.201) see Marsden and Ratiu 1994.
[6] For another proof, see section 4.3.2.

where we represent elements $R \in SO(3)$ by 3×3 orthogonal matrices. Since $\mathbf{M}_b(T) = \mathbf{M}_b(0)$, one obtains

$$R(T)^{-1}\mathbf{M}_s = R(0)^{-1}\mathbf{M}_s \ . \tag{3.203}$$

However, $\mathbf{M}_s$ is a constant vector, say $\mathbf{J}$, and hence

$$R(T) \cdot R(0)^{-1}\mathbf{J} = \mathbf{J} \ , \tag{3.204}$$

which means that

$$R_{\Delta\theta} := R(T)^{-1} \cdot R(0) \tag{3.205}$$

is a rotation about the $\mathbf{J}$ axis by the angle $\Delta\theta$ given explicitly by the formula (3.201), as we now shall prove.

Suppose that for $t = 0$, $R(0) = \mathbb{1}$, which means that $\mathbf{M}_b(0) = \mathbf{J}$. The phase space trajectory is represented by the curve $z(t) := (R(t), \mathbf{M}_b(t)) \in T^*SO(3) \cong SO(3) \times \mathbb{R}^3$, starting at $z(0) = (\mathbb{1}, \mathbf{J})$. Consider the following two curves in $T^*SO(3)$, both beginning at $z(0)$: the first one, C_1, is simply the trajectory of the body, i.e.,

$$C_1(t) = z(t) \ \text{for} \ 0 \le t \le T \ ,$$

and the second one,

$$C_2(\theta) = (R_\theta, \mathbf{J}) \ \text{for} \ 0 \le \theta \le \Delta\theta \ ,$$

denotes a counterclockwise spatial rotation of the body about the $\mathbf{J}$ axis by an angle θ. Note that

$$C_1(T) = C_2(\Delta\theta) = (R_{\Delta\theta}, \mathbf{J}) \ , \tag{3.206}$$

and, therefore, the curve

$$C := C_1 - C_2$$

is closed on $T^*SO(3)$.

To prove formula (3.201) we shall compute the line integral along the closed curve C from the canonical one-form "pdq" on $T^*SO(3)$ (cf. Example 3.1.3). The canonical one-form reads

$$\Theta = pdq = \sum_k \mathbf{p}_k \cdot d\mathbf{x}_k \ ,$$

where the sum runs over all points of the body. Now, by the very definition of rotation, $d\mathbf{x}_k$ is defined by

$$d\mathbf{x}_k = d\boldsymbol{\alpha} \times \mathbf{x}_k \ ,$$

where $d\boldsymbol{\alpha} = \widehat{\boldsymbol{\alpha}}\, d\alpha$ represents rotation about the axis $\widehat{\boldsymbol{\alpha}}$ (with $|\widehat{\boldsymbol{\alpha}}| = 1$) by an angle $d\alpha$. Hence

$$\Theta = \mathbf{M}_s \cdot d\boldsymbol{\alpha} \tag{3.207}$$

where

$$\mathbf{M}_s = \sum_k \mathbf{x}_k \times \mathbf{p}_k \,,$$

denotes the angular momentum in the inertial frame. Now, along the curve C_1,

$$d\boldsymbol{\alpha} = \boldsymbol{\omega}\, dt \,,$$

where $\boldsymbol{\omega}$ is the angular velocity of the body with respect to the inertial frame (that is, $\mathbf{M}_s = \widehat{\mathbf{I}}_s\boldsymbol{\omega}$, where $\widehat{\mathbf{I}}_s$ denotes the body's inertia tensor with respect to an inertial frame). Hence, along C_1 one has

$$\Theta = \mathbf{M}_s \cdot \boldsymbol{\omega}\, dt = \boldsymbol{\omega} \cdot (\widehat{\mathbf{I}}_s\boldsymbol{\omega})\, dt = 2E\, dt \,. \tag{3.208}$$

Along the second curve, we instead have

$$d\boldsymbol{\alpha} = \boldsymbol{\alpha}\, d\theta = \frac{\mathbf{J}}{J}\, d\theta \,,$$

and, therefore,

$$\Theta = \mathbf{M}_s \cdot \frac{\mathbf{J}}{J}\, d\theta = J\, d\theta \,, \tag{3.209}$$

since $\mathbf{M}_s = \mathbf{J}$. At this point we apply the Stokes theorem:

$$\int_{\partial\Sigma = C} d\Theta = \oint_C \Theta \,.$$

Using the definition of $C = C_1 - C_2$ and the formulae (3.208) and (3.209), one obtains

$$\int_{\partial\Sigma = C} d\Theta = \int_{C_1} \Theta - \int_{C_2} \Theta = 2ET - J\Delta\theta \,. \tag{3.210}$$

The last step of our calculation is to show that

$$\int_{\partial\Sigma = C} d\Theta = J\Omega_b \,. \tag{3.211}$$

To prove (3.211) let us parametrize $SO(3)$ by Euler angles $(\varphi, \vartheta, \psi)$, as follows:

$$R(\varphi, \vartheta, \psi) = R_3(\varphi) \cdot R_2(\vartheta) \cdot R_3(\psi) \,, \tag{3.212}$$

where $R_k(\alpha)$ denotes the counterclockwise rotation about the kth axis by an angle α (see formulae in footnote 3 in section 2.2.5). We choose the (inertial) coordinate system so that

$$\mathbf{J} = J\mathbf{e}_3 \ .$$

Now, any rotation $R \in SO(3)$ relates the coordinates $\mathbf{x}$ in the inertial frame to the body coordinates $\mathbf{X}$ by $\mathbf{x} = R\mathbf{X}$. Viewed from the body frame, $\mathbf{X}$ is constant and, therefore,

$$d\mathbf{x} = (dR)\,\mathbf{X} \ .$$

Moreover using, $d\mathbf{x} = d\boldsymbol{\alpha} \times \mathbf{x}$, so that

$$(dR)\mathbf{X} = d\boldsymbol{\alpha} \times \mathbf{x} \ ,$$

or, equivalently,

$$(dR) \cdot R^{-1}\mathbf{x} = d\boldsymbol{\alpha} \times \mathbf{x} \ . \tag{3.213}$$

Let us differentiate formula (3.212):

$$\begin{aligned}
dR(\varphi, \vartheta, \psi) = {} & [dR_3(\varphi)] \cdot R_2(\vartheta) \cdot R_3(\psi) \\
& + \ R_3(\varphi) \cdot [dR_2(\vartheta)] \cdot R_3(\psi) + R_3(\varphi) \cdot R_2(\vartheta) \cdot [dR_3(\psi)] \ .
\end{aligned} \tag{3.214}$$

Multiplying dR by R^{-1} from the right one obtains

$$\begin{aligned}
(dR) \cdot R^{-1} = {} & [dR_3(\varphi)] \cdot R_3(\varphi)^{-1} + R_3(\varphi) \cdot [dR_2(\vartheta)] \cdot R_2(\vartheta)^{-1} \cdot R_3(\varphi)^{-1} \\
& + \ R_3(\varphi) \cdot R_2(\vartheta) \cdot [dR_3(\psi)] \cdot R_3(\psi)^{-1} \cdot R_2(\vartheta)^{-1} \cdot R_3(\varphi)^{-1} \ .
\end{aligned} \tag{3.215}$$

Therefore, using equation (3.213), one gets

$$\begin{aligned}
(dR) \cdot R^{-1}\mathbf{x} = {} & d\varphi\,\mathbf{e}_3 \times \mathbf{x} + d\vartheta\,R_3(\varphi) \left[\mathbf{e}_2 \times R_3(\varphi)^{-1}\mathbf{x} \right] \\
& + \ d\psi\,R_3(\varphi) \cdot R_2(\vartheta) \left\{ \mathbf{e}_3 \times [R_3(\varphi) \cdot R_2(\vartheta)]^{-1}\mathbf{x} \right\} \ .
\end{aligned} \tag{3.216}$$

Taking into account the following obvious property of rotations:

$$R(\mathbf{v} \times \mathbf{w}) = R\mathbf{v} \times R\mathbf{w} \ ,$$

one obtains

$$(dR) \cdot R^{-1}\mathbf{x} = [d\varphi\,\mathbf{e}_3 + d\vartheta\,R_3(\varphi)\mathbf{e}_2 + d\psi\,R_3(\varphi) \cdot R_2(\theta)\mathbf{e}_3] \times \mathbf{x} \ . \tag{3.217}$$

Comparing the above formula with equation (3.213), one concludes that

$$d\boldsymbol{\alpha} = d\varphi\,\mathbf{e}_3 + d\vartheta\,R_3(\varphi)\mathbf{e}_2 + d\psi\,R_3(\varphi) \cdot R_2(\theta)\mathbf{e}_3 \ . \tag{3.218}$$

Therefore, the formula for the canonical one-form Θ reads

$$\Theta = \mathbf{M} \cdot d\boldsymbol{\alpha} = J\mathbf{e}_3 \cdot d\boldsymbol{\alpha} = J\{d\varphi + [\mathbf{e}_3 \cdot R_2(\vartheta)\,\mathbf{e}_3]\,d\psi\}\,, \tag{3.219}$$

and observing that

$$R_2(\vartheta)\mathbf{e}_3 = \cos\vartheta\,\mathbf{e}_3 + \sin\vartheta\,\mathbf{e}_1\,,$$

one ends up with

$$\Theta = J\,[d\varphi + \cos\vartheta\,d\psi]\,. \tag{3.220}$$

Hence,

$$d\Theta = -J\sin\vartheta\,d\vartheta \wedge d\psi = -J\,d\Omega_s\,. \tag{3.221}$$

To finish the proof of equation (3.211), and hence that of equation (3.201), we have to relate the solid angle element $d\Omega_s$ on the two-sphere of the space frame angular momentum $|\mathbf{M}_s| = J$ to the solid angle element $d\Omega_b$ on the two-sphere of the body frame angular momentum $|\mathbf{M}_b| = J$. Let us show that, in fact,

$$d\Omega_s = -d\Omega_b\,. \tag{3.222}$$

Indeed, since $\mathbf{M}_s = R\mathbf{M}_b$ and $\mathbf{M}_s = \mathbf{J} = J\mathbf{e}_3$, we have, on the one hand,

$$\begin{aligned}
\mathbf{M}_b &= R^{-1}\mathbf{J} = JR_3(\psi)^{-1} \cdot R_2(\vartheta)^{-1} \cdot R_3(\varphi)^{-1}\,\mathbf{e}_3 = JR_3(\psi)^{-1} \cdot R_2(\vartheta)^{-1}\,\mathbf{e}_3 \\
&= JR_3(\psi)^{-1}\,[\cos\vartheta\,\mathbf{e}_3 - \sin\vartheta\,\mathbf{e}_1] = J\left[\cos\vartheta\,\mathbf{e}_3 - \sin\vartheta\,R_3(\psi)^{-1}\mathbf{e}_1\right] \\
&= J\,[\cos\vartheta\,\mathbf{e}_3 + \sin\vartheta\,\sin\psi\,\mathbf{e}_2 - \sin\vartheta\,\cos\psi\,\mathbf{e}_1]\,. \tag{3.223}
\end{aligned}$$

On the other hand, by the very definition of the spherical angles (ϑ_b, ψ_b) in the $\mathbf{M}_b$ space, we have

$$\mathbf{M}_b = J\,[\cos\vartheta_b\,\mathbf{e}_3 + \sin\vartheta_b\,\sin\psi_b\,\mathbf{e}_2 + \sin\vartheta_b\,\cos\psi_b\,\mathbf{e}_1]\,, \tag{3.224}$$

and, therefore, $\psi = -\psi_b$ and $\vartheta = -\vartheta_b$, which implies formula (3.222), finally proves equation (3.211), and, hence, equation (3.201).

Further reading

Section 3.1. A detailed exposition of geometric approaches to classical mechanics may be found in Abraham and Marsden 1978; Arnold 1989; Marsden and Ratiu 1999; and Thirring 1978.

Section 3.2. The adiabatic theorem in classical mechanics is treated in Arnold 1989 (see also Landau and Lifshitz 1976 for an interesting exposition). For the derivation of Hannay angles see Hannay 1985; Berry 1985a; and Berry and Hannay 1988b.

For details of the semiclassical approach in quantum mechanics the reader may consult Berry 1983, and Maslov and Fedoruk 1981. The classical limit of a quantum adiabatic phase is discussed in Berry 1885a, and Gozzi and Thacker 1987a,b. Several authors use the method of coherent states to show a direct correspondence between quantum and classical geometric phases. We refer the reader to papers of Maamache, Provost and Vallee (1990 and 1991) (see also Benedict and Schleich 1993; Aravind 1999). For the non-adiabatic approach to classical phases see Berry and Hannay 1988b. The geometric phase was generalized to classical chaotic systems in Robbins and Berry 1992.

Section 3.3. Simple experiments demonstrating the appearance of geometric phases in noninertial frames were proposed by Kugler (1989) and Kugler and Shtrikman (1988). For other aspects of the classical geometric phase see also Hannay 1998a.

Problems

3.1. Show that a symplectic manifold has to be even dimensional.

3.2. Prove the following properties of the Poisson bracket: the Jacobi identity

$$\{\{F, G\}, H\} + \{\{H, F\}, G\} + \{\{G, H\}, F\} = 0 ;$$

and the Leibnitz rule

$$\{FG, H\} = \{F, H\} G + F \{G, H\} ,$$

for any $F, G, H \in C^\infty(\mathcal{P})$.

3.3. Prove the following, for any functions $F, G \in C^\infty(\mathcal{P})$:

$$[X_F, X_G] = -X_{\{F,G\}} .$$

3.4. Let $(\mathcal{P}, \Omega)$ be a symplectic manifold. Show that a set of canonical transformations $\varphi : \mathcal{P} \longrightarrow \mathcal{P}$ form a group.

3.5. Find the standard action variables for elliptic motion in the Kepler problem in $\mathbb{R}^3$.

3.6. Show that the definition of the standard action does not depend on the particular choice of basic cycles of T^n.

3.7. Show that the coordinates q and p, defined by formula (3.106), are canonical on S^2.

3.8. Derive the formula (3.107) for $d_\mathrm{B} p \wedge d_\mathrm{B} q$.

3.9. Show that $s(t)$, defined by the formula (3.148), does solve the equation

$$\ddot{s} = \dot{\theta}^2 \mathbf{q} \cdot \mathbf{t} - \ddot{\theta} q \sin \alpha .$$

3.10. Using the Serret–Frenet formulae derive

$$\ddot{\mathbf{N}} = -\dot{\kappa}\mathbf{S} - (\kappa^2 + \tau^2)\mathbf{N} + \dot{\tau}\mathbf{B},$$
$$\ddot{\mathbf{B}} = -\dot{\tau}\mathbf{N} + \kappa\tau\mathbf{S} - \tau^2\mathbf{B}.$$

3.11. Show that if $(\mathbf{U}_1, \mathbf{U}_2)$ is parallel transported along a curve on a two-dimensional manifold M, then

$$\ddot{\mathbf{U}}_1 = (\kappa\tau\sin\beta - \dot{\kappa}\cos\beta)\mathbf{S} - \kappa^2\cos\beta\mathbf{N},$$
$$\ddot{\mathbf{U}}_2 = -(\kappa\tau\cos\beta + \dot{\kappa}\sin\beta)\mathbf{S} - \kappa^2\sin\beta\mathbf{N}.$$

4

Geometric Approach to Classical Phases

4.1 Hamiltonian systems with symmetries

4.1.1 Hamiltonian actions and momentum maps

Suppose that $(\mathcal{P}, \Omega)$ is a symplectic manifold and let G be a Lie group acting from the left on $\mathcal{P}$ by canonical transformations. That is, there is a mapping

$$\Phi : G \times \mathcal{P} \longrightarrow \mathcal{P} ,$$

such that for any $g \in G$,

$$\Phi_g : \mathcal{P} \longrightarrow \mathcal{P} ,$$

defined by $\Phi_g = \Phi(g, \cdot)$, is a canonical transformation:

$$\Phi_g^* \Omega = \Omega .$$

Denote by $\mathfrak{g}$ the corresponding Lie algebra of G. For any $\xi \in \mathfrak{g}$, let $\mathbf{X}_\xi$ denote an infinitesimal generator of the above action (cf. Definition 1.2.3):

$$\mathbf{X}_\xi(x) = \frac{d}{dt} \Phi(\exp(t\xi), x) \Big|_{t=0} . \tag{4.1}$$

Definition 4.1.1 *Suppose there is a linear map* $J : \mathfrak{g} \longrightarrow C^\infty(\mathcal{P})$*, such that for all* $\xi \in \mathfrak{g}$,

$$X_{J(\xi)} = \mathbf{X}_\xi , \tag{4.2}$$

where $X_{J(\xi)}$ is a hamiltonian vector field for $J(\xi) \in C^\infty(P)$. The map $\mathbf{J} : P \longrightarrow \mathfrak{g}^*$, defined by

$$\langle \mathbf{J}(x), \xi \rangle := J(\xi)(x), \tag{4.3}$$

for all $\xi \in \mathfrak{g}$ and $x \in P$, is called a momentum map of the action Φ.

This means that each infinitesimal generator $\mathbf{X}_\xi$ admits its own hamiltonian function $X_{J(\xi)}$.

Consider now a hamiltonian system living on (P, Ω) and let $H \in C^\infty(P)$ be the corresponding Hamiltonian. We call a Lie group G a *symmetry group* of our system iff

$$\Phi_g^* H = H , \tag{4.4}$$

for any $g \in G$, that is, each transformation Φ_g leaves the Hamiltonian H invariant. Suppose that the action of G on P admits a momentum map $\mathbf{J}$. Then one proves the classical

Theorem 4.1.1 (Noether theorem) *If $\mathbf{J}$ is a momentum map of the action of a Lie group G on a symplectic manifold (P, Ω), then*

$$\{J(\xi), H\} = 0 , \tag{4.5}$$

i.e., for all $\xi \in \mathfrak{g}$, the function $J(\xi) \in C^\infty(P)$ defines a conserved quantity.

The name momentum map comes from the fact that it recovers the standard definitions of linear and angular momenta.

Example 4.1.1 (Linear momentum) Consider a system of N particles in $\mathbb{R}^3$. Then the configuration space $Q = \mathbb{R}^{3N}$ and the corresponding phase space $P = T^*Q \cong \mathbb{R}^{6N}$. Define the action of $G = \mathbb{R}^3$ on P as follows: For any $\mathbf{x} \in \mathbb{R}^3$,

$$\Phi_{\mathbf{x}} (\mathbf{q}^1, \ldots, \mathbf{q}^N, \mathbf{p}_1, \ldots, \mathbf{p}_N) := (\mathbf{q}^1 + \mathbf{x}, \ldots, \mathbf{q}^N + \mathbf{x}, \mathbf{p}_1, \ldots, \mathbf{p}_N) , \tag{4.6}$$

that is, it translates the position of each particle by $\mathbf{x}$. Clearly, $\mathbb{R}^3$ is an abelian group:

$$\Phi_{\mathbf{x}} \circ \Phi_{\mathbf{y}} = \Phi_{\mathbf{y}} \circ \Phi_{\mathbf{x}} = \Phi_{\mathbf{x}+\mathbf{y}} .$$

One can easily prove that this is a hamiltonian action with respect to the standard Poisson bracket in $\mathbb{R}^{6N}$, i.e.,

$$\{F, G\} = \sum_{i=1}^{N} \left(\frac{\partial F}{\partial \mathbf{q}^i} \cdot \frac{\partial G}{\partial \mathbf{p}_i} - \frac{\partial G}{\partial \mathbf{q}^i} \cdot \frac{\partial F}{\partial \mathbf{p}_i} \right) . \tag{4.7}$$

Let $\xi \in \mathfrak{g} \cong \mathbb{R}^3$. The infinitesimal generator $\mathbf{X}_\xi$ at a point

$$(\mathbf{q}^1, \ldots, \mathbf{q}^N, \mathbf{p}_1, \ldots, \mathbf{p}_N) \in \mathbb{R}^{6N}$$

is defined by differentiating formula (4.6) with respect to $\mathbf{x}$, in the direction $\boldsymbol{\xi}$, i.e.,

$$\mathbf{X}_{\boldsymbol{\xi}}(\mathbf{q}^1, \ldots, \mathbf{q}^N, \mathbf{p}_1, \ldots, \mathbf{p}_N) := (\boldsymbol{\xi}, \ldots, \boldsymbol{\xi}, 0, \ldots, 0) . \tag{4.8}$$

Now, the hamiltonian vector field $X_{J(\boldsymbol{\xi})}$ defined with respect to (4.7) is given by

$$X_{J(\boldsymbol{\xi})}(\mathbf{q}^1, \ldots, \mathbf{q}^N, \mathbf{p}_1, \ldots, \mathbf{p}_N) = \left(\frac{\partial J(\boldsymbol{\xi})}{\partial \mathbf{p}_1}, \ldots, \frac{\partial J(\boldsymbol{\xi})}{\partial \mathbf{p}_N}, -\frac{\partial J(\boldsymbol{\xi})}{\partial \mathbf{q}^1}, \ldots, -\frac{\partial J(\boldsymbol{\xi})}{\partial \mathbf{q}^N} \right) . \tag{4.9}$$

Therefore, the defining condition $X_{J(\boldsymbol{\xi})} = \mathbf{X}_{\boldsymbol{\xi}}$ implies that

$$\frac{\partial J(\boldsymbol{\xi})}{\partial \mathbf{p}_i} = \boldsymbol{\xi} , \qquad \frac{\partial J(\boldsymbol{\xi})}{\partial \mathbf{q}^i} = 0 , \quad i = 1, \ldots, N . \tag{4.10}$$

It is evident that the linear map $J : \mathfrak{g} \to C^\infty(\mathcal{P})$ is given by

$$J(\boldsymbol{\xi})(\mathbf{q}^1, \ldots, \mathbf{q}^N, \mathbf{p}_1, \ldots, \mathbf{p}_N) = \left(\sum_{i=1}^N \mathbf{p}_i \right) \cdot \boldsymbol{\xi} , \tag{4.11}$$

and, hence,

$$\mathbf{J}(\mathbf{q}^1, \ldots, \mathbf{q}^N, \mathbf{p}_1, \ldots, \mathbf{p}_N) = \sum_{i=1}^N \mathbf{p}_i =: \mathbf{P} , \tag{4.12}$$

with $\mathbf{P}$ being the total linear momentum of the N particle system. $\diamond$

Example 4.1.2 (Angular momentum) Consider a single particle with a configuration space $Q = \mathbb{R}^3$ and the corresponding phase space $\mathcal{P} = T^*Q \cong \mathbb{R}^6$. Define the action of $SO(3)$ on $\mathcal{P}$ by

$$\Phi_A(\mathbf{q}, \mathbf{p}) = (A\mathbf{q}, A\mathbf{p}) , \tag{4.13}$$

for any $A \in SO(3)$ and $(\mathbf{q}, \mathbf{p}) \in \mathbb{R}^6$. Now, the Lie algebra $so(3)$ consists of 3×3 antisymmetric matrices and it is isomorphic to $\mathbb{R}^3$ (cf. Example 1.2.11). Take any $\boldsymbol{\xi} \in \mathbb{R}^3$, and let $\hat{\boldsymbol{\xi}}$ be the corresponding element from $so(3)$. The infinitesimal generator $\mathbf{X}_{\hat{\boldsymbol{\xi}}}$ is defined by

$$\mathbf{X}_{\hat{\boldsymbol{\xi}}}(\mathbf{q}, \mathbf{p}) = (\hat{\boldsymbol{\xi}}\mathbf{q}, \hat{\boldsymbol{\xi}}\mathbf{p}) = (\boldsymbol{\xi} \times \mathbf{q}, \boldsymbol{\xi} \times \mathbf{p}) , \tag{4.14}$$

where we have used the canonical isomorphism between $\mathbb{R}^3$ and $so(3)$. To find the momentum map one solves the Hamilton equations:

$$\frac{\partial J(\boldsymbol{\xi})}{\partial \mathbf{p}} = \hat{\boldsymbol{\xi}}\mathbf{q} , \qquad \frac{\partial J(\boldsymbol{\xi})}{\partial \mathbf{q}} = -\hat{\boldsymbol{\xi}}\mathbf{p} . \tag{4.15}$$

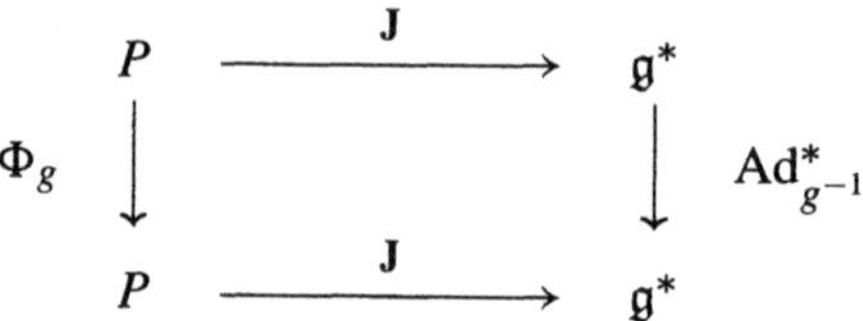

Figure 4.1: Equivariance of the momentum map

A solution $J(\xi)$, which is linear in ξ, is given by

$$J(\xi)(\mathbf{q}, \mathbf{p}) = (\hat{\xi}\mathbf{q}) \cdot \mathbf{p} = (\xi \times \mathbf{q}) \cdot \mathbf{p} = (\mathbf{q} \times \mathbf{p}) \cdot \xi \,, \qquad (4.16)$$

so that

$$\mathbf{J}(\mathbf{q}, \mathbf{p}) = \mathbf{q} \times \mathbf{p} \,, \qquad (4.17)$$

which is the standard formula for the angular momentum of a single particle.   ◇

It turns out that in physical applications one usually considers momentum maps satisfying the following additional property:

$$J([\xi, \eta]) = \{J(\xi), J(\eta)\} \,, \qquad (4.18)$$

for any $\xi, \eta \in \mathfrak{g}$. A momentum map satisfying (4.18) is called *equivariant*. Hence, an equivariant momentum map defines a homomorphism between a Lie algebra $\mathfrak{g}$ and a Lie algebra of classical observables $(C^\infty(\mathcal{P}), \{\, , \,\})$.

Proposition 4.1.2 *A momentum map* $\mathbf{J} : \mathcal{P} \longrightarrow \mathfrak{g}^*$ *is equivariant iff*

$$\mathrm{Ad}^*_{g^{-1}} \circ \mathbf{J} = \mathbf{J} \circ \Phi_g \,, \qquad (4.19)$$

or, equivalently,

$$J(\mathrm{Ad}_g\xi)(\Phi_g(x)) = J(\xi)(x) \qquad (4.20)$$

for $x \in P$, $g \in G$ and $\xi \in \mathfrak{g}$, i.e., the diagram in Fig. 4.1 commutes.

Equipped with the notion of an equivariant momentum map we end this section with the following

Definition 4.1.2 *An action of a Lie group G on a symplectic manifold $(\mathcal{P}, \Omega)$ admitting an equivariant momentum map is called hamiltonian.*

4.1.2   Reduced phase space

Let $(\mathcal{P}, \Omega)$ be a symplectic manifold and suppose that there is a hamiltonian action of a Lie group G on $\mathcal{P}$. Denote by $\mathbf{J}$ the corresponding (equivariant) momentum map

$$\mathbf{J} : \mathcal{P} \longrightarrow \mathfrak{g}^* \,.$$

Let us take $\mu \in \mathfrak{g}^*$ and consider the level set $\mathbf{J}^{-1}(\mu) \subset \mathcal{P}$, i.e.,

$$\mathbf{J}^{-1}(\mu) = \{ x \in \mathcal{P} \mid \mathbf{J}(x) = \mu \} \,.$$

In general, $\mathbf{J}^{-1}(\mu)$ does not define a differentiable manifold. It does so, however, when, for example, μ is a regular value[1] of $\mathbf{J}$. Let us therefore suppose that μ is a regular value of $\mathbf{J}$, and let G_μ be an isotropy subgroup of the coadjoint action of G on $\mathfrak{g}^*$, i.e.,

$$G_\mu := \{ g \in G \mid \mathrm{Ad}^*_g \mu = \mu \} \,. \tag{4.21}$$

Note that, due to the equivariance property of the momentum map, G_μ leaves $\mathbf{J}^{-1}(\mu)$ invariant, i.e., for any $g \in G_\mu$ and $x \in \mathbf{J}^{-1}(\mu)$ one has $\Phi_g(x) \in \mathbf{J}^{-1}(\mu)$. Suppose that the action of the isotropy subgroup G_μ, restricted to $\mathbf{J}^{-1}(\mu)$, is free, i.e., it has no fixed points (see section 1.2.1). This construction gives rise to the following G_μ-bundle:

$$\mathbf{J}^{-1}(\mu) \longrightarrow \mathcal{P}_\mu \,,$$

with the base space

$$\mathcal{P}_\mu := \mathbf{J}^{-1}(\mu)/G_\mu \,. \tag{4.22}$$

$\mathcal{P}_\mu$ is called a *reduced phase space*. It is the space of orbits of the action of G_μ on $\mathbf{J}^{-1}(\mu)$. Now, we are going to equip $\mathcal{P}_\mu$ with the canonical two-form Ω_μ. This form is defined as follows: Let us take an arbitrary point $a \in \mathcal{P}_\mu$ and two arbitrary vectors $v, w \in T_a\mathcal{P}_\mu$. A point $a \in \mathcal{P}_\mu$ corresponds to an orbit $\mathcal{O}_a$ of G_μ contained in $\mathbf{J}^{-1}(\mu)$ — see Fig. 4.2. Let $x \in \mathcal{O}_a$ and take two arbitrary vectors $\tilde{v}, \tilde{w} \in T_x\mathbf{J}^{-1}(\mu)$, such that

$$T_x\pi_\mu(\tilde{v}) = v \,, \quad \text{and} \quad T_x\pi(\tilde{w}) = w \,, \tag{4.23}$$

where $\pi_\mu : \mathbf{J}^{-1}(\mu) \longrightarrow G_\mu$ denotes the canonical bundle projection. Finally, define

$$\Omega_\mu(a)(v, w) := (i_\mu^* \Omega)(x)(\tilde{v}, \tilde{w}) \,, \tag{4.24}$$

where $i_\mu : \mathbf{J}^{-1}(\mu) \hookrightarrow \mathcal{P}$ denotes the canonical embedding as follows:

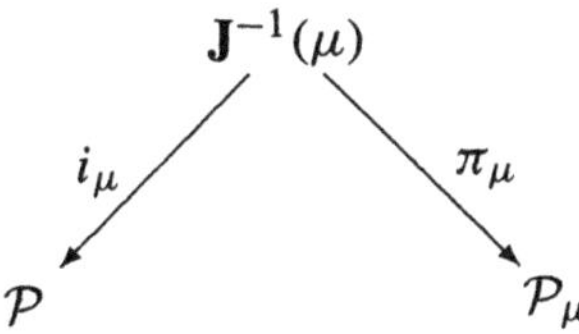

It is easy to see that the above definition is correct, i.e., it does not depend upon the choice of $x \in \mathcal{O}_a$ or the vectors $\tilde{v}, \tilde{w}$, provided that the condition (4.23) holds. One can then prove the celebrated

---

[1] Let $f : X \to Y$. A point $y \in Y$ is a regular value of f iff

$$(T_x f)(T_x X) = T_y Y \,,$$

for all $x \in f^{-1}(y) \subset X$.

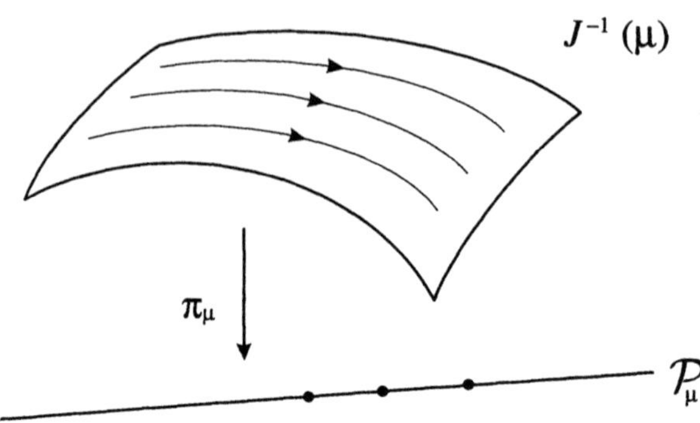

Figure 4.2: Reduced phase space.

Theorem 4.1.3 (Marsden–Weinsten) *A pair* $(\mathcal{P}_\mu, \Omega_\mu)$ *defines a symplectic manifold.*

By its construction, Ω_μ is a unique two-form on $\mathcal{P}_\mu$, such that

$$i_\mu^* \Omega = \pi_\mu^* \Omega_\mu$$

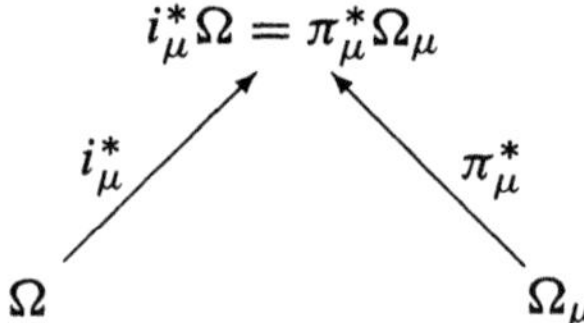

Example 4.1.3 Consider the following action of $SO(2)$ on $\mathbb{R}^{2n}$ parametrized by $(\mathbf{x}, \mathbf{y}) = (x^1, \ldots, x^n, y^1, \ldots, y^n)$:

$$
\begin{aligned}
x^k &\longrightarrow \cos\lambda\, x^k + \sin\lambda\, y^k\,, \\
y^k &\longrightarrow -\sin\lambda\, x^k + \cos\lambda\, y^k\,.
\end{aligned}
$$

This action is canonical with respect to the following symplectic form on $\mathbb{R}^{2n}$:

$$\Omega = d\mathbf{x} \cdot \wedge\, d\mathbf{y} = \sum_{k=1}^{n} dx^k \wedge dy^k\,. \tag{4.25}$$

The corresponding momentum map $\mathbf{J} : \mathbb{R}^{2n} \longrightarrow so(2)^* \cong \mathbb{R}$ reads

$$\mathbf{J}(\mathbf{x}, \mathbf{y}) = \frac{1}{2}(\mathbf{x}^2 + \mathbf{y}^2)\,, \tag{4.26}$$

that is, it defines the Hamiltonian of the harmonic oscillator. Note that $\mathbf{J}^{-1}(\mu) \cong S^{2n-1}$. Moreover, the isotropy subgroup $G_\mu = SO(2)$ for any $\mu \in so(2)^*$. Therefore, the corresponding reduced space reads

$$\mathcal{P}_\mu := S^{2n-1}/SO(2)\,. \tag{4.27}$$

Using complex coordinates

$$\mathbf{z} := \mathbf{x} + i\mathbf{y} , \tag{4.28}$$

one easily finds that the following $U(1) \cong SO(2)$ action on $\mathbb{C}^n \cong \mathbb{R}^{2n}$:

$$\mathbf{z} \longrightarrow e^{i\lambda}\mathbf{z} , \tag{4.29}$$

leaves the symplectic form

$$\Omega = \frac{1}{2} \operatorname{Im}(d\bar{\mathbf{z}} \cdot \wedge \, d\mathbf{z}) \tag{4.30}$$

invariant. Clearly, $\mathbf{J}(\mathbf{z}) = \frac{1}{2}|\mathbf{z}|^2$. The reduced phase space

$$\mathcal{P}_\mu := S^{2n-1}/U(1) \cong \mathbb{C}P^{n-1} \tag{4.31}$$

defines a complex projective space (cf. Example 1.2.14). Therefore, due to the Marsden–Weinstein theorem, $\mathbb{C}P^{n-1}$ is a symplectic manifold. $\diamond$

4.2 Geometric approach to adiabatic phases

4.2.1 Families of hamiltonian actions

Consider a hamiltonian system defined on a symplectic manifold $(\mathcal{P}, \Omega)$. Suppose that the hamiltonian of the system depends upon some external parameters, which parametrize a manifold M, that is,

$$H : \mathcal{P} \times M \longrightarrow \mathbb{R} . \tag{4.32}$$

Clearly, the total space $E := \mathcal{P} \times M$ defines a trivial fibre over M. Let us denote by π_M and π_P the following canonical projections:

$$\pi_M \; : \; \mathcal{P} \times M \longrightarrow M ,$$
$$\pi_P \; : \; \mathcal{P} \times M \longrightarrow \mathcal{P} .$$

Definition 4.2.1 *Let G be a Lie group. A family of hamiltonian G-actions on E is a smooth (left) action of G on E, such that*

- *each fibre $\pi_M^{-1}(x) = \mathcal{P}$ is invariant under the action,*

- *the action restricted to the fibre $\pi_M^{-1}(x)$ is symplectic,*

- *it admits a smooth family of momentum maps*

$$\mathbf{J} : \mathcal{P} \times M \longrightarrow \mathfrak{g}^* , \tag{4.33}$$

  i.e., for any $x \in M$, the map $\mathbf{J}(\,\cdot\,, x) : \mathcal{P} \longrightarrow \mathfrak{g}^*$ defines a momentum map in the usual sense.

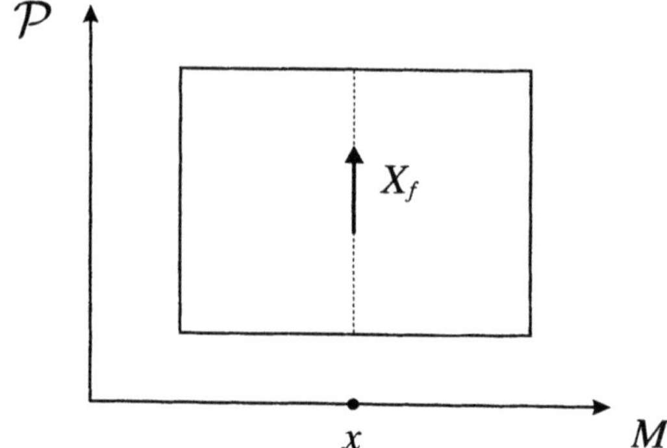

Figure 4.3: Fibrewise hamiltonian vector field on $\mathcal{P} \times M$

To clarify the last property let us note that the pull-back $\pi_{\mathrm{P}}^* \Omega$ of the symplectic form Ω on $\mathcal{P}$ defines a two-form on $P \times M$. We shall use the suggestive notation of Montgomery 1988 and write

$$\pi_{\mathrm{P}}^* \Omega =: \Omega \oplus 0 . \tag{4.34}$$

It is evident that the restriction of $\pi_{\mathrm{P}}^* \Omega$ to any fibre $\pi_{\mathrm{M}}^{-1}(x)$ gives a symplectic form on this fibre. Take any function $f \in C^\infty(\mathcal{P} \times M)$, and define the *fibrewise* hamiltonian vector field X_f corresponding to f by

$$i_{X_f} (\pi_{\mathrm{P}}^* \Omega) = d_{\mathrm{P}} f , \tag{4.35}$$

where d_{P} denotes an exterior derivative in the "$\mathcal{P}$-direction," that is, the total exterior derivative d on $\mathcal{P} \times M$ splits into d_{M} and d_{P}, as follows:

$$df = d_{\mathrm{P}} f + d_{\mathrm{M}} f ,$$

meaning that, if $(y^1, \dots, y^{2n})$ are local coordinates on $\mathcal{P}$ and $(x^1, \dots, x^m)$ are local coordinates on M, then

$$d_{\mathrm{P}} f = \sum_{i=1}^{2n} \frac{\partial f}{\partial y^i} \, dy^i , \qquad d_{\mathrm{M}} f = \sum_{i=1}^{m} \frac{\partial f}{\partial x^i} \, dx^i .$$

Note that X_f is tangent to each fibre $\pi_{\mathrm{M}}^{-1}(x)$ and hence defines a hamiltonian vector field on $\pi_{\mathrm{M}}^{-1}(x)$, in the usual sense — see Fig. 4.3. Let

$$J : \mathfrak{g} \longrightarrow C^\infty(\mathcal{P} \times M) ,$$

be the map associated with a family of momentum maps $\mathbf{J} : \mathcal{P} \times M \longrightarrow \mathfrak{g}^*$, such that

$$\langle \mathbf{J}(p, x), \xi \rangle =: J(\xi)(p, x) ,$$

for any $(p, x) \in \mathcal{P} \times M$ and $\xi \in \mathfrak{g}$. To define this map we proceed as follows: Let $\xi \in \mathfrak{g}$ and denote by $\mathbf{X}_\xi$ the corresponding infinitesimal generator of G-action on $E = \mathcal{P} \times M$. We define $J(\xi)$ such that the fibrewise hamiltonian vector field $X_{J(\xi)}$ satisfies

$$\mathbf{X}_\xi = X_{J(\xi)} , \tag{4.36}$$

on $\mathcal{P} \times M$.

Example 4.2.1 (A family of completely integrable systems) Suppose that for each $x \in M$, the Hamiltonian $H(\,\cdot\,, x)$ defines a completely integrable system on $\mathcal{P}$. We then call (4.32) a *family of completely integrable systems*. Due to the Liouville theorem (cf. Section 3.1.3) there exists a set of local x-dependent action variables

$$I_1(\,\cdot\,; x), \ldots , I_n(\,\cdot\,; x) ,$$

with $2n = \dim \mathcal{P}$. If this system is globally defined on $\mathcal{P} \times M$, then one can define the family of momentum maps

$$\mathbf{J} := (I_1, \ldots , I_n) : \mathcal{P} \times M \longrightarrow \mathbb{R}^n , \tag{4.37}$$

corresponding to the family of G-actions, where G is an abelian n-torus T^n, and $\mathbb{R}^n$ its (dual) Lie algebra. $\diamondsuit$

4.2.2   Hannay's angles and the Hannay–Berry connection

A trivial bundle $E = \mathcal{P} \times M$ is equipped with a natural connection, and hence it gives rise to the following horizontal lift:

$$\mathrm{h}_0\,(Z) = 0 \oplus Z , \tag{4.38}$$

where $0 \oplus Z$ denotes the following vector field on $\mathcal{P} \times M$:

$$\mathcal{P} \times M \ni (p, x) \longrightarrow (0, Z) \in T_p\mathcal{P} \times T_x M . \tag{4.39}$$

Let us assume that a Lie group G is compact and connected. If $\mathfrak{T}$ is an arbitrary tensor field on $\mathcal{P} \times M$, then its average is a G-invariant tensor field $\langle \mathfrak{T} \rangle$ defined by

$$\langle \mathfrak{T} \rangle := \frac{1}{|G|} \int_G \Phi_g^* \mathfrak{T}\, dg , \tag{4.40}$$

where dg is an invariant volume form on G, and $|G|$ denotes the total volume of G, i.e., $|G| = \int_G dg$. Now, averaging the natural connection on $\mathcal{P} \times M$, we are led to the

Definition 4.2.2 *A Hannay–Berry connection in a trivial bundle $\mathcal{P} \times M$ is a connection defined by the following horizontal lift:*

$$\mathrm{h}(Z) = \langle \mathrm{h}_0(Z) \rangle = \langle 0 \oplus Z \rangle . \tag{4.41}$$

For the proof that the above formula does indeed define a connection see Marsden, Montgomery and Ratiu 1990. The Hannay–Berry connection is a connection of a general type (or Ehresmann connection), as defined in section 1.3.3. We stress that, in general, $E = \mathcal{P} \times M \longrightarrow M$ is neither a principal nor a vector bundle.

Introducing local canonical coordinates $(q^1, \ldots, q^n, p_1, \ldots, p_n)$ on $\mathcal{P}$ and $(x^1, \ldots, x^m)$ on M, we may describe the family of hamiltonian G-actions as follows:

$$\Phi_g \; : \; \mathcal{P} \times M \; \longrightarrow \; \mathcal{P} \times M \, , \tag{4.42}$$

with (in the obvious notation)

$$\Phi_g(q, p; x) = (Q(q, p, g; x), P(q, p, g; x); x) \, , \tag{4.43}$$

for any $g \in G$. Let

$$X = \sum_{\alpha=1}^{m} X^\alpha \frac{\partial}{\partial x^\alpha} \, ,$$

be a vector field on the m-dimensional parameter space M. Then the horizontal lift $h(X)$ with respect to the Hannay–Berry connection is given by

$$\begin{aligned} h(X) \;\; &= \;\; \langle 0 \oplus X \rangle \\ &= \;\; \sum_{\alpha=1}^{m} X^\alpha \left[\frac{\partial}{\partial x^\alpha} + \sum_{i=1}^{n} \left\langle \frac{\partial Q^i}{\partial x^\alpha} \right\rangle \frac{\partial}{\partial q^i} + \sum_{i=1}^{n} \left\langle \frac{\partial P^i}{\partial x^\alpha} \right\rangle \frac{\partial}{\partial p^i} \right] . \end{aligned} \tag{4.44}$$

Consider a family of completely integrable system, as introduced in Example 4.2.1. Fix an arbitrary point in the parameter space, $x \in M$, and take any regular value $\mu \in \mathbb{R}^n$ of the momentum map

$$\mathbf{J}(\,\cdot\,, x) \; : \; \mathcal{P} \; \longrightarrow \; \mathbb{R}^n \, .$$

The following subset of $\mathcal{P}$:

$$E_x^\mu \; := \; \mathbf{J}^{-1}(\mu) \cap \pi_M^{-1}(x) \, , \tag{4.45}$$

defines the Arnold n-torus T^n. In a neighborhood of any such torus there exist local angle variables $(\varphi_1, \ldots, \varphi_n)$, and hence we may construct parameter-dependent action-angle variables on $\mathcal{P}$, as follows:

$$I_i = I_i(q, p; x) \quad \text{and} \quad \varphi_i = \varphi_i(q, p; x) \, , \quad i = 1, \ldots, n \, . \tag{4.46}$$

The set $(I_i, \varphi_i, x^\alpha)$ defines local coordinates on the bundle space $\mathcal{P} \times M$. Using these coordinates, the x-dependent hamiltonian vector field has the following form:

$$X_H = \sum_{i=1}^{n} \omega_i(I; x) \frac{\partial}{\partial \varphi_i} \, , \tag{4.47}$$

where the x-dependent frequencies are defined by

$$\omega_i(I; x) = \frac{\partial H(I; x)}{\partial I_i} \, . \tag{4.48}$$

It is clear that the formula (4.44) for the horizontal lift with respect to the Hannay–Berry connection may be rewritten in the action-angle variables as follows:

$$\begin{aligned} h(X) &= \langle 0 \oplus X \rangle \\ &= \sum_{\alpha=1}^{m} X^\alpha \left[\frac{\partial}{\partial x^\alpha} + \sum_{i=1}^{n} \left\langle \frac{\partial I_i}{\partial x^\alpha} \right\rangle \frac{\partial}{\partial I_i} + \sum_{i=1}^{n} \left\langle \frac{\partial \varphi_i}{\partial x^\alpha} \right\rangle \frac{\partial}{\partial \varphi_i} \right] \, . \end{aligned} \tag{4.49}$$

However, as we showed in section 3.2.2, the standard actions I_i define adiabatic invariants, and hence

$$\left\langle \frac{\partial I_i}{\partial x^\alpha} \right\rangle = 0 \, , \tag{4.50}$$

or, equivalently,

$$\langle d_M I_i \rangle = 0 \, , \qquad i = 1, \ldots, n. \tag{4.51}$$

Therefore, formula (4.49) reduces to

$$h(X) = \sum_{\alpha=1}^{m} X^\alpha \left[\frac{\partial}{\partial x^\alpha} + \sum_{i=1}^{n} \left\langle \frac{\partial \varphi_i}{\partial x^\alpha} \right\rangle \frac{\partial}{\partial \varphi_i} \right] \, , \tag{4.52}$$

for any vector field X on a parameter manifold M. Hence, if C is a curve in M, and $\dot{x}$ is a corresponding velocity vector along C, then the above formula implies

$$h(\dot{x}) = \sum_{\alpha=1}^{m} \dot{x}^\alpha \left[\frac{\partial}{\partial x^\alpha} \oplus \sum_{i=1}^{n} \left(\langle d_M \varphi_i \rangle \cdot \frac{\partial}{\partial x^\alpha} \right) \frac{\partial}{\partial \varphi_i} \right] \, . \tag{4.53}$$

Now we shall construct a principal torus bundle over M. Let $\mu \in \mathbb{R}^n$ be a regular value of the momentum map, and define a subset $E^\mu \subset \mathcal{P} \times M$ by

$$E^\mu := \left\{ (p, x) \in \mathcal{P} \times M \,\middle|\, \mathbf{J}(p, x) = \mu \right\} \, . \tag{4.54}$$

It is easy to see that the projection

$$\pi_\mu := \pi_M \Big|_{E^\mu}$$

defines a principal bundle

$$\pi_\mu : E^\mu \longrightarrow M \, ,$$

with a typical fibre $F = T^n$. Clearly, a fibre at $x \in M$ reads

$$\pi_\mu^{-1}(x) = E_x^\mu \, , \tag{4.55}$$

where E_x^μ is the Arnold torus introduced in (4.45).

Corollary 4.2.1 *A bundle $\pi_\mu : E^\mu \longrightarrow M$ is a principal-torus bundle and the restriction of the Hannay–Berry connection to E^μ defines a connection on a principal fibre bundle.*

Note that if C is a closed curve in M, then formula (4.53) implies that the corresponding holonomy element from T^n is given by

$$\Delta\varphi_i(C) \;=\; -\oint_C \langle d_M\varphi_i(x) \rangle , \qquad i = 1,\ldots,n . \tag{4.56}$$

This formula reproduces the formula for the Hannay angles (3.88); one has

$$\frac{\partial}{\partial I_j} \sum_{i=1}^n \langle p_i d_M q^i \rangle = \frac{\partial}{\partial I_j} \sum_{i=1}^n \langle I_i d_M\varphi_i \rangle = \langle d_M\varphi_j \rangle , \tag{4.57}$$

and formula (4.56) follows. Hence, we may summarize that

$$\text{Hannay's angles} \;=\; \text{holonomy of the Hannay–Berry connection} .$$

4.3   Reduction, reconstruction and phases

4.3.1   Reconstruction of dynamics

Consider now a hamiltonian system defined on a symplectic manifold $(\mathcal{P}, \Omega)$, together with a Hamiltonian function $H \in C^\infty(\mathcal{P})$. Let a Lie group G act on $\mathcal{P}$ (on the left) by canonical transformations, and let

$$\mathbf{J} : \mathcal{P} \longrightarrow \mathfrak{g}^* \tag{4.58}$$

be the corresponding momentum map. The Marsden–Weinstein reduction theorem 4.1.3 implies that for a regular value $\mu \in \mathfrak{g}^*$, the reduced phase

$$\mathcal{P}_\mu := \mathbf{J}^{-1}(\mu)/G_\mu \tag{4.59}$$

defines a symplectic manifold with a symplectic form Ω_μ given by (4.24). If π_μ denotes the canonical projection

$$\pi_\mu : \mathbf{J}^{-1}(\mu) \longrightarrow \mathcal{P}_\mu , \tag{4.60}$$

then the reduced Hamiltonian function H_μ on $\mathcal{P}_\mu$ satisfies

$$\pi_\mu^* H_\mu = H_\mu \circ \pi_\mu = H . \tag{4.61}$$

Suppose, now, that we found a solution of the reduced dynamics. It means that we know a solution, say

$$t \longrightarrow c_\mu(t) \in \mathcal{P}_\mu ,$$

of the Hamilton equations on $\mathcal{P}_\mu$, such that

$$\frac{d}{dt} c_\mu(t) = X_{H_\mu}(c_\mu) . \tag{4.62}$$

The *reconstruction problem* is as follows: Knowing $c_\mu(t)$, find the trajectory of the original system in $\mathcal{P}$, i.e., the solution

$$t \longrightarrow c(t) \in \mathcal{P}$$

to the Hamilton equations, such that

$$\frac{d}{dt} c(t) = X_H(c) . \tag{4.63}$$

Recall from section 4.1.2 that a map π_μ, given by (4.60), defines a principal G_μ-bundle. Therefore, a curve $c(t)$ is nothing but a lift of the reduced trajectory $c_\mu(t)$ in the bundle $\mathbf{J}^{-1}(\mu) \longrightarrow \mathcal{P}_\mu$, and hence

$$\pi_\mu(c(t)) = c_\mu(t) .$$

Let $\mathcal{A}$ denote an arbitrary connection one-form on $\mathbf{J}^{-1}(\mu)$:

$$\mathcal{A} \in \Lambda^1(\mathbf{J}^{-1}(\mu)) \otimes \mathfrak{g}_\mu .$$

Take a point $p_0 \in \mathbf{J}^{-1}(\mu) \subset \mathcal{P}$, and let

$$t \longrightarrow d(t)$$

be a horizontal lift of $c_\mu(t)$ passing through a point p_0. Finally, let

$$t \longrightarrow \xi(t) \in \mathfrak{g}_\mu$$

be a curve in the Lie algebra $\mathfrak{g}_\mu$, defined by

$$\xi(t) := \mathcal{A}(X_H(d(t))) . \tag{4.64}$$

With this notation, one has

Theorem 4.3.1 (Reconstruction theorem) *An integral curve $c(t)$ of the hamiltonian system on $\mathcal{P}$ passing through a point p_0 is given by*

$$c(t) := g(t)d(t) , \tag{4.65}$$

where $g(t) \in G_\mu$ satisfies the following equation:

$$\dot{g}(t) = g(t)\xi(t) , \tag{4.66}$$

with $\xi(t)$ defined in (4.64).

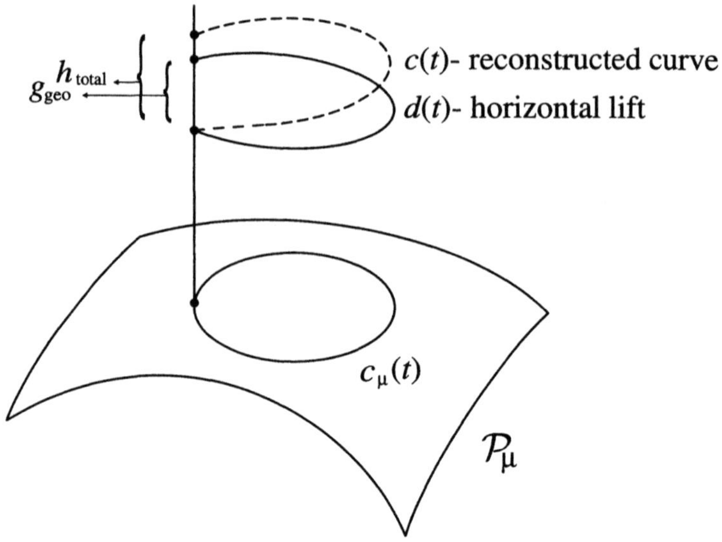

Figure 4.4: Reconstruction of the closed curve in $\mathcal{P}_\mu$ and its holonomy.

For the proof, see Marsden, Montgomery and Ratiu 1990. If the reduced curve $c_\mu(t)$ is closed, i.e,

$$c_\mu(T) = c_\mu(0) \ , \tag{4.67}$$

for some $T > 0$, then

$$d(T) = g_{\mathrm{geo}} \cdot d(0) \ , \tag{4.68}$$

and $g_{\mathrm{geo}} \in G_\mu$ defines the holonomy of the loop $c_\mu(t)$ with respect to the connection $\mathcal{A}$ (cf. Fig. 4.4). Note that

$$c(T) = h_{\mathrm{total}} \cdot c(0) \ , \tag{4.69}$$

with

$$h_{\mathrm{total}} = g(T) \cdot g_{\mathrm{geo}} \cdot g(0)^{-1} \ . \tag{4.70}$$

Remark 4.3.1 It is clear that the horizontal lift $d = d(t)$ *does* depend on a particular connection $\mathcal{A}$ in the bundle $\mathbf{J}^{-1}(\mu) \longrightarrow \mathcal{P}_\mu$. Hence, the *geometric phase factor*, or holonomy, g_{geo} also depends on the chosen connection. However, neither the reconstructed trajectory $c = c(t)$, nor the *total phase factor* h_{total} depend on $\mathcal{A}$. Therefore, to reconstruct $c(t)$ from $c_\mu(t)$ one may choose an arbitrary connection. $\diamond$

Example 4.3.1 Let Q be the configuration space of some mechanical system and let $\mathcal{P} = T^*Q$ be the corresponding phase space. Suppose there is a (left) action of a Lie group G on Q, i.e.,

$$\Phi_g^Q : Q \longrightarrow Q \ ,$$

with $g \in G$. An action of G on Q may, in a natural way, be lifted to an action on $\mathcal{P}$, as follows:

$$\Phi_g^P(q, \alpha_q) := \left(\Phi_g^Q(q), (T_q \Phi_g^Q)^* \cdot \alpha_q \right) , \tag{4.71}$$

for any $\alpha_q \in T_q^* Q$. The reader can easily show that Φ^P acts on $\mathcal{P}$ by canonical transformations, and gives rise to the momentum map

$$\mathbf{J} : \mathcal{P} \longrightarrow \mathfrak{g} ,$$

defined by

$$\langle \mathbf{J}(\alpha_q), \xi \rangle := (i_{\mathbf{X}_\xi^P} \Theta)(\alpha_q) = \alpha_q \left(\mathbf{X}_\xi^Q(q) \right) = \langle \alpha_q, \mathbf{X}_\xi^Q(q) \rangle , \tag{4.72}$$

where Θ is a canonical one-form on $T^* Q$ (cf. Example 3.1.3), and $\mathbf{X}_\xi^Q$ and $\mathbf{X}_\xi^P$ stand for the infinitesimal generators of Φ^Q and Φ^P, respectively, corresponding to $\xi \in \mathfrak{g}$. That is, $\mathbf{X}_\xi^Q \in \mathfrak{X}(Q)$ and $\mathbf{X}_\xi^P \in \mathfrak{X}(P)$.

Suppose that G_μ is one-dimensional and let $\zeta = \mu/|\mu|$ denote the generator of $\mathfrak{g}_\mu$ ($|\mu|$ stands for the length of $\mu \in \mathfrak{g}^* \cong \mathfrak{g}$).[2] We may identify $\mathfrak{g}_\mu$, which is one-dimensional, with $\mathbb{R}$ via

$$\mathbb{R} \ni a \longrightarrow a\zeta \in \mathfrak{g}_\mu .$$

Let Θ_μ and Ω_μ be the reduced canonical forms on $\mathcal{P}_\mu$, such that

$$\Theta = \pi_\mu^* \Theta_\mu , \qquad \Omega = \pi_\mu^* \Omega_\mu , \tag{4.73}$$

and $\Omega = -d\Theta$ is a symplectic form in $T^* Q$.

Proposition 4.3.2 *The following objects, defined on $\mathcal{P}_\mu$:*

$$A = \frac{1}{|\mu|} \Theta_\mu \otimes \zeta , \tag{4.74}$$

and

$$F = \frac{-1}{|\mu|} \Omega_\mu \otimes \zeta , \tag{4.75}$$

define a local connection and curvature in a principal G_μ-bundle $\mathbf{J}^{-1}(\mu) \longrightarrow \mathcal{P}_\mu$.

---

[2]A Lie algebra is equipped with the canonical metric form — the so-called Cartan tensor — defined by

$$h_{ij} := f_{il}^k f_{jk}^l .$$

For example, for $so(3)$ one has

$$h_{ij} = \epsilon_{il}^k \epsilon_{jk}^l = 2\delta_{ij} ,$$

that is, one reproduces the standard euclidean metric in $\mathbb{R}^3$.

Note that the holonomy of the closed curve $c_\mu(t)$ is defined by

$$g_{\text{geo}} = \exp(\Delta\theta_{\text{geo}}\zeta) ,$$

where

$$\Delta\theta_{\text{geo}} = \frac{-1}{|\mu|} \oint_{c_\mu} \Theta_\mu = \frac{1}{|\mu|} \int_\Sigma \Omega_\mu , \qquad (4.76)$$

with Σ being any two-dimensional region in $\mathcal{P}_\mu$, such that $\partial\Sigma = c_\mu$. $\diamond$

Example 4.3.2 (Mechanical connection) Let the configuration space Q be a Riemannian manifold. Suppose there is a (left) action of a Lie group G on Q:

$$\Phi_g^Q : Q \longrightarrow Q ,$$

and let $\mathbf{X}_\xi$ denote an infinitesimal generator corresponding to $\xi \in \mathfrak{g}$. For any $q \in Q$ and $\mu \in \mathfrak{g}^*$, let us define

$$I_\mu(q) : \mathfrak{g}_\mu \longrightarrow \mathfrak{g}_\mu^* \qquad (4.77)$$

by

$$\langle\, I_\mu(q)(\xi), \eta \,\rangle := g_q(\mathbf{X}_\xi, \mathbf{X}_\eta) , \qquad (4.78)$$

with $\xi, \eta \in \mathfrak{g}_\mu$. In the above formula g_q stands for the riemannian metric at the point $q \in Q$. Moreover, let

$$\mathbf{J} : \mathcal{P} \longrightarrow \mathfrak{g}^*$$

denote a momentum map of the lifted action Φ^P on $\mathcal{P}$ (cf. previous example). Now we are ready to define a connection in a principal G_μ-bundle,

$$Q \longrightarrow Q_\mu := Q/G_\mu ,$$

that is, we define a one-form on Q taking values in $\mathfrak{g}_\mu$. Take any $v_q \in T_q Q$, and denote by $v_q^\flat$ an element from $T_q^* Q$ defined by

$$v_q^\flat(u) := g_q(u, v_q) ,$$

for any vector $u \in T_q Q$. One defines a so-called *mechanical connection* by

$$A_{\text{mech}}(v_q) := (I_\mu(q))^{-1}(\mathbf{J}(v_q^\flat)) . \qquad (4.79)$$

Note that, due to the definition of $I_\mu(q)$, $A_{\text{mech}}(v_q)$ does belong to $\mathfrak{g}_\mu$. The reader can easily show that the above formula indeed defines a connection form in a principal bundle. $\diamond$

Remark 4.3.2 Note that if $G = SO(3)$, then the above construction shows that angular momentum (i.e., the momentum map for the action of $SO(3)$) defines a connection. This observation was made by several authors. This connection is closely related to the so-called *Cat's problem* (see, e.g., Montgomery 1990). For example, Shapere and Wilczek (1989a,b) used this connection to describe the dynamics of deformable bodies. It was also investigated by Iwai (1987a, 1987b, 1988). Guichardet (1984) applied this connection to molecular systems. $\diamond$

4.3.2 Rigid bodies

Using the general scheme of the previous section, we now show how to reconstruct the dynamics of the free rigid body in $\mathbb{R}^3$. Recall that a configuration space for a free rigid body with one fixed point in $\mathbb{R}^3$ is the group of rotations in $\mathbb{R}^3$, i.e., $G = SO(3)$. Each element of $SO(3)$ uniquely describes the orientation of the body in space. The motion of the body is therefore described by the trajectory $t \longrightarrow g(t) \in G$, and the velocity $\dot{g}$ is a vector tangent to G at g, i.e., $\dot{g} \in T_g G$. This tangent vector may be mapped to the Lie algebra $\mathfrak{g}$ by left and right translations, as follows:

$$\boldsymbol{\omega}_b \quad := \quad T_g L_{g^{-1}} \dot{g} \in \mathfrak{g}\,, \tag{4.80}$$

$$\boldsymbol{\omega}_s \quad := \quad T_g R_{g^{-1}} \dot{g} \in \mathfrak{g}\,. \tag{4.81}$$

These two vectors correspond to the angular velocity in the body frame and in the space frame, respectively. In this way we define two natural isomorphisms:

$$(\text{Body}) \quad G \times \mathfrak{g} \xleftarrow{\;L\;} TG \xrightarrow{\;R\;} G \times \mathfrak{g} \quad (\text{Space})\,, \tag{4.82}$$

and $\mathfrak{g}$ serves as the space of angular velocities. Now, the dual space $\mathfrak{g}^*$ is the space of angular momenta. The analog of the above diagram reads

$$(\text{Body}) \quad G \times \mathfrak{g}^* \xleftarrow{\;L\;} T^*G \xrightarrow{\;R\;} G \times \mathfrak{g}^* \quad (\text{Space})\,, \tag{4.83}$$

that is, for any $\mathbf{M} \in T_g^* G$, one has

$$\mathbf{M}_b \quad := \quad (T_e L_g)^*(\mathbf{M}) \in \mathfrak{g}^*\,, \tag{4.84}$$

$$\mathbf{M}_s \quad := \quad (T_e R_g)^*(\mathbf{M}) \in \mathfrak{g}^*\,, \tag{4.85}$$

and $\mathbf{M}_b$ and $\mathbf{M}_s$ denote the angular momenta in the body frame and in the space frame, respectively. In the above formulae, $(T_e L_g)^*$ $((T_e R_g)^*)$ denotes the operator adjoint to $T_e L_g$ $(T_e R_g)$, that is,

$$T_e L_g \; : \; \mathfrak{g} \longrightarrow T_g G,$$

and the adjoint operator

$$(T_e L_g)^* \; : \; T_g^* G \longrightarrow \mathfrak{g}^*$$

is defined by

$$\langle (T_e L_g)^*(\mathbf{M}), \xi \rangle := \langle \mathbf{M}, (T_e L_g)(\xi) \rangle\,,$$

for any $\mathbf{M} \in T_g^* G$ and $\xi \in \mathfrak{g}$. In the above formula $\langle\,,\,\rangle$ is a natural pairing between $\mathfrak{g}^*$ and $\mathfrak{g}$.

The properties of the rigid body are encoded into the inertia tensor (a symmetric, positively defined operator)

$$\hat{\mathbf{I}} \; : \; \mathfrak{g} \longrightarrow \mathfrak{g}^*\,, \tag{4.86}$$

which defines the correspondence between angular velocities and angular momenta, according to

$$\mathbf{M}_b = \hat{\mathbf{I}}\boldsymbol{\omega}_b \, . \tag{4.87}$$

Now, the kinetic energy of the body is defined by

$$T(\boldsymbol{\omega}_b) := \frac{1}{2} \langle \hat{\mathbf{I}}\boldsymbol{\omega}_b, \boldsymbol{\omega}_b \rangle \, . \tag{4.88}$$

Note that $\hat{\mathbf{I}}$ defines a scalar product $(\,,\,)$ in $\mathfrak{g}$: for any $\xi, \eta \in \mathfrak{g}$

$$(\xi, \eta) := \langle \hat{\mathbf{I}}\xi, \eta \rangle \, . \tag{4.89}$$

Using left translations one may define the corresponding scalar product on each tangent space $T_g G$, and, therefore, one obtains a left-invariant riemannian metric on G. It is well known that the motion of the free rigid body is described by the geodesics of this metric on $G = SO(3)$.

This well-known picture may be generalized in the obvious way: Instead of $SO(3)$, take an arbitrary Lie group G as a configuration space of the dynamical system, and let its motion be described by the left-invariant riemannian metric on G. One calls such a system a *generalized rigid body*. The corresponding equation reads

$$\dot{\mathbf{M}}_s = 0 \, , \tag{4.90}$$

i.e., the angular momentum in the space is conserved in time (for a free rigid body). Now, to find the equation for the angular momentum in the body frame let us define the following operation in the Lie algebra $\mathfrak{g}$ of G: For $\xi \in \mathfrak{g}$, let

$$\mathrm{ad}_\xi \, : \, \mathfrak{g} \longrightarrow \mathfrak{g}$$

be defined by

$$\mathrm{ad}_\xi \, \eta := [\xi, \eta] \, .$$

The assignment $\xi \longrightarrow \mathrm{ad}_\xi$ is called the adjoint representation of $\mathfrak{g}$. Now, let ad_ξ^* denote the adjoint operation, i.e., $\mathrm{ad}_\xi : \mathfrak{g}^* \longrightarrow \mathfrak{g}^*$ such that

$$\langle \mathrm{ad}_\xi^* \alpha, \eta \rangle := \langle \alpha, \mathrm{ad}_\xi \eta \rangle \, , \qquad \eta \in \mathfrak{g}, \, \alpha \in \mathfrak{g}^* \, .$$

It turns out that the analog of the standard Euler equation in $\mathbb{R}^3$ is given by

$$\dot{\mathbf{M}}_b = \mathrm{ad}_{\omega_b}^* \mathbf{M}_b \, . \tag{4.91}$$

with $\boldsymbol{\omega}_b = \hat{\mathbf{I}}^{-1}\mathbf{M}_b$.

Consider now the standard rigid body in $\mathbb{R}^3$ with a configuration space $Q = SO(3)$. The corresponding phase space is, therefore, a cotangent bundle $\mathcal{P} = T^*SO(3)$. The standard left action of $SO(3)$ on itself, i.e.,

$$SO(3) \ni g \longrightarrow \Phi_g^Q(h) := L_g h = g \cdot h \, , \tag{4.92}$$

may be lifted to the action Φ^P on $T^*SO(3)$:

$$\Phi_g^P(h, \alpha_g) := \left(gh, (T_e L_g)^*(\alpha_h)\right), \tag{4.93}$$

for any $\alpha_h \in T_h^* SO(3)$. The corresponding momentum map

$$\mathbf{J} : T^*SO(3) \longrightarrow so(3)^*$$

is given by

$$\mathbf{J}(\alpha_g) = (T_e L_g)^*(\alpha_g). \tag{4.94}$$

The cotangent bundle $T^*SO(3)$ may be canonically identified with

$$T^*SO(3) \cong SO(3) \times so(3)^*.$$

Moreover, the Lie algebra $so(3)$ and its dual, are canonically isomorphic, and, as we have already observed in Example 1.2.11, $so(3)$ may be identified with $\mathbb{R}^3$ via

$$\mathbb{R}^3 \ni \mathbf{x} \longrightarrow \hat{\mathbf{x}} \in so(3), \tag{4.95}$$

such that, for any $\mathbf{y} \in \mathbb{R}^3$,

$$\hat{\mathbf{x}}\mathbf{y} := \mathbf{x} \times \mathbf{y}.$$

Using formula (4.84), the lifted action Φ^P on $SO(3) \times so(3)^*$ may be rewritten as follows:

$$\Phi_g^P(h, \boldsymbol{\mu}) := (gh, \mathrm{Ad}^*_{g^{-1}}\boldsymbol{\mu}) = (gh, g^{-1} \cdot \boldsymbol{\mu}), \tag{4.96}$$

for any $g, h \in SO(3)$, and $\boldsymbol{\mu} \in so(3)^* \cong \mathbb{R}^3$. It is easy to see that the corresponding momentum map reads

$$\mathbf{J}(h, \boldsymbol{\mu}) = \boldsymbol{\mu}. \tag{4.97}$$

Clearly, for any $\boldsymbol{\mu} \in \mathbb{R}^3$,

$$\mathbf{J}^{-1}(\boldsymbol{\mu}) \cong SO(3). \tag{4.98}$$

Moreover, the isotropy subgroup $G_\mu \subset SO(3)$ consists of all $SO(2)$ rotations around an axis defined by $\boldsymbol{\mu} \in \mathbb{R}^3$. Therefore, the reduced phase space

$$\mathcal{P}_\mu = \mathbf{J}^{-1}(\boldsymbol{\mu})/G_\mu \cong S_\mu^2 \tag{4.99}$$

may be identified with a two-dimensional sphere S_μ^2, of radius $\mu = |\boldsymbol{\mu}|$, equipped with a symplectic form

$$\Omega_\mu = -\frac{dS_\mu}{\mu}, \tag{4.100}$$

with dS_μ being the standard volume two-form on a sphere of radius μ.

Consider a closed trajectory of the reduced system on S_μ^2 corresponding to a periodic motion of the body, i.e., we have a closed curve in the space of body frame angular momenta:

$$t \longrightarrow c_\mu(t) \equiv \mathbf{M}_b(t) \in S_\mu^2 , \qquad (4.101)$$

with $\mathbf{M}_b(T) = \mathbf{M}_b(0)$. This is our reduced trajectory from the previous section; more precisely,

$$c_\mu(t) = (\text{fixed orientation in the body frame}, \mathbf{M}_b(t)) .$$

The corresponding reconstructed trajectory $t \longrightarrow c(t) \in \mathcal{P} = T^*SO(3)$ has the following form:

$$c(t) = (R(t), \mathbf{M}_b(t)) \in SO(3) \times \mathbb{R}^3 , \qquad (4.102)$$

with

$$R(t)\mathbf{M}_b(t) = \mathbf{M}_s , \qquad (4.103)$$

where $\mathbf{M}_s$ is the (constant) value of the angular momentum in the space frame, and $R(t)$ defines a curve in $SO(3)$, i.e., a one-parameter subgroup of rotations. Now, after a period T the body angular momentum $\mathbf{M}_b$ coincides with its initial value $\mathbf{M}_b(0)$, but in general the body has performed a nontrivial rotation, i.e.,

$$R(T) \neq R(0) .$$

Denote by

$$R_{\Delta\theta} := R(T) \cdot R(0)^{-1} \qquad (4.104)$$

the net rotation corresponding to the change in orientation of the body between $t = 0$ and $t = T$ (cf. section 3.3.6). Clearly, $R_{\Delta\theta}$ defines an element from G_μ, i.e.,

$$R_{\Delta\theta} = \exp(\Delta\theta\,\boldsymbol{\zeta}) , \qquad (4.105)$$

with $\boldsymbol{\zeta} = \boldsymbol{\mu}/\mu$ being a generator in $\mathfrak{g}_\mu$. Take a connection of the sort defined in (4.74) and let $t \longrightarrow d(t) \in \mathbf{J}^{-1}(\mu)$ be a horizontal lift of c_μ. Since the connection (4.74) corresponds to the natural connection in a monopole bundle (cf. section 1.4.4), the corresponding holonomy element g_{geo}, such that $d(T) = g_{\text{geo}} \cdot d(0)$, is, due to (4.76), given by

$$g_{\text{geo}} = \exp\left[\Delta\theta_{\text{geo}}\zeta\right] , \qquad (4.106)$$

with

$$\Delta\theta_{\text{geo}} = \frac{1}{\mu} \int_{S^2} \Omega_\mu = -\Omega(c_\mu) , \qquad (4.107)$$

where $\Omega(c_\mu)$ denotes the solid angle subtended by the closed curve c_μ on S_μ^2.

Now, the horizontal lift $d(t)$, such that $d(0) = (\mathbb{1}, \mathbf{M}_s)$, is given by

$$d(t) = (u(t), \mathbf{M}_b(t)) \in SO(3) \times \mathbb{R}^3 . \tag{4.108}$$

with some $u(t) \in SO(3)$. Actually, to find the reconstructed trajectory $c(t)$ we do not need the explicit form of $u(t)$; the interested reader can find the construction of $u(t)$ in section 6.1.1. Clearly, $u(T) = g_{\text{geo}}$. To use our reconstruction algorithm, let us define

$$\xi(t) := \mathcal{A}\Big(X_H(d(t))\Big) \in \mathbb{R}^3 . \tag{4.109}$$

Using the definition of connection (4.74), we then obtain

$$
\begin{aligned}
\mathcal{A}(X_H) &= \mathcal{A}(X_{H_\mu}) = \frac{1}{\mu} \sum_{i=1}^{3} p_i dq^i \left[\sum_{k=1}^{3} \left(p^k \frac{\partial}{\partial q^k} - q^k \frac{\partial}{\partial p^k} \right) \right] \otimes \zeta \\
&= \frac{1}{\mu} \sum_{i=1}^{3} p_i p^i \otimes \zeta = \frac{2E_\mu}{\mu} \otimes \zeta ,
\end{aligned}
\tag{4.110}
$$

with E_μ standing for the energy for a free rigid body of the unit mass. The reconstructed trajectory is defined by

$$c(t) = (g(t)u(t), \mathbf{M}_b(t)) , \tag{4.111}$$

with $g(t)$ satisfying

$$\dot{g}(t) = g(t) \cdot \xi(t) = \frac{2E_\mu}{\mu} g(t) \cdot \zeta . \tag{4.112}$$

The solution to (4.112) reads

$$g(t) = \exp\left[\frac{2E_\mu t}{\mu} \zeta \right] , \tag{4.113}$$

which corresponds to rotation about the $\boldsymbol{\mu}$-axis with angular velocity $2E_\mu/\mu$. Finally, the total rotation is, due to (4.70), defined by

$$h_{\text{total}} := g(T) \cdot g_{\text{geo}} \cdot g(0) ,$$

and it corresponds to a rotation about the $\boldsymbol{\mu}$-axis by a total angle

$$\Delta\theta = \Delta\theta_{\text{dyn}} + \Delta\theta_{\text{geo}} , \tag{4.114}$$

where the geometric phase $\Delta\theta_{\text{geo}}$ is given by (4.107), and the dynamical phase $\Delta\theta_{\text{dyn}}$ reads as follows:

$$\Delta\theta_{\text{dyn}} = \frac{2E_\mu T}{\mu} . \tag{4.115}$$

The above formula agrees with (3.201), which was derived without the use of the reconstruction theorem.

Further reading

Section 4.1. A detailed exposition of the geometric approach to classical Hamiltonian mechanics may be found in Abraham and Marsden 1978; Arnold 1989; Marsden and Ratiu 1999; and Guillemin and Sternberg 1984.

Section 4.2. For more information about the geometric properties of Hannay–Berry connections the reader is referred to Montgomery 1988; Golin, Knauf and Marmi 1989; and Marsden, Montgomery and Ratiu 1990 (see also Golin 1989; Golin and Marmi 1990; and Koiler 1989).

Section 4.3. For a more detailed discussion of the reconstruction of Hamiltonian dynamics in classical dynamics we refer the reader to Marsden, Montgomery and Ratiu 1990.

Problems

4.1. Let $\mathbf{J}$ be a momentum map corresponding to a hamiltonian action of a Lie group G on $\mathcal{P}$. Show that, for any $\xi, \eta \in \mathfrak{g}$,

$$X_{J([\xi,\eta])} = X_{\{J(\xi), J(\eta)\}} \, .$$

4.2. Prove Proposition 4.1.2.

4.3. Show that linear and angular momenta, defined in examples 4.1.1 and 4.1.2, respectively, are equivariant momentum maps.

4.4. Show that the definition of the reduced symplectic form Ω_μ (see (4.24)) does not depend on the particular choice of vectors $\tilde{u}$ and $\tilde{v}$ provided the condition (4.23) holds.

4.5. Check that the geometric phase factor g_{geo} in the reconstruction theorem does not depend on a particular connection $\mathcal{A}$.

4.6. Show that the lifted action defined in (4.71) is hamiltonian.

4.7. Check that a mechanical connection, defined in (4.79), does define a connection form in a principal bundle.

5
Geometry of Quantum Evolution

5.1 Geometrical formulation of quantum mechanics

Usually one uses completely different mathematical descriptions to formulate classical and quantum mechanics. Classical theory may be nicely formulated in terms of symplectic geometry, and the quantum one in terms of algebraic objects related to a complex Hilbert space. However, it turns out that standard, nonrelativistic quantum mechanics possesses natural geometric structure that is even richer than that found in classical mechanics. This section reveals the beauty of the geometric approach to quantum theory and stands as a basis for the elegant geometrical ideas of Pancharatnam and, later on, of Aharonov and Anandan.

5.1.1 Hilbert space as a Kähler manifold

Let us begin with the standard Hilbert space formulation of nonrelativistic quantum mechanics. Denote by $\mathcal{H}$ a complex Hilbert space and decompose the hermitian scalar product in $\mathcal{H}$,

$$\langle \cdot \,|\, \cdot \rangle \; : \; \mathcal{H} \times \mathcal{H} \; \longrightarrow \; \mathcal{H} \, ,$$

into real and imaginary parts, as follows:

$$\langle \psi | \phi \rangle =: G(\psi, \phi) + i\Omega(\psi, \phi) \, . \tag{5.1}$$

One easily finds that G satisfies

$$G(\psi, \phi) = G(\phi, \psi) \tag{5.2}$$

and, therefore, defines a positive, real scalar product, whereas Ω defines a symplectic two-form, i.e.,

$$\Omega(\psi, \phi) = -\Omega(\phi, \psi) \, . \tag{5.3}$$

Moreover, they satisfy the following relation:

$$\Omega(\psi, \phi) = G(\psi, i\phi) \, , \tag{5.4}$$

for any $\psi, \phi \in \mathcal{H}$. Formula (5.4) is a defining relation of the so-called *Kähler space*. In order to introduce a precise definition of a Kähler space we shall proceed as follows: Let us consider a real, m-dimensional vector space V. A linear map

$$J \, : \, V \, \longrightarrow \, V \, ,$$

such that $J^2 = -\mathbb{1}_m$, is called a *complex structure* on V. Note, that the real dimension of V is necessarily even: $\det(J^2) = (\det J)^2 = (-1)^m$, and hence we have $m = 2n$. A real vector space V endowed with a complex structure J becomes a complex vector space of complex dimension $n = \frac{m}{2}$. Indeed, for any $\alpha = a + ib \in \mathbb{C}$ with $a, b \in \mathbb{R}$, and $v \in V$, we define

$$\alpha v := av + bJv \, .$$

Conversely, if W is an n-dimensional complex vector space, then W is a $2n$-dimensional real vector space endowed with a complex structure defined by

$$Jw := iw,$$

for any $w \in W$.

Example 5.1.1 Consider $\mathbb{C}^n = \{ (z^1, \dots, z^n) \mid z^k \in \mathbb{C} \}$. Introducing real coordinates

$$z^k = x^k + iy^k \, , \quad x^k, y^k \in \mathbb{R} \, , \quad k = 1, \dots, n \, ,$$

we may identify $\mathbb{C}^n$ with the $2n$-dimensional real vector space $\mathbb{R}^{2n}$. Now, $\mathbb{R}^{2n}$ is endowed with a canonical complex structure J_0 defined as follows: For $v = (x^1, \dots, x^n, y^1, \dots, y^n) \in \mathbb{R}^{2n}$,

$$J_0 v := (y^1, \dots, y^n, -x^1, \dots, -x^n) \, ,$$

that is, in the standard basis in $\mathbb{R}^{2n}$, J_0 is given by the following matrix:

$$J_0 = \begin{pmatrix} 0 & \mathbb{1}_n \\ -\mathbb{1}_n & 0 \end{pmatrix} \, . \tag{5.5}$$

Clearly, $J_0^2 = -\mathbb{1}_{2n}$. $\diamond$

Now let us turn to manifolds. A real, $2n$-dimensional manifold M is called a *complex manifold* if it admits an atlas (U_i, φ_i) such that all transition functions

$$\varphi_{ij} := \varphi_j \circ \varphi_i^{-1} \; : \; U_i \cap U_j \;\longrightarrow\; \mathbb{C}^n \, ,$$

are holomorphic (complex analytic). The number n is called the complex dimension of M. Note that every complex manifold admits a field of endomorphisms

$$J_x \; : T_x M \;\longrightarrow\; T_x M \, ,$$

such that $J_x^2 = -\mathbb{1}_{2n}$, called a complex structure on M.[1] Clearly, in the case of a complex vector space $\mathbb{C}^n$ this field $x \longrightarrow J_x$ is constant, i.e., J_x reproduces J_0 from Example 5.1.1.

Let M be a complex manifold. A *hermitian metric h* on M is a smooth assignment of a hermitian scalar product on each tangent space. That is,

$$h \; : \; TM \times TM \;\longrightarrow\; \mathbb{C} \, ,$$

such that

$$h_x(u, v) := \langle u, v \rangle_x \, , \tag{5.6}$$

with $\langle u, v \rangle_x$ being a hermitian scalar product on $T_x M$, for any $u, v \in T_x M$. A complex manifold endowed with a hermitian metric is called a *hermitian manifold*. Introducing local complex coordinates $(z^1, \ldots, z^n)$ on M, any hermitian metric can be written as:

$$h = 2 g_{i\bar{j}} \, dz^i \otimes d\bar{z}^j \, , \tag{5.7}$$

with $g_{i\bar{j}}$ being a hermitian matrix, i.e., $\overline{g_{i\bar{j}}} = g_{j\bar{i}}$, where a bar denotes complex conjugation. In what follows we shall use the standard complex notation: A complexified tangent space is spanned by $2n$ vectors:

$$\partial/\partial z^1, \ldots, \partial/\partial z^n, \partial/\partial \bar{z}^1, \ldots, \partial/\partial \bar{z}^n \, ,$$

whereas the corresponding cotangent space is spanned by $2n$ covectors:

$$dz^1, \ldots, dz^n, d\bar{z}^1, \ldots, d\bar{z}^n \, .$$

One often writes ∂_k for $\partial/\partial z^k$ and $\bar{\partial}_k$ for $\partial/\partial \bar{z}^k$. Having any tensor field on M the components corresponding to the holomorphic quantities ∂_k and dz^k are denoted by a simple index k, whereas those corresponding to the anti-holomorphic objects $\bar{\partial}_k$ and $d\bar{z}^k$ are denoted by $\bar{k}$.

---

[1]It should be stressed that the existence of a field $x \to J_x$ is only a necessary condition for a manifold to be complex. For example, among even-dimensional spheres, S^{2n}, only S^2 defines a complex manifold. However, also S^6 admits $x \to J_x$ with $J_x^2 = -\mathbb{1}_6$.

Note that any hermitian metric h is invariant under the action of the complex structure J, i.e.,

$$h(Ju, Jv) = h(u, v) \ . \tag{5.8}$$

Actually, this condition may be used as an equivalent definition of h. Any hermitian metric on M gives rise to the tensor field

$$K \ : \ \mathfrak{X}(M) \times \mathfrak{X}(M) \ \longrightarrow \ \mathbb{C} \ ,$$

defined by

$$K(u, v) := h(u, Jv) \ . \tag{5.9}$$

Proposition 5.1.1 *K satisfies the following properties:*

- $K(u, v) = -K(v, u) \ , \ \ i.e., \ K \in \Lambda^2(M),$

- $K(u, v) = \overline{K(u, v)} \ , \ \ i.e., \ K \ is \ real,$

- $K(Ju, Jv) = K(u, v) \ , \ \ i.e., \ K \ is \ J\text{-}invariant.$

In local complex coordinates one can then write:

$$K = i g_{i\bar{j}} \, dz^i \wedge d\bar{z}^j \ . \tag{5.10}$$

Definition 5.1.1 *If K is closed, then one calls K a Kähler form, h a Kähler metric, and the corresponding hermitian manifold M a Kähler manifold.*

Corollary 5.1.1 *Recall that a two-dimensional complex manifold is called a Riemann surface. Now, on a two-dimensional manifold any two-form K is necessarily closed (since dK, being a three-form, has to vanish) and hence any hermitian metric is also Kähler. This implies that each Riemann surface is a Kähler manifold.*

Now, it is easy to show that a hermitian form K is nondegenerate and hence a Kähler two-form, being nondegenerate, and closed, defines a symplectic structure on M. In this way, any Kähler manifold is endowed with a (Kähler) metric h and a symplectic form K, which are related by the defining formula (5.9). This observation may be used as another equivalent definition of a Kähler manifold. Actually, one can prove the following

Proposition 5.1.2 *Let M be a real manifold endowed with a complex structure J and a riemannian J-invariant metric g. Define the symplectic form ω as follows:*

$$\omega(u, v) = g(u, Jv) \ . \tag{5.11}$$

Then M is a Kähler manifold if and only if $\nabla J = 0$ (or equivalently, $\nabla \omega = 0$), where ∇ denotes the covariant derivative with respect to the riemannian connection.[2]

---

[2]Recall, that for the $(1, 1)$-tensor field t on M the covariant derivative is defined as follows:

$$\nabla_k t^i{}_j := \partial_k t^i{}_j + \Gamma^i_{kl} t^l{}_j - \Gamma^m_{kj} t^i{}_m \ ,$$

Example 5.1.2 Consider the simplest complex manifold, i.e., let M be the complex plane $\mathbb{C}$. Taking $z = x + iy$, one has

$$g = dx \otimes dx + dy \otimes dy = (dx + idy) \otimes (dx - idy) = dz \otimes d\bar{z} , \qquad (5.12)$$

and, therefore, the Kähler form

$$K = \frac{i}{2} dz \wedge d\bar{z} = \frac{i}{2}(dx + idy) \wedge (dx - idy) = dx \wedge dy , \qquad (5.13)$$

where we have used $g_{1\bar{1}} = 1/2$. Note that K is obviously closed, and defines the standard symplectic form in $\mathbb{R}^2$. This example may be immediately generalized to $\mathbb{C}^n$. One easily finds that

$$g = \sum_{\alpha=1}^{n} (dx_\alpha \otimes dx_\alpha + dy_\alpha \otimes dy_\alpha) = \sum_{\alpha=1}^{n} dz_\alpha \otimes d\bar{z}_\alpha , \qquad (5.14)$$

for the hermitian metric, and

$$K = \frac{i}{2} \sum_{\alpha=1}^{n} dz_\alpha \wedge d\bar{z}_\alpha = \sum_{\alpha=1}^{n} dx_\alpha \wedge dy_\alpha , \qquad (5.15)$$

for the symplectic form. $\diamond$

Example 5.1.3 Let $M = S^2$. First of all, it is well known that S^2 defines a complex manifold — the celebrated Riemann sphere, and hence, due to Corollary 5.1.1, it is a Kähler manifold. Consider S^2 as a real manifold and parametrize it locally by the spherical angles (θ, φ). One then has for the metric,[3]

$$ds^2 = d\theta^2 + \sin^2 \theta \, d\varphi^2 ,$$

and for the symplectic structure,

$$\Omega = \sin \theta \, d\theta \wedge d\varphi .$$

Defining a complex structure J on S^2 by

$$J = \begin{pmatrix} 0 & -\sin \theta \\ (\sin \theta)^{-1} & 0 \end{pmatrix} , \qquad (5.16)$$

---

where the connection coefficients Γ^i_{kl}, so-called Christoffel symbols, are defined by

$$\Gamma^i_{kl} := \frac{1}{2} g^{im} (g_{mk,l} + g_{ml,k} - g_{kl,m}) .$$

The generalization to arbitrary tensor fields is straightforward, see e.g., Kobayashi and Nomizu 1969; Choquet-Bruhat and DeWitt–Morette 1982; and Nash and Sen 1983.

[3] From now on we shall use the convention preferred by physicists: $dx^2 := dx \otimes dx$, and the line element $ds^2 := g_{\alpha\beta} dx^\alpha \otimes dx^\beta$.

one immediately finds that formula (5.11) holds. Now, to show that J is covariantly constant let us note that the Christoffel symbols have the following form:

$$\Gamma^{\varphi}_{\varphi\theta} = \cot\theta\,, \qquad \Gamma^{\theta}_{\varphi\varphi} = -\cos\theta\sin\theta\,,$$

and the remaining components vanish. The reader will easily show that

$$\nabla_i J^k{}_j = 0\,,$$

for any $i,\,j,\,k = \theta,\,\varphi$, which proves that S^2 is a Kähler manifold. $\diamond$

5.1.2   The quantum phase space

In the previous section, we have shown that a Hilbert space $\mathcal{H}$ corresponding to any quantum system carries a structure of a Kähler manifold. Note, however, that the Hilbert space $\mathcal{H}$ is not the quantum analog of a classical phase space. Indeed, any two vectors $\psi,\,\phi \in \mathcal{H}$, such that

$$\psi = c\phi\,, \quad c \in \mathbb{C}\,,$$

are physically equivalent ($\psi \sim \phi$), that is, they define the same physical state. Therefore, the proper phase space of a quantum system is the space of *rays* in $\mathcal{H}$:

$$\mathcal{P}(\mathcal{H}) := \mathcal{H}/\sim\,, \tag{5.17}$$

called a *projective Hilbert space*. Define a canonical projection

$$\Pi\ :\ \mathcal{H}\ \longrightarrow\ \mathcal{P}(\mathcal{H})\,, \tag{5.18}$$

and denote $[\psi] := \Pi(\psi)$, so that $[\psi]$ corresponds to a ray in $\mathcal{H}$ passing through ψ. Note that the above construction defines a vector bundle over $\mathcal{P}(\mathcal{H})$ with a typical fibre $F = \mathbb{C}$, and a structure group $G = GL(1,\mathbb{C}) \cong \mathbb{C}^* := \mathbb{C} - \{0\}$. Clearly, the fibres

$$\Pi^{-1}([\psi]) = \text{complex line (ray) passing through } \psi$$

are one-dimensional, and one calls the above bundle a *complex line bundle*. Now define a unit sphere in $\mathcal{H}$:

$$S(\mathcal{H}) := \left\{\ \psi \in \mathcal{H}\ \middle|\ \langle\psi|\psi\rangle = 1\ \right\} \subset \mathcal{H}\,.$$

Any two points $\psi,\,\phi \in S(\mathcal{H})$ are physically equivalent ($\psi \sim \phi$), if they differ by a phase factor, i.e., $\psi = e^{i\lambda}\phi$. Therefore, one has, equivalently,

$$\mathcal{P}(\mathcal{H}) = S(\mathcal{H})/\sim\,.$$

Clearly, if $\psi \in S(\mathcal{H})$, then the corresponding equivalence class $[\psi]$ may be identified with the one-dimensional projector

$$P_\psi := |\psi\rangle\langle\psi|\,.$$

Hence,

$$\mathcal{P}(\mathcal{H}) \cong \text{space of one-dimensional projectors in } \mathcal{H} .$$

Note that this construction defines a principal $U(1)$-bundle

$$\pi \; : \; S(\mathcal{H}) \longrightarrow \mathcal{P}(\mathcal{H}) ,$$

over $\mathcal{P}(\mathcal{H})$. Indeed, for any vector $\psi \in \mathcal{H}$ the corresponding fibre

$$\pi^{-1}([\psi]) = \left\{ \frac{e^{i\lambda} \psi}{\langle \psi | \psi \rangle} \in S(\mathcal{H}) \right\} ,$$

may be identified with the Lie group $U(1)$.

Theorem 5.1.3 *A quantum phase space — projective Hilbert space $\mathcal{P}(\mathcal{H})$ — is a Kähler manifold.*

Proof. In order to prove this theorem we shall construct a hermitian scalar product on each tangent space $T_p\mathcal{P}(\mathcal{H})$, i.e.,

$$\langle \cdot | \cdot \rangle_p \; : \; T_p\mathcal{P}(\mathcal{H}) \times T_p\mathcal{P}(\mathcal{H}) \longrightarrow \mathbb{C} .$$

Any tangent vector $\xi \in T_p\mathcal{P}(\mathcal{H})$ may be represented as a projection of some vector $X \in T_\psi\mathcal{H}$, as follows:

$$\xi = T_\psi \Pi(X) , \tag{5.19}$$

with $\psi \in \Pi^{-1}(p)$. A tangent space $T_\psi\mathcal{H}$, being a linear space, may be identified with $\mathcal{H}$ itself, that is $T_\psi\mathcal{H} \cong \mathcal{H}$. Evidently, there are infinitely many vectors $X \in \mathcal{H}$ projecting to a given vector $\xi \in T_p\mathcal{P}(\mathcal{H})$; note that $X + \alpha\psi$, with $\alpha \in \mathbb{C}$, has the same projection as X. Let us observe that any vector $X \in \mathcal{H}$ may be uniquely represented by

$$X = \lambda\psi + X^\perp , \tag{5.20}$$

where

$$X^\perp \in (\mathbb{C}\psi)^\perp := \{ \phi \in \mathcal{H} \mid \langle \psi | \phi \rangle = 0 \} . \tag{5.21}$$

It is clear from (5.20) that $\lambda = \langle \psi | X \rangle / \langle \psi | \psi \rangle$. Therefore,

$$\xi = T_\psi \Pi(X) = T_\psi \Pi(X^\perp) , \tag{5.22}$$

where X_1 and X_2 are arbitrary vectors from $T_{\psi_0}\mathcal{H}$ projecting to ξ_1 and ξ_2, respectively. Let $\xi_1, \xi_2 \in T_p\mathcal{P}(\mathcal{H})$ and take any element $\psi_0 \in S(\mathcal{H})$ belonging to a fibre $\Pi^{-1}(p)$. We define

$$\begin{aligned}
\langle \xi_1 | \xi_2 \rangle_p \; &:= \; \langle X_1^\perp | X_2^\perp \rangle \\
&= \; \langle X_1 | X_2 \rangle - \langle X_1 | \psi_0 \rangle \langle \psi_0 | X_2 \rangle .
\end{aligned} \tag{5.23}$$

It is clear that the above definition does not depend on $\psi_0 \in \Pi^{-1}(p)$ provided ψ_0 is of unit norm.

Now let $\psi \in \mathcal{H}$ be an arbitrary element from $\Pi^{-1}(p)$, i.e. $\psi = c\psi_0$ for some $c \in \mathbb{C}$. Using the definition of the tangent map, i.e.,

$$T_\psi \Pi(X) = \frac{d}{dt}\bigg|_{t=0} \Pi(\psi + tX) \,,$$

and recalling that $T_\psi \Pi$ is a linear map, we have

$$
\begin{aligned}
T_{c\psi_0} \Pi(X) &= \frac{d}{dt}\bigg|_{t=0} \Pi(c\psi_0 + tX) = \frac{d}{dt}\bigg|_{t=0} \Pi\left(c(\psi_0 + tc^{-1}X)\right) \\
&= T_{\psi_0}\Pi(c^{-1}X) = c^{-1} T_{\psi_0}\Pi(X) \,,
\end{aligned}
$$

where we have used the basic property of Π:

$$\Pi(c\phi) = \Pi(\phi) \,. \tag{5.24}$$

Therefore,

$$\langle\, T_\psi \Pi(X_1) \,|\, T_\psi \Pi(X_2) \,\rangle_p = \frac{1}{c^2}\langle\, T_{\psi_0}\Pi(X_1) \,|\, T_{\psi_0}\Pi(X_2)\,\rangle_p = \frac{\langle\, X_1^\perp | X_2^\perp \,\rangle}{\langle\, \psi|\psi\,\rangle} \,, \tag{5.25}$$

and hence, using (5.23), we obtain the following formula for the hermitian scalar product in $T_p \mathcal{P}(\mathcal{H})$:

$$\langle \xi_1 | \xi_2 \rangle_p = \frac{\langle X_1 | X_2 \rangle \langle \psi|\psi\rangle - \langle X_1|\psi\rangle\langle\psi|X_2\rangle}{\langle\psi|\psi\rangle^2} \,, \tag{5.26}$$

where X_1, X_2 are arbitrary vectors in $\mathcal{H}$, provided

$$\xi_1 = T_\psi \Pi(X_1) \,, \quad \xi_2 = T_\psi \Pi(X_2) \,. \tag{5.27}$$

Now it is clear how to define the metric g and symplectic form ω on $\mathcal{P}(\mathcal{H})$: We write

$$g_{[\psi]}(\xi_1, \xi_2) := \mathrm{Re}\,\langle \xi_1|\xi_2\rangle_p \,, \tag{5.28}$$

and

$$\omega_{[\psi]}(\xi_1, \xi_2) := \mathrm{Im}\,\langle \xi_1|\xi_2\rangle_p \,. \tag{5.29}$$

Evidently,

$$\omega_{[\psi]}(\xi_1, \xi_2) = g_{[\psi]}(\xi_1, i\xi_2) \,, \tag{5.30}$$

for any $\xi_1, \xi_2 \in \mathfrak{X}(\mathcal{P}(\mathcal{H}))$, which proves that $\mathcal{P}(\mathcal{H})$ is a Kähler space. $\qquad\square$

A riemannian metric g as defined in (5.28), is called a *Fubini–Study* metric on $\mathcal{P}(\mathcal{H})$ (Fubini 1903, Study 1905). The above formulae for g and ω are compatible with the Marsden–Weinstein reduction procedure presented in section 4.1.2. Denote by j a canonical embedding

$$j : S(\mathcal{H}) \hookrightarrow \mathcal{H} .$$

Then there exists a unique symplectic form ω on $\mathcal{P}(\mathcal{H})$, such that

$$\pi^* \omega = j^* \Omega ,$$

that is,

$$\omega_p(\xi_1, \xi_2) = \Omega(X_1, X_2) , \tag{5.31}$$

where X_1, X_2 are arbitrary vectors in $T_\psi S(\mathcal{H})$ and ψ is an arbitrary element from $\pi^{-1}(p)$, provided $\xi_k = T_\psi \pi(X_k)$. Clearly, the formula (5.29) reduces to (5.31) on $S(\mathcal{H})$, since any vector $X \in T_\psi S(\mathcal{H})$ is orthogonal to ψ, i.e.,

$$T_\psi S(\mathcal{H}) \cong (\mathbb{C}\psi)^\perp ,$$

and hence $X^\perp = X$, implying that

$$\omega_p(\xi_1, \xi_2) = \mathrm{Im} \, \langle X_1 | X_2 \rangle = \Omega(X_1, X_2) . \tag{5.32}$$

Analogously, one has $\pi^* g = j^* G$, and hence

$$g_p(\xi_1, \xi_2) = G(X_1, X_2) , \tag{5.33}$$

where X_1, X_2 are arbitrary vectors from $(\mathbb{C}\psi)^\perp$ projecting to ξ_1 and ξ_2, respectively, and $\psi \in \pi^{-1}(p)$.

5.1.3   Example: geometry of $\mathbb{C}P^n$

Consider $(n + 1)$-dimensional Hilbert space $\mathcal{H} \cong \mathbb{C}^{n+1}$. The corresponding quantum phase space $\mathcal{P}(\mathcal{H})$ is a *complex projective space*

$$\mathbb{C}P^n := S^{2n+1}/U(1) , \tag{5.34}$$

where S^{2n+1} is a unit sphere in $\mathbb{C}^{n+1} \cong \mathbb{R}^{2n+2}$. This space is the most important example of a Kähler manifold. Recall that the map

$$S^{2n+1} \longrightarrow \mathbb{C}P^n$$

defines a celebrated Hopf fibration (see Example 1.3.5). The complex projective space $\mathbb{C}P^n$ may be parametrized as follows: Choose complex coordinates $(z^0, z^1, \ldots, z^n)$ in $\mathbb{C}^{n+1}$. For $z^0 \neq 0$, we may parametrize $\mathbb{C}P^n$ using n complex variables

$$w^k = \frac{z^k}{z^0} , \quad k = 1, 2, \ldots, n . \tag{5.35}$$

Define a sphere S^{2n+1} of radius r in $\mathbb{R}^{2n+2}$ by

$$|z^0|^2 + |z^1|^2 + \ldots + |z^n|^2 = r^2 .$$

The above equations imply the following formula for z^0:

$$z^0 = \frac{re^{i\varphi}}{(1 + \overline{w}_k w^k)^{1/2}} , \tag{5.36}$$

where φ is an arbitrary phase. (Note that we assume summation over k from 1 to n.) In this way we established the following sets of coordinates:

$$
\begin{array}{lcl}
r, \varphi, w_k, \overline{w}_k & \longleftrightarrow & 2n+2 \ \text{coordinates in } \mathbb{R}^{2n+2} \cong \mathbb{C}^{n+1}, \\
\varphi, w_k, \overline{w}_k & \longleftrightarrow & 2n+1 \ \text{coordinates in } S^{2n+1}, \\
w_k, \overline{w}_k & \longleftrightarrow & 2n \ \text{coordinates in } \mathbb{C}P^n .
\end{array}
$$

In the above coordinates the principal $U(1)$-bundle $\pi : S^{2n+1} \longrightarrow \mathbb{C}P^n$ is defined locally by

$$\pi(\varphi, w, \overline{w}) = (w, \overline{w}) , \tag{5.37}$$

i.e., the fibres are parametrized by $\varphi \in [0, 2\pi)$, and therefore the vertical vectors $X \sim \partial_\varphi$.

Now, let us look for the Fubini–Study metric on $\mathbb{C}P^n$.[4] First of all let us observe that the standard euclidean metric in $\mathbb{C}^{n+1}$,

$$ds^2(\mathbb{C}^{n+1}) := \delta_{\alpha\beta} dz^\alpha \, d\overline{z}^\beta ,$$

with $\alpha, \beta = 1, \ldots, n+1$, induces the following metric $ds^2(S^{2n+1})$ on a sphere S^{2n+1}:

$$ds^2(\mathbb{C}^{n+1}) = dr^2 + r^2 ds^2(S^{2n+1}) . \tag{5.38}$$

We define the metric on $\mathbb{C}P^n$ such that the projection $\pi : S^{2n+1} \longrightarrow \mathbb{C}P^n$ is a riemannian submersion, i.e., it is an isometry when restricted to the orthogonal complement of the kernel of $T\pi$. Now, the kernel of $T\pi$ consists of all vertical vector fields. Therefore, the metric on S^{2n+1} has the following structure:

$$ds^2(S^{2n+1}) = (d\varphi - \Theta)^2 + ds^2(\mathbb{C}P^n) , \tag{5.39}$$

where the first term, $d\varphi - \Theta$, is degenerate, that is, it vanishes along the $2n$-dimensional subspaces orthogonal to the fibres of the bundle $S^{2n+1} \longrightarrow \mathbb{C}P^n$. Using (5.35), (5.36) and (5.38) it is easy to show that

$$\Theta = \frac{i}{2} \frac{\overline{w}_k dw^k - w^k d\overline{w}_k}{1 + \overline{w}_k w^k} , \tag{5.40}$$

---

[4]We follow Page 1987.

and

$$ds^2(\mathbb{C}P^n) = \frac{(1 + \overline{w}_k w^k)\delta_{ij} - \overline{w}_i w_j}{(1 + \overline{w}_k w^k)^2} \, dw^i \, d\overline{w}^j \, . \tag{5.41}$$

Therefore, the (holomorphic–antiholomorphic) components of the Fubini–Study metric on $\mathbb{C}P^n$ read

$$g_{i\overline{j}} = \frac{1}{2} \frac{(1 + \overline{w}_k w^k)\delta_{ij} - \overline{w}_i w_j}{(1 + \overline{w}_k w^k)^2} \, . \tag{5.42}$$

The reader can show that the formula for $g_{i\overline{j}}$ agrees with (5.28). Let us note that

$$g_{i\overline{j}} = \frac{\partial^2 \mathcal{K}}{\partial w^i \partial \overline{w}^j} \, , \tag{5.43}$$

where

$$\mathcal{K} := \ln \sqrt{1 + \overline{w}_k w^k} \, , \tag{5.44}$$

is a function called a Kähler potential. Actually, one can prove that any Kähler metric admits a Kähler potential. Finally, the Kähler two-form reads as follows:

$$K = \frac{i}{2} \frac{(1 + \overline{w}_k w^k)\delta_{ij} - \overline{w}_i w_j}{(1 + \overline{w}_k w^k)^2} \, dw^i \wedge d\overline{w}^j \, . \tag{5.45}$$

Note that

$$K = -\frac{1}{2} d\Theta \, , \tag{5.46}$$

with Θ defined in (5.40). It should be stressed that the above formula holds only locally (Θ is a locally defined object) and hence it does not imply that K is an exact form.

Example 5.1.4 Consider the simplest case, namely the complex projective line $\mathbb{C}P^1 \cong S^2$. The hermitian metric g is given by

$$g = \frac{dw \, d\overline{w}}{(1 + \overline{w}w)^2} \, , \tag{5.47}$$

whereas the symplectic form $\omega = K$ reads

$$\omega = \frac{i}{2} \frac{dw \wedge d\overline{w}}{(1 + \overline{w}w)^2} \, . \tag{5.48}$$

Introducing a real parametrization

$$w = x + iy \, , \quad \overline{w} = x - iy \, ,$$

one obtains

$$g = \frac{dx^2 + dy^2}{(1 + x^2 + y^2)^2} = \frac{dr^2 + r^2 d\phi^2}{(1 + r^2)^2} \,, \tag{5.49}$$

and

$$\omega = \frac{dx \wedge dy}{(1 + x^2 + y^2)^2} = \frac{r dr \wedge d\phi}{(1 + r^2)^2} \,, \tag{5.50}$$

where (r, ϕ) are standard polar coordinates in the xy-plane. In particular, the surface area of S^2, computed with respect to ω,

$$\int_{S^2} \omega = 2\pi \int_0^\infty \frac{r dr}{(1 + r^2)^2} = \pi \,, \tag{5.51}$$

is 4 times smaller than the standard one. $\diamond$

5.1.4   Symplectic structure and quantum dynamics

The standard Hilbert space approach to nonrelativistic quantum mechanics is based on the Schrödinger equation for a state vector ψ:

$$i\hbar \frac{d}{dt}\psi = \hat{H}\psi \,, \tag{5.52}$$

where $\hat{H}$ denotes the quantum Hamiltonian, which is a self-adjoint operator on $\mathcal{H}$. Let us recall that in classical mechanics the phase space $\mathcal{P}$ is equipped with a symplectic structure ω and the dynamics is governed by the hamiltonian vector field X_H, defined by

$$dH(Y) = \omega(X_H, Y) \,, \tag{5.53}$$

for any vector field $Y \in \mathfrak{X}(\mathcal{P})$. In the quantum case the Hilbert space of the system, being a Kähler space, is also equipped with a symplectic form Ω. Does the Schrödinger dynamics (5.52) on $\mathcal{H}$ correspond to Hamiltonian dynamics defined by Ω? It turns out that the answer to the above question is in the affirmative. To see this, define the following vector field on $\mathcal{H}$:

$$X_{\hat{H}}(\psi) := -\frac{i}{\hbar}\hat{H}\psi \,. \tag{5.54}$$

Since $\mathcal{H}$ is a linear space, each tangent space $T_\psi \mathcal{H}$ may be identified with $\mathcal{H}$ itself. Therefore, the action of the vector field can be regarded as the following assignment:

$$\mathcal{H} \ni \psi \longrightarrow X_{\hat{H}}(\psi) \in \mathcal{H} \,.$$

Clearly, $X_{\hat{H}}(\psi)$ is the vector field generating the Schrödinger dynamics. Now, to any self-adjoint operator $\hat{A}$ we may associate a function A on $\mathcal{H}$:

$$\hat{A} \longrightarrow A \in C^\infty(\mathcal{H})$$

defined by

$$A(\psi) := \langle \psi | \hat{A}\psi \rangle = G(\psi, \hat{A}\psi) . \tag{5.55}$$

Let us call a function A corresponding to a self-adjoint operator $\hat{A}$ an *evaluation function*. Recall, that the *expectation value function* $\langle \hat{A} \rangle$ corresponding to $\hat{A}$ is defined by

$$\langle \hat{A} \rangle_\psi := \frac{\langle \psi | \hat{A}\psi \rangle}{\langle \psi | \psi \rangle} . \tag{5.56}$$

If ψ is normalized, then

$$A(\psi) = \langle \hat{A} \rangle_\psi .$$

Now, since $\mathcal{H}$ is equipped with a symplectic structure Ω, we may compute the hamiltonian vector field X_H corresponding to a function H.

Theorem 5.1.4 *The Schrödinger vector field $X_{\hat{H}}$ is hamiltonian, and*

$$2\hbar X_{\hat{H}} = X_H ;$$

that is, the Schrödinger equation defines a classical hamiltonian system on $\mathcal{H}$.

Proof. Fix an element $\psi \in \mathcal{H}$, and let $Y \in T_\psi \mathcal{H} \cong \mathcal{H}$. Performing a standard computation, one finds

$$
\begin{aligned}
dH(\psi)(Y) &= \left. \frac{d}{dt} H(\psi + tY) \right|_{t=0} = \left. \frac{d}{dt} \langle \psi + tY | \hat{H}(\psi + tY) \rangle \right|_{t=0} \\
&= \langle Y | \hat{H}\psi \rangle + \langle \psi | \hat{H}Y \rangle = 2G(Y, \hat{H}\psi) = 2\hbar\, \Omega\left(Y, \frac{i}{\hbar}\hat{H}\psi\right) \\
&= 2\hbar\, \Omega(X_{\hat{H}}, Y)(\psi) ,
\end{aligned}
\tag{5.57}
$$

which shows that $2\hbar X_{\hat{H}}$ is Hamiltonian with respect to Ω. $\qquad\square$

Example 5.1.5 Let $\mathcal{H} = \mathbb{C}^n$ and denote by $(e_1, \ldots, e_n)$ the orthonormal base in $\mathbb{C}^n$, i.e., such that:

$$\langle e_\alpha | e_\beta \rangle = \delta_{\alpha\beta} .$$

For any $z \in \mathbb{C}^n$ one has

$$z = \sum_{\alpha=1}^{n} z_\alpha e_\alpha ,$$

and the Schrödinger equation implies the following equations for the complex coefficients z_α:

$$i\hbar \dot{z}_\alpha = \sum_{\beta=1}^{n} H_{\alpha\beta} z_\beta , \tag{5.58}$$

with $H_{\alpha\beta} = \langle e_\alpha|\hat{H}|e_\beta\rangle$. The corresponding evaluation function H on $\mathbb{C}^n$ reads

$$H(z) = \langle z|\hat{H}|z\rangle = \sum_{\alpha,\beta=1}^{n} H_{\alpha\beta}\bar{z}_\alpha z_\beta \ .$$

Hence,

$$\dot{z}_\alpha = \frac{1}{i\hbar}\frac{\partial H(z)}{\partial\bar{z}_\alpha}\ , \qquad \dot{\bar{z}}_\alpha = \frac{-1}{i\hbar}\frac{\partial H(z)}{\partial z_\alpha}\ , \tag{5.59}$$

which shows that the Schrödinger equation defines a hamiltonian system on $\mathbb{C}^n$ with respect to the symplectic structure

$$i\hbar\sum_{\alpha=1}^{n} dz_\alpha \wedge d\bar{z}_\alpha = 2\hbar\Omega\ ,$$

where Ω stands for the canonical symplectic form on $\mathbb{C}^n$ (cf. (5.15)). Clearly,

$$\dot{z}_\alpha = \frac{1}{2\hbar}\{z_\alpha, H(z)\}_\Omega\ , \tag{5.60}$$

where $\{\ ,\ \}_\Omega$ denotes the Poisson bracket corresponding to Ω. One can easily show that

$$\{A, B\}_\Omega = \frac{2}{i}\sum_{\alpha=1}^{n}\left(\frac{\partial A}{\partial z_\alpha}\frac{\partial B}{\partial\bar{z}_\alpha} - \frac{\partial B}{\partial z_\alpha}\frac{\partial A}{\partial\bar{z}_\alpha}\right)\ , \tag{5.61}$$

for any $A, B \in C^\infty(\mathcal{H})$. $\diamond$

Let us note that there is a close relation between the Poisson bracket $\{\ ,\ \}_\Omega$ defined by Ω on a Hilbert space $\mathcal{H}$ and the commutator of quantum observables (self-adjoint operators on $\mathcal{H}$). Let $\hat{A}$ and $\hat{B}$ be two quantum observables and A and B the corresponding evaluation functions on $\mathcal{H}$, i.e., such that

$$\hat{A}, \hat{B} \longrightarrow A, B \in C^\infty(\mathcal{H})\ .$$

The hamiltonian vector field corresponding to A reads

$$X_A(\psi) = 2\hbar X_{\hat{A}}(\psi) = -2i\hat{A}\psi\ . \tag{5.62}$$

Therefore, using the definition of Ω, one obtains

$$\begin{aligned}
\{A, B\}_\Omega(\psi) &= \Omega_\psi(X_A, X_B) = \frac{1}{2i}\Big(\langle X_A(\psi)|X_B(\psi)\rangle - \langle X_B(\psi)|X_A(\psi)\rangle\Big)\\
&= \frac{2}{i}\langle\psi|\hat{A}\hat{B} - \hat{B}\hat{A}|\psi\rangle = \frac{2}{i}\langle\psi|[\hat{A}, \hat{B}]|\psi\rangle\ ,
\end{aligned} \tag{5.63}$$

for any $\psi \in \mathcal{H}$. This means that the commutator of two quantum observables corresponds to the Poisson bracket of their evaluation functions. Hence, for $\psi \in S(\mathcal{H})$ we have

$$\{A, B\}_\Omega(\psi) = \frac{2}{i} \langle [\hat{A}, \hat{B}] \rangle_\psi \ .$$

Consider now the dynamics on the projective Hilbert space $\mathcal{P}(\mathcal{H})$. Note that each point in $\mathcal{P}(\mathcal{H})$ corresponds to a one-dimensional projector operator in $\mathcal{H}$, as follows:

$$S(\mathcal{H}) \ni \psi \longrightarrow [\psi] = P_\psi := |\psi\rangle\langle\psi| \in \mathcal{P}(\mathcal{H}) \ .$$

One also calls P_ψ the density operator (or density matrix) corresponding to the pure state ψ. If ψ satisfies the Schrödinger equation (5.52), then P_ψ satisfies the von Neumann equation:

$$i\hbar \frac{d}{dt} P_\psi = [\hat{H}, P_\psi] \ . \tag{5.64}$$

The solution of the von Neumann equation defines a curve in a quantum phase space $\mathcal{P}(\mathcal{H})$. Now, each quantum observable $\hat{A}$ in $\mathcal{H}$ gives rise to a function $a \in C^\infty(\mathcal{P}(\mathcal{H}))$, defined according to

$$\mathcal{P}(\mathcal{H}) \ni P_\psi \longrightarrow a(P_\psi) := \mathrm{Tr}(\hat{A} P_\psi) \ ; \tag{5.65}$$

equivalently,

$$a(P_\psi) = \langle \hat{A} \rangle_\psi \ . \tag{5.66}$$

Recall that the projective Hilbert space $\mathcal{P}(\mathcal{H})$, being a Kähler manifold, is equipped with a symplectic structure ω. Clearly, the Hamiltonian dynamics on $\mathcal{H}$, given by the Schrödinger equation, projects to the Hamiltonian dynamics on $(\mathcal{P}(\mathcal{H}), \omega)$ with a hamiltonian $h \in C^\infty(\mathcal{P}(\mathcal{H}))$ defined by

$$h(P_\psi) = H(\psi) = \langle \hat{H} \rangle_\psi \ .$$

Hence, the von Neumann equation (5.64) may be rewritten as follows:

$$\frac{d}{dt} P_\psi = \frac{1}{2\hbar} \{h, P_\psi\}_\omega \ , \tag{5.67}$$

where $\{\ ,\ \}_\omega$ denotes the Poisson bracket corresponding to ω. Hence, the von Neumann equation defines a hamiltonian system with respect to $2\hbar\omega$.

5.1.5 Metric structure and uncertainty relation

Let us turn to the metric structure of $\mathcal{H}$ encoded in G. In general, the classical phase space is not equipped with a riemannian metric; therefore, the information encoded in

G does not have a classical analog. Actually, as we shall see, the metric structure of $\mathcal{H}$ is closely related to the Heisenberg uncertainty principle. Recall that the uncertainty of the observable $\hat{A}$ in the state corresponding to a normalized state vector ψ is given by

$$(\Delta\hat{A})^2_\psi := \langle \hat{A}^2 \rangle_\psi - \langle \hat{A} \rangle^2_\psi \,. \tag{5.68}$$

Consider two quantum observables $\hat{A}$ and $\hat{B}$, and let

$$\hat{A}_\perp := \hat{A} - \mathbb{1} \cdot A \,, \tag{5.69}$$

where A is the evaluation function of $\hat{A}$, and similarly for $\hat{B}$. Clearly, $(\Delta\hat{A})^2_\psi = \langle \hat{A}^2_\perp \rangle_\psi$, and hence

$$(\Delta\hat{A})^2_\psi (\Delta\hat{B})^2_\psi = \langle \hat{A}^2_\perp \rangle_\psi \langle \hat{B}^2_\perp \rangle_\psi = \langle \psi | \hat{A}^2_\perp | \psi \rangle \langle \psi | \hat{B}^2_\perp | \psi \rangle \,.$$

Now, using the Schwartz inequality

$$\langle \psi | \hat{A}^2_\perp | \psi \rangle \langle \psi | \hat{B}^2_\perp | \psi \rangle \geq |\langle \psi | \hat{A}_\perp \hat{B}_\perp | \psi \rangle|^2 \,,$$

and

$$\hat{A}_\perp \hat{B}_\perp = \frac{1}{2}[\hat{A}_\perp, \hat{B}_\perp] + \frac{1}{2}[\hat{A}_\perp, \hat{B}_\perp]_+ \,,$$

where $[\hat{X}, \hat{Y}]_+ = \hat{X}\hat{Y} + \hat{Y}\hat{X}$, one obtains

$$(\Delta\hat{A})^2_\psi (\Delta\hat{B})^2_\psi \geq \frac{1}{4}\Big(|\langle [\hat{A}_\perp, \hat{B}_\perp]_+ \rangle_\psi|^2 + |\langle [\hat{A}_\perp, \hat{B}_\perp] \rangle_\psi|^2\Big) \,.$$

Finally, noting that $[\hat{A}_\perp, \hat{B}_\perp] = [\hat{A}, \hat{B}]$, and that

$$|\langle [\hat{A}_\perp, \hat{B}_\perp]_+ \rangle_\psi|^2 = \langle [\hat{A}_\perp, \hat{B}_\perp] \rangle^2_\psi \,, \qquad |\langle [\hat{A}, \hat{B}] \rangle_\psi|^2 = -\langle [\hat{A}, \hat{B}] \rangle^2_\psi \,,$$

one finds

$$(\Delta\hat{A})^2_\psi (\Delta\hat{B})^2_\psi \geq \frac{1}{4}\left(\langle [\hat{A}_\perp, \hat{B}_\perp]_+ \rangle^2_\psi - \langle [\hat{A}, \hat{B}] \rangle^2_\psi\right) \,, \tag{5.70}$$

which is the standard form of the uncertainty relation for two quantum observables $\hat{A}$ and $\hat{B}$ in the Hilbert space formulation. Let us rewrite the above formula using the canonical geometric structures of $\mathcal{H}$, i.e., the symplectic structure Ω and the metric structure G. By analogy with formula (5.63) one obtains

$$\begin{aligned}
G_\psi(X_A, X_B) &= \frac{1}{2}\Big(\langle X_A(\psi) | X_B(\psi) \rangle + \langle X_B(\psi) | X_A(\psi) \rangle\Big) \\
&= 2\langle \psi | \hat{A}\hat{B} + \hat{B}\hat{A} | \psi \rangle = 2\langle \psi | [\hat{A}, \hat{B}]_+ | \psi \rangle \,, \tag{5.71}
\end{aligned}$$

where we have used the formula (5.62) for X_A and X_B. Moreover, the fact that

$$[\hat{A}_\perp, \hat{B}_\perp]_+ = [\hat{A}, \hat{B}]_+ + 2(AB\mathbb{1} - A\hat{B} - B\hat{A}) \,,$$

leads to

$$\langle [\hat{A}_\perp, \hat{B}_\perp]_+ \rangle_\psi = \frac{1}{2} G_\psi(X_A, X_B) - 2(AB)(\psi) .$$

Therefore, we may rewrite the Heisenberg uncertainty relation (5.70), without reference to a particular state vector, as follows:

$$(\Delta A)^2 (\Delta B)^2 \geq \left(\frac{1}{4} \Omega(X_A, X_B) \right)^2 + \left(\frac{1}{4} G(X_A, X_B) - AB \right)^2 , \qquad (5.72)$$

where $(\Delta A)^2$ is a function on $\mathcal{H}$ defined by $(\Delta A)^2(\psi) := (\Delta \hat{A})^2_\psi$.

Finally, let us see how the Heisenberg relation is encoded on the level of the quantum phase space $\mathcal{P}(\mathcal{H})$. Let $\hat{A}$ and $\hat{B}$ be two quantum observables, and denote by a and b the corresponding functions on $\mathcal{P}(\mathcal{H})$, i.e.,

$$a \circ \pi = \langle \hat{A} \rangle = A|_{S(\mathcal{H})} , \qquad b \circ \pi = \langle \hat{B} \rangle = B|_{S(\mathcal{H})} ,$$

where π denotes the canonical projection $S(\mathcal{H}) \longrightarrow \mathcal{P}(\mathcal{H})$. Denote by g and ω the corresponding metric tensor and symplectic form on $\mathcal{P}(\mathcal{H})$, respectively. By analogy with the Poisson bracket

$$\{a, b\}_\omega := \omega(X_a, X_b) , \qquad (5.73)$$

let us define the so-called *Riemann bracket*

$$(a, b)_g := g(X_a, X_b) . \qquad (5.74)$$

Using local coordinates on $\mathcal{P}(\mathcal{H})$, one finds that

$$\{a, b\}_\omega = \omega^{\alpha\beta} \partial_\alpha a \partial_\beta b ,$$

and

$$(a, b)_g = g_{\alpha\beta}(X_a)^\alpha (X_b)^\beta = g_{\alpha\beta} \omega^{\alpha\gamma} \partial_\gamma a \, \omega^{\beta\delta} \partial_\delta b = g^{\alpha\beta} \partial_\alpha a \, \partial_\beta b .$$

Now, recall (see section 5.1.2) that if $\xi, \eta \in T_p \mathcal{P}(\mathcal{H})$, then

$$g_p(\xi, \eta) = G_\psi(X^\perp, Y^\perp) ,$$

and

$$\omega_p(\xi, \eta) = \Omega_\psi(X^\perp, Y^\perp) ,$$

where ψ is any vector from $\mathcal{H}$ projecting to $p \in \mathcal{P}(\mathcal{H})$, X and Y are arbitrary vectors in $\mathcal{H}$, and

$$X^\perp = X - X^{\parallel} = X - \frac{\langle \psi | X \rangle}{\langle \psi | \psi \rangle} \psi ,$$

and similarly for $Y^{\perp}$. In particular, one has

$$g_p(X_a, X_b) = G_\psi(X_A^{\perp}, X_B^{\perp}) , \qquad \omega_p(X_a, X_b) = \Omega_\psi(X_A^{\perp}, X_B^{\perp}) ,$$

and can establish the following relations:

$$\Omega_\psi(X_A^{\perp}, X_B^{\perp}) = \Omega_\psi(X_A, X_B) , \tag{5.75}$$

and

$$G_\psi(X_A^{\perp}, X_B^{\perp}) = G_\psi(X_A, X_B) + 4(AB)(\psi) . \tag{5.76}$$

Therefore, the Heisenberg relation (5.72) may be rewritten as the following inequality between objects defined on $\mathcal{P}(\mathcal{H})$:

$$\begin{aligned}
(\Delta a)^2 (\Delta b)^2 &\geq \frac{1}{16} \left(\omega(X_a, X_b)^2 + g(X_a, X_b)^2 \right) \\
&= \frac{1}{16} \left(\{a, b\}_\omega^2 + (a, b)_g^2 \right) ,
\end{aligned} \tag{5.77}$$

where $(\Delta a)^2 (P_\psi) := (\Delta A)^2 (\psi)$. In particular,

$$(\Delta a)^2 = \frac{1}{4} (a, a)_g .$$

This leads to a nice geometrical interpretation of quantum mechanical uncertainty. For example, the uncertainty of the energy,

$$(\Delta h)^2 = \frac{1}{4} g(X_h, X_h) , \tag{5.78}$$

is (up to a constant $1/4$) equal to the length of the hamiltonian vector field X_h. Thus the energy uncertainty measures the speed at which the quantum system travels through the quantum state space $\mathcal{P}(\mathcal{H})$ (Anandan and Aharonov 1990).

5.2   Aharonov–Anandan phase

5.2.1   Standard derivation

The solution of the Schrödinger equation

$$i\hbar \frac{d}{dt} \psi = \hat{H} \psi , \qquad \psi(0) = \psi_0 , \tag{5.79}$$

defines a trajectory

$$t \longrightarrow \psi(t) \in \mathcal{H} ,$$

in the Hilbert space $\mathcal{H}$. If the initial state vector $\psi_0 \in S(\mathcal{H})$, then the solution $\psi(t)$ remains in $S(\mathcal{H})$ for any $t \in \mathbb{R}$. Such a trajectory on $S(\mathcal{H})$ projects onto a trajectory in the quantum phase space:

$$t \;\longrightarrow\; P(t) \in \mathcal{P}(\mathcal{H}) \,,$$

that is,

$$P(t) := \pi(\psi(t)) \,, \tag{5.80}$$

where $\pi : S(\mathcal{H}) \longrightarrow \mathcal{P}(\mathcal{H})$. This defines a solution to the von Neumann equation, i.e.,:

$$i\hbar \frac{d}{dt} P = [\hat{H}, P] \,, \qquad P(0) = P_0 := |\psi_0\rangle\langle\psi_0| \,. \tag{5.81}$$

Suppose that a trajectory $P = P(t)$ is closed, i.e., $P(T) = P(0)$ for some $T > 0$. We call such an evolution cyclic. We stress that we do not make any assumption about the Hamiltonian $\hat{H}$ of our system. It is not even important whether or not it depends on time. Since $\psi(T)$ and $\psi(0)$ define the same physical state they may differ by a phase factor only, i.e.,

$$\psi(T) = e^{i\varphi} \psi(0) \,, \tag{5.82}$$

for some $\varphi \in [0, 2\pi)$. Our task in this section is to find the phase shift φ knowing the system Hamiltonian $\hat{H}$ and a closed trajectory $P(t)$ in $\mathcal{P}(\mathcal{H})$.

First of all, let us note that we may make certain changes to $\hat{H}$ without affecting $P(t)$. It is evident from the commutator structure of the von Neumann equation that the following transformation:

$$\hat{H}' = \hat{H} + \mathbb{1}\,a(t) \,, \tag{5.83}$$

where $\mathbb{1}$ is an identity in $\mathcal{H}$ and $a(t)$ is any real function of time, leaves the solution $P(t)$ invariant. The corresponding solution to the Schrödinger equation changes as follows:

$$\psi'(t) = \exp\left(\frac{1}{i\hbar}\int_0^t a(\tau)d\tau\right) \psi(t) \,, \tag{5.84}$$

and hence

$$\psi'(T) = e^{i\varphi'} \psi(0) \,, \tag{5.85}$$

with

$$\varphi' = \varphi - \frac{1}{\hbar}\int_0^T a(\tau)d\tau \,. \tag{5.86}$$

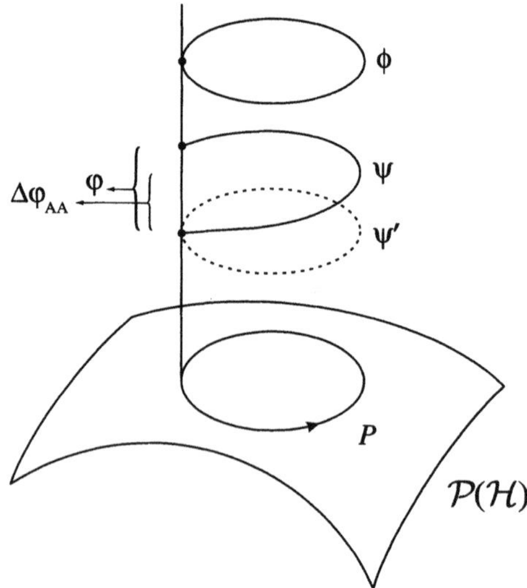

Figure 5.1: Three lifts of the closed curve P in $\mathcal{P}(\mathcal{H})$: ψ — solution of the original Schrödinger equation; ψ' — solution of the Schrödinger equation corresponding to the transformed Hamiltonian $\hat{H}'$; and ϕ — an arbitrary closed curve.

Therefore, by performing a trivial change of the Hamiltonian (5.83) we may change the corresponding phase φ completely arbitrarily. However, as was shown by Aharonov and Anandan (Aharonov and Anandan 1987), the total phase φ may be naturally divided into two parts, as follows:

$$\varphi = \varphi_{\mathrm{dyn}} + \varphi_{\mathrm{geo}} , \tag{5.87}$$

such that the *geometric phase* φ_{geo} is invariant under the transformation (5.83) and depends only on the closed curve $P(t)$ in the quantum phase space. To see this let us take a function $a = a(t)$ in (5.83) such that $\varphi' = 0$, which means that the curve $\psi'(t)$ is closed, i.e., $\psi'(T) = \psi'(0)$ (see Fig. 5.1). Note that $a(t)$ then satisfies

$$\frac{1}{\hbar} \int_0^T a(t)dt = \varphi , \tag{5.88}$$

where φ is defined in (5.82). The new function $\psi'(t)$ solves the Schrödinger equation

$$i\hbar \frac{d}{dt} \psi'(t) = \left(\hat{H} + \mathbb{1}\, a(t) \right) \psi'(t) , \tag{5.89}$$

and hence, taking a scalar product with $\psi'(t)$ and integrating over time from 0 to T, one obtains

$$\int_0^T \left\langle \psi'(t) \middle| i \frac{d}{dt} \psi'(t) \right\rangle dt = \frac{1}{\hbar} \int_0^T \langle \psi'(t) | \hat{H} | \psi'(t) \rangle \, dt + \frac{1}{\hbar} \int_0^T a(t)dt . \tag{5.90}$$

Therefore, using (5.88), we find the following formula for the phase shift φ:

$$\varphi = \int_0^T \left\langle \psi'(t) \middle| i\frac{d}{dt}\psi'(t) \right\rangle dt - \frac{1}{\hbar}\int_0^T \langle \psi'(t)|\hat{H}|\psi'(t)\rangle dt \ . \tag{5.91}$$

In this way the total phase shift φ is divided into the following two parts:

- *the dynamical phase*

$$\varphi_{\mathrm{dyn}} := -\frac{1}{\hbar}\int_0^T \langle \psi'(t)|\hat{H}|\psi'(t)\rangle \, dt = -\frac{1}{\hbar}\int_0^T \langle \psi(t)|\hat{H}|\psi(t)\rangle \, dt \ , \tag{5.92}$$

  which manifestly depends upon the system hamiltonian $\hat{H}$, and

- *the geometric phase*

$$\varphi_{\mathrm{geo}} = \int_0^T \left\langle \psi'(t) \middle| i\frac{d}{dt}\psi'(t) \right\rangle dt \ , \tag{5.93}$$

  which, as we shall see, depends only on the closed curve $P(t)$ in $\mathcal{P}(\mathcal{H})$ and not on a particular choice of the Hamiltonian.

Now we show that φ_{geo} depends only on the geometry of the projected curve

$$P(t) = |\psi(t)\rangle\langle\psi(t)| = |\psi'(t)\rangle\langle\psi'(t)| \ , \tag{5.94}$$

in $\mathcal{P}(\mathcal{H})$. Let $\phi = \phi(t)$ be an arbitrary closed curve in $S(\mathcal{H})$ projecting onto $P(t)$, cf. Fig. 5.1. Clearly, $\psi'(t)$ and $\phi(t)$ differ by a time-dependent phase factor, i.e.,

$$\phi(t) = e^{if(t)} \, \psi'(t) \ , \tag{5.95}$$

such that $f(T) = f(0)$. The easy computation

$$
\begin{aligned}
\int_0^T \left\langle \phi(t) \middle| i\frac{d}{dt}\phi(t) \right\rangle dt &= \int_0^T \left[-\left\langle \psi'(t) \middle| \frac{df}{dt}\psi'(t) \right\rangle + \left\langle \psi'(t) \middle| i\frac{d}{dt}\psi'(t) \right\rangle \right] dt \\
&= \left(f(0) - f(T) \right) + \int_0^T \left\langle \psi'(t) \middle| i\frac{d}{dt}\psi'(t) \right\rangle dt \ , \tag{5.96}
\end{aligned}
$$

leads to the following conclusion:

$$\int_0^T \left\langle \phi(t) \middle| i\frac{d}{dt}\phi(t) \right\rangle dt = \varphi_{\mathrm{geo}} \ . \tag{5.97}$$

This proves that φ_{geo} is a characteristic geometric feature of the closed curve in $\mathcal{P}(\mathcal{H})$. It is called the *Aharonov–Anandan phase*:

$$\Delta\varphi_{\mathrm{AA}} := \varphi_{\mathrm{geo}} = \text{Aharonov–Anandan phase} \ .$$

Hence, the total phase shift corresponding to a cyclic evolution is given by

$$\varphi = \varphi_{\mathrm{dyn}} + \Delta\varphi_{\mathrm{AA}} \ , \tag{5.98}$$

and, as we have just shown, the geometric part $\Delta\varphi_{\mathrm{AA}}$ does not change under the transformation (5.83).

Remark 5.2.1 The definition of the Aharonov–Anandan phase for a cyclic evolution in $\mathcal{P}(\mathcal{H})$ is a special case of the reconstruction procedure from section 4.3.1. Projective Hilbert space defines a reduced phase space for the Schrödinger dynamics. Hence, we have a direct correspondence between Fig. 4.4 and Fig. 5.1.   ◇

5.2.2   Example: spin-half in a magnetic field

Consider a spin-half particle interacting with a constant magnetic field **B**. Choose a coordinate system such that $\mathbf{B} = (0, 0, B)$. The corresponding quantum Hamiltonian $\hat{H}$ reads (cf. Example 2.2.1)

$$\hat{H} = \frac{1}{2}\mathbf{B}\cdot\sigma = \frac{\hbar}{2}B\begin{pmatrix} 1 & 0 \\ 0 & -1 \end{pmatrix}, \tag{5.99}$$

where, for simplicity, we set the giromagnetic ratio μ equal to one. Let $|\pm\rangle \in \mathcal{H} \cong \mathbb{C}^2$ denote two eigenvectors of σ_3, i.e., such that

$$\sigma_3|\pm\rangle = \pm|\pm\rangle .$$

Taking as an initial state vector

$$\psi_0 = \cos\left(\frac{\theta_0}{2}\right)|+\rangle + \sin\left(\frac{\theta_0}{2}\right)e^{i\varphi_0}|-\rangle , \tag{5.100}$$

one easily finds the following solution of the corresponding Schrödinger equation:

$$\begin{aligned}
\psi(t) &= \exp\left(\frac{-iBt}{2\hbar}\right)\cos\left(\frac{\theta_0}{2}\right)|+\rangle + \exp\left(\frac{iBt}{2\hbar}\right)\sin\left(\frac{\theta_0}{2}\right)e^{i\varphi_0}|-\rangle \\
&= \exp\left(\frac{-iBt}{2\hbar}\right)\left[\cos\left(\frac{\theta_0}{2}\right)|+\rangle + \exp\left(\frac{iBt}{\hbar}\right)\sin\left(\frac{\theta_0}{2}\right)e^{i\varphi_0}|-\rangle\right] . \tag{5.101}
\end{aligned}$$

Note that for $T = \frac{2\pi\hbar}{B} = \frac{h}{B}$, one obtains

$$\psi(T) = e^{-i\pi}\left(\cos\left(\frac{\theta_0}{2}\right)|+\rangle + \sin\left(\frac{\theta_0}{2}\right)e^{i\varphi_0}|-\rangle\right) = e^{-i\pi}\psi(0) , \tag{5.102}$$

which means that the evolution is cyclic. The total phase shift $\varphi = -\pi$ decomposes into dynamical and geometric parts:

$$-\pi = \varphi_{\text{dyn}} + \varphi_{\text{geo}} .$$

The dynamical phase φ_{dyn} is easy to calculate. One has

$$\begin{aligned}
\varphi_{\text{dyn}} &= -\frac{1}{\hbar}\int_0^T \langle\psi(t)|\hat{H}|\psi(t)\rangle\, dt = -\frac{B}{2\hbar}\int_0^T \langle\psi(t)|\sigma_3|\psi(t)\rangle\, dt \\
&= -\frac{BT}{2\hbar}\left(\cos^2\left(\frac{\theta_0}{2}\right) - \sin^2\left(\frac{\theta_0}{2}\right)\right) = -\pi\cos\theta_0 , \tag{5.103}
\end{aligned}$$

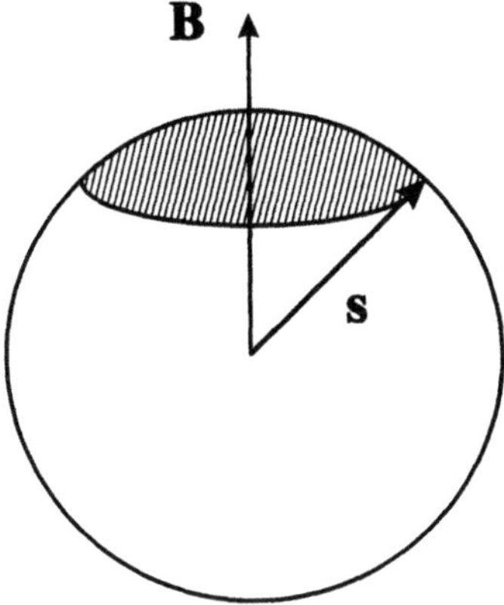

Figure 5.2: Cyclic evolution of the polarization vector s.

and hence, the geometric Aharonov–Anandan phase $\Delta\varphi_{AA}$ reads

$$\Delta\varphi_{AA} = -\pi - \varphi_{dyn} = \pi(\cos\theta_0 - 1) \qquad \text{mod } 2\pi . \qquad (5.104)$$

Let us interpret this result in terms of the geometry of a quantum phase space

$$\mathcal{P}(\mathcal{H}) = \mathbb{C}P^1 \cong S^2 .$$

Define a *polarization vector*, as follows:

$$\mathbf{s} := \langle \psi | \boldsymbol{\sigma} | \psi \rangle . \qquad (5.105)$$

Note, that if $\langle \psi | \psi \rangle = 1$, then $|\mathbf{s}| = 1$, and hence we may define a map

$$S(\mathbb{C}^2) \equiv S^3 \ni \psi \longrightarrow \mathbf{s} \in S^2 ,$$

which is the celebrated Hopf map (cf. section 1.4.4). It is easy to see that the corresponding evolution of $\mathbf{s}$ on a quantum phase space S^2 is given by

$$\frac{d}{dt}\mathbf{s} = \mathbf{s} \times \mathbf{B} , \qquad (5.106)$$

which describes precession of $\mathbf{s}$ about the direction of $\mathbf{B}$ with an angular velocity B. Hence, the solution $\mathbf{s} = \mathbf{s}(t)$, satisfying the initial condition

$$\mathbf{s}_0 = \langle \psi_0 | \boldsymbol{\sigma} | \psi_0 \rangle = (\sin\theta_0 \cos\varphi_0, \sin\theta_0 \sin\varphi_0, \cos\theta_0) , \qquad (5.107)$$

defines a closed curve C on S^2, i.e, a circle (see Fig. 5.2). It is given by

$$\mathbf{s}(t) = (\sin\theta_0 \cos(\varphi_0 + Bt), \sin\theta_0 \sin(\varphi_0 + Bt), \cos\theta_0) , \qquad (5.108)$$

and satisfies

$$\mathbf{s}(T = h/B) = \mathbf{s}(0) . \qquad (5.109)$$

Recall that the solid angle $\Omega(C)$ subtended by C is given by

$$\Omega(C) = 2\pi(1 - \cos\theta_0) \, ,$$

and, therefore, the Aharonov–Anandan phase may be rewritten as

$$\Delta\varphi_{AA} = -\frac{1}{2}\Omega(C) \qquad \mathrm{mod}\ 2\pi \, . \tag{5.110}$$

Hence, $\Delta\varphi_{AA}$ is defined entirely in terms of geometric structures living on a quantum phase space $\mathcal{P}(\mathcal{H}) \cong S^2$.

5.2.3   Fibre bundle approach

Evidently, the Aharonov–Anandan phase $\Delta\varphi_{AA}$ corresponds to a holonomy of an appropriate connection in a principal $U(1)$-bundle

$$S(\mathcal{H}) \longrightarrow \mathcal{P}(\mathcal{H}) \, ,$$

or its associated line bundle $\mathcal{H} \longrightarrow \mathcal{P}(\mathcal{H})$. It turns out that there is a natural connection giving rise to the Aharonov-Anandan phase. Recall that to introduce a connection we have to define a subspace of horizontal vectors. To do so, let $\psi \in S(\mathcal{H})$. The tangent space $T_\psi S(\mathcal{H})$ may be identified as a linear subspace in $\mathcal{H}$. Hence, the subspaces of vertical and horizontal vectors, which are related to $T_\psi S(\mathcal{H})$ as follows:

$$\mathcal{H} \supset T_\psi S(\mathcal{H}) = V_\psi + H_\psi \, , \tag{5.111}$$

are linear subspaces in $\mathcal{H}$. A fibre $\pi^{-1}(\psi)$ consists of all vectors of the form $e^{i\lambda}\psi$. Therefore, the vertical subspace V_ψ is defined by

$$V_\psi = \left\{ i\lambda\,\psi \,\middle|\, \lambda \in \mathbb{R} \right\} , \tag{5.112}$$

and hence it may be identified with $u(1) \cong i\mathbb{R}$. To define a *natural connection* we use a hermitian scalar product in $\mathcal{H}$. Let X be a vector tangent to $S(\mathcal{H})$ at a point ψ. We identify X as a vector in $\mathcal{H}$ (using the same letter for $X \in \mathcal{H}$), and we call X a horizontal vector w.r.t. a natural connection if

$$\langle \psi | X \rangle = 0 \, . \tag{5.113}$$

Thus, the space of horizontal vectors at ψ consists of all vectors orthogonal to ψ, i.e.,

$$H_\psi = \left\{ X \in \mathcal{H} \,\middle|\, \langle \psi | X \rangle = 0 \right\} . \tag{5.114}$$

A curve $t \longrightarrow \psi(t) \in S(\mathcal{H})$ is horizontal if

$$\langle \psi(t) | \dot{\psi}(t) \rangle = 0 \, , \tag{5.115}$$

for any t. Now, since $\langle \psi | \psi \rangle = 1$, one has

$$\mathrm{Re}\, \langle \psi(t) | \dot{\psi}(t) \rangle \equiv 0 \, ,$$

and hence the horizontality condition reduces to

$$\mathrm{Im}\, \langle \psi(t) | \dot{\psi}(t) \rangle = 0 \, . \tag{5.116}$$

Recall that a connection one-form $\mathcal{A}$ in a principal $U(1)$-bundle $S(\mathcal{H}) \longrightarrow P(\mathcal{H})$ is a $u(1)$-valued one-form on $S(\mathcal{H})$. Take $X \in T_\psi S(\mathcal{H}) \subset \mathcal{H}$, and define

$$\mathcal{A}_\psi(X) := i \, \mathrm{Im}\, \langle \psi | X \rangle \, \in \, u(1) \, . \tag{5.117}$$

It is clear that X is a horizontal vector at a point $\psi \in S(\mathcal{H})$ if $\mathcal{A}_\psi(X) = 0$. Consider now a local connection form A on a quantum phase space $P(\mathcal{H})$. Let

$$\psi \, : \, P(\mathcal{H}) \longrightarrow S(\mathcal{H})$$

be a local section. The pull-back

$$A := i \psi^* \mathcal{A} \tag{5.118}$$

defines a local connection one-form on $P(\mathcal{H})$ (in a gauge ψ).[5] The definition of $\mathcal{A}$ implies the following formula for the local connection A (in a gauge ψ):

$$A = i \langle \psi | d\psi \rangle \, . \tag{5.119}$$

By performing a gauge transformation, as follows:

$$\psi \longrightarrow \psi' = e^{-if} \, \psi \, , \tag{5.120}$$

we obtain a gauge-transformed connection A'

$$A' = A + df \, , \tag{5.121}$$

which agrees with the transformation law for a connection form. Having defined a connection we may compute the corresponding holonomy element

$$\Phi(C) := \exp\left(i \oint_C A \right) \, , \tag{5.122}$$

where C is a closed curve in $P(\mathcal{H})$. Evidently, $\Phi(C)$ reproduces the Aharonov–Anadan phase factor, that is

$$\Phi(C) = e^{i \Delta \varphi_{AA}} \, . \tag{5.123}$$

In summary we have proved the following

Theorem 5.2.1 *Let C be a closed curve in a quantum phase space $P(\mathcal{H})$, corresponding to a cyclic evolution of a quantum system. Then the corresponding Aharonov–Anandan phase factor $\exp(i\Delta\varphi_{AA})$ defines the holonomy of C with respect to the natural connection in a principal $U(1)$-bundle $S(\mathcal{H}) \longrightarrow P(\mathcal{H})$.*

---

[5]Due to the factor of i, A is $\mathbb{R}$-valued.

5.2.4 Geometry of $\mathbb{C}P^n$ and Aharonov–Anandan phase

Consider, now, an $(n+1)$-level quantum system living in a finite-dimensional Hilbert space $\mathcal{H} \cong \mathbb{C}^{n+1}$ (cf. section 5.1.3). Taking cartesian coordinates $(z^0, z^1, \ldots, z^n)$ in $\mathbb{C}^{n+1}$, we may express a natural connection one-form in a principal $U(1)$-bundle $S^{2n+1} \longrightarrow \mathbb{C}P^n$ as follows:

$$A = i\,\mathrm{Im}\,(\bar{z}_\alpha dz^\alpha) \;=\; \frac{1}{2}\,(\bar{z}_\alpha dz^\alpha - z^\alpha d\bar{z}_\alpha)\,. \tag{5.124}$$

Using local coordinates $\{(\varphi, w^k, \bar{w}_k)\,, k = 1, \ldots, n\}$ in S^{2n+1} we obtain the following formula for A:

$$A = id\varphi + \frac{1}{2}\,\frac{\bar{w}_k dw^k - w^k d\bar{w}_k}{1 + \bar{w}_k w^k} = i(d\varphi - \Theta)\,, \tag{5.125}$$

where Θ was already defined in (5.40). Hence, it is clear that a local connection one-form on $\mathbb{C}P^n$ is given by

$$A = -\Theta\,, \tag{5.126}$$

where we take A to be real-valued. The corresponding (real-valued) local curvature reads

$$F = dA = -d\Theta\,. \tag{5.127}$$

Note that, due to (5.46), we have

$$F = 2\omega\,. \tag{5.128}$$

This shows that the first Chern class of the Hopf bundle $S^{2n+1} \longrightarrow \mathbb{C}P^n$ is entirely determined by the symplectic form ω, as follows:

$$c_1(F) = \frac{i}{2\pi}\,(-iF) = \frac{\omega}{\pi}\,. \tag{5.129}$$

Example 5.2.1 Consider once again the complex projective line $\mathbb{C}P^1 \cong S^2$. Using the symplectic form ω derived in Example 5.1.4, i.e.,

$$\omega = \frac{i}{2}\,\frac{dw \wedge d\bar{w}}{(1 + \bar{w}w)^2}\,,$$

one easily finds that the

$$\text{Chern number of a complex line bundle} = \int_{S^2} c_1(F) = \frac{1}{\pi}\int_{S^2} \omega = 1\,,$$

and, hence, that the $U(1)$-bundle $S(\mathcal{H}) \longrightarrow P(\mathcal{H})$ reproduces the monopole bundle $S^3 \longrightarrow S^2$ with a monopole charge $g = 1$. $\diamond$

Finally, the Aharonov–Anandan phase $\Delta\varphi_{\text{AA}}$ is given by the following formula:

$$\Delta\varphi_{\text{AA}} = \oint_C A = -\frac{1}{2}\oint_C \frac{\overline{w}_k dw^k - w^k d\overline{w}_k}{1 + \overline{w}_k w^k} \qquad \text{mod } 2\pi \ . \qquad (5.130)$$

Applying the Stokes theorem one finds

$$\Delta\varphi_{\text{AA}} = \oint_C A = \int_\Sigma F = 2\int_\Sigma \omega \qquad \text{mod } 2\pi \ , \qquad (5.131)$$

where Σ is any two-dimensional submanifold in $\mathbb{C}P^n$, such that $C = \partial\Sigma$. This shows that the Aharonov–Anandan phase corresponding to a closed curve C in $\mathcal{P}(\mathcal{H})$ equals twice the symplectic area of Σ. In particular, for a two-level system one has (cf. Example 5.1.4) $\omega = \frac{1}{4}\omega_0$, where ω_0 is the standard volume form on S^2, i.e., $\omega_0 = \sin\theta d\theta \wedge d\varphi$. Hence,

$$\Delta\varphi_{\text{AA}} = \frac{1}{2}\int_\Sigma \omega_0 = \frac{1}{2}\Omega(C) \qquad \text{mod } 2\pi \ , \qquad (5.132)$$

where, as usual, $\Omega(C)$ is the solid angle subtended by the closed curve C. Clearly, this result agrees, as it should, with formula (5.110) for the Aharonov–Anandan phase for a spin-half particle.

5.3   Quantum measurement and Pancharatnam phase

5.3.1   Geodesics in quantum phase space

The projective Hilbert space $\mathcal{P}(\mathcal{H})$, being a Kähler manifold, is equipped with a canonical Fubini–Study metric. This metric enables us to measure the distance between quantum states, i.e., points in $\mathcal{P}(\mathcal{H})$, and the length of curves in $\mathcal{P}(\mathcal{H})$. Consider a curve

$$C : \quad [0, 1] \ni t \longrightarrow p(t) \in \mathcal{P}(\mathcal{H}) \ ,$$

and let

$$\widetilde{C} : \quad [0, 1] \ni t \longrightarrow \psi(t) \in S(\mathcal{H}) \ ,$$

be a lift of C in the bundle $S(\mathcal{H}) \longrightarrow \mathcal{P}(\mathcal{H})$. The length of $\widetilde{C}$ may be computed in terms of the Hilbert space structure as follows:

$$L(\widetilde{C}) = \int_0^1 \sqrt{\langle \dot{\psi} | \dot{\psi} \rangle} \, dt \ . \qquad (5.133)$$

Evidently, $L(\widetilde{C})$ does depend on a particular lift, and hence the length of a lifted curve does not define a gauge-invariant quantity. If $\widetilde{C}'$ is another lift of C, such that $\psi' = e^{i\varphi}\psi$, then

$$L(\widetilde{C}') = \int_0^1 \sqrt{\langle \dot{\psi}' | \dot{\psi}' \rangle} \, dt = \int_0^1 \sqrt{\langle \dot{\psi} | \dot{\psi} \rangle + \dot{\varphi}^2 + 2\dot{\varphi} \, \text{Im}\langle \psi | \dot{\psi} \rangle} \, dt \ . \qquad (5.134)$$

This poses a natural question: which lift of C has minimal length? Let us call such a lift a *minimal lift*.

Theorem 5.3.1 *A lift is minimal if and only if it is a horizontal lift w.r.t. the natural connection. Moreover, the length of C with respect to the Fubini–Study metric on $\mathcal{P}(\mathcal{H})$ is minimal, i.e.,*

$$L_{\mathrm{FS}}(C) \; = \; \text{length of its horizontal lift} \,.$$

Proof. The proof is based on the following simple inequality:

$$\langle \dot{\psi} | \dot{\psi} \rangle \geq \langle \dot{\psi} | \dot{\psi} \rangle - |\langle \dot{\psi} | \psi \rangle|^2 = \langle \dot{\psi} | \dot{\psi} \rangle - \langle \dot{\psi} | \psi \rangle \langle \psi | \dot{\psi} \rangle \,. \tag{5.135}$$

Note that the r.h.s. of the above formula is exactly the scalar product computed in terms of the Fubini–Study metric (cf. (5.23)). Hence, the above inequality may be rewritten as follows:

$$\langle \dot{\psi} | \dot{\psi} \rangle \geq (\pi^* g)_\psi (\dot{\psi}, \dot{\psi}) \,. \tag{5.136}$$

Therefore a lift is minimal if and only if the lifted curve satisfies

$$\langle \psi | \dot{\psi} \rangle = 0 \,, \tag{5.137}$$

which is exactly the condition for a lift to be horizontal with respect to the natural connection in a bundle $S(\mathcal{H}) \longrightarrow \mathcal{P}(\mathcal{H})$. Now, if $\psi(t)$ is a horizontal lift, then

$$\langle \dot{\psi} | \dot{\psi} \rangle = (\pi^* g)_\psi (\dot{\psi}, \dot{\psi}) \,, \tag{5.138}$$

and hence the second part of the theorem follows. $\qquad\qquad\square$

Knowing how to measure length of a curve in $\mathcal{P}(\mathcal{H})$ lets us turn to curves with minimal possible length, i.e., geodesics with respect to a Fubini–Study metric. Consider two points $p_1, p_2 \in \mathcal{P}(\mathcal{H})$. Let ψ_1 and ψ_2 be two arbitrary nonorthogonal state vectors from $S(\mathcal{H})$ projecting to p_1 and p_2, respectively. Define a *real plane* in $\mathcal{H}$ spanned by ψ_1 and ψ_2, as follows:

$$\left\{ \psi = \lambda_1 \psi_1 + \lambda_2 \psi_2 \,\middle|\, \lambda_1, \lambda_2 \in \mathbb{R} \right\} \subset \mathcal{H} \,.$$

The intersection of any real plane with the unit sphere $S(\mathcal{H})$ is a great circle which defines a geodesic on $S(\mathcal{H})$ with respect to the metric induced from $\mathcal{H}$. Clearly, its length equals 2π. Now, a geodesic on $S(\mathcal{H})$ projects to a geodesic on $\mathcal{P}(\mathcal{H})$ and hence each geodesic on $\mathcal{P}(\mathcal{H})$ is a closed curve, since it is a projection of a closed curve (a circle) on $S(\mathcal{H})$.

Let us parametrize a geodesic joining ψ_1 and ψ_2 on $S(\mathcal{H})$, i.e., an arc of a great circle passing through ψ_1 and ψ_2, by an angle $\vartheta \in [0, 2\pi)$,[6] i.e., such that

$$\psi(\vartheta) \; = \; \lambda_1(\vartheta) \, \psi_1 + \lambda_2(\vartheta) \, \psi_2 \,, \tag{5.139}$$

---

[6]We follow Uhlmann 1987.

and let us introduce the following real parameter

$$a := \mathrm{Re}\, \langle\, \psi_1 | \psi_2 \,\rangle \,. \tag{5.140}$$

Suppose, that $a > 0$ (if not we take a pair ψ_1 and $-\psi_2$). The normalization condition $\langle\, \psi(\vartheta) | \psi(\vartheta) \,\rangle = 1$ implies the following formulae for λ's

$$\lambda_1(\vartheta) \;=\; \cos\vartheta - \frac{a}{\sqrt{1-a^2}}\, \sin\vartheta \,, \tag{5.141}$$

$$\lambda_2(\vartheta) \;=\; \frac{\sin\vartheta}{\sqrt{1-a^2}} \,. \tag{5.142}$$

Note, that

$$\psi(0) = \psi_1 \,, \qquad \psi(\vartheta_0) = \psi_2 \,, \tag{5.143}$$

where the angle ϑ_0 is defined by

$$\cos\vartheta_0 = a \,, \tag{5.144}$$

with $\vartheta_0 \in [0, \pi/2)$. Now, one may easily show that

$$\langle\, \partial_\vartheta \psi | \partial_\vartheta \psi \,\rangle = 1 \,, \tag{5.145}$$

and hence, a length of the geodesic arc between ψ_1 and ψ_2 reads

$$L = \int_0^{\vartheta_0} \sqrt{\langle\, \partial_\vartheta \psi | \partial_\vartheta \psi \,\rangle}\, d\vartheta = \vartheta_0 \,. \tag{5.146}$$

It is clear that this length attains its minimum if we choose ψ_1 and ψ_2 such that the parameter a defined in (5.140) is maximized

$$a = \mathrm{Re}\, \langle\, \psi_1 | \psi_2 \,\rangle \;=\; \max \,. \tag{5.147}$$

Now, since

$$\langle\, \psi_1 | \psi_2 \,\rangle = r e^{i\lambda} = r\cos\lambda + i\sin\lambda \,,$$

it is clear that $\mathrm{Re}\, \langle\, \psi_1 | \psi_2 \,\rangle$ attains its maximum if and only if $\langle\, \psi_1 | \psi_2 \,\rangle$ is real and positive, in which case $a = |\langle\, \psi_1 | \psi_2 \,\rangle|$. Uhlmann (Uhlmann 1986) proposed the terminology *horizontal plane* for real plane spanned by such vectors. Now, let γ be the shortest geodesic joining p_1 with p_2,[7] and let

$$\sigma(p_1, p_2) := \text{ length of } \gamma \,.$$

The above discussion may be summarized in the following

---

[7] Since any geodesic on $\mathcal{P}(\mathcal{H})$ is closed, any pair of point in $\mathcal{P}(\mathcal{H})$ defines two arcs of the closed geodesic passing through them.

Theorem 5.3.2 *The length of the shortest geodesic joining p_1 and p_2 in $\mathcal{P}(\mathcal{H})$ is given by the following formula:*

$$\cos[\sigma(p_1, p_2)] = |\langle \psi_1 | \psi_2 \rangle|,$$

with ψ_1 and ψ_2 being arbitrary elements from the corresponding fibres $\pi^{-1}(p_1)$ and $\pi^{-1}(p_2)$, respectively. In particular, choosing ψ_1 and ψ_2 such that

$$\langle \psi_1 | \psi_2 \rangle \text{ is real and positive},$$

one has

$$\cos[\sigma(p_1, p_2)] = \langle \psi_1 | \psi_2 \rangle.$$

Corollary 5.3.1 *Each geodesic on $\mathcal{P}(\mathcal{H})$ is closed and its length equals π.*

Proof. A closed geodesic on $S(\mathcal{H})$ (a circle) passing through ψ_1 and ψ_2 is defined by (5.139), where the λ's are given by (5.141)–(5.142) and $\vartheta \in [0, 2\pi)$. Note, however, that

$$\psi(\vartheta) = -\psi(\vartheta + \pi),$$

and hence

$$|\psi(\vartheta)\rangle\langle\psi(\vartheta)| = |\psi(\vartheta + \pi)\rangle\langle\psi(\vartheta + \pi)|,$$

i.e., $\psi(\vartheta)$ and $\psi(\vartheta + \pi)$ project to the same point on the projected geodesic in $\mathcal{P}(\mathcal{H})$. Therefore, the length of the closed geodesic on $\mathcal{P}(\mathcal{H})$ is half that of the closed geodesic in $S(\mathcal{H})$. Recalling that the length of any closed geodesic in $S(\mathcal{H})$ equals 2π, the corollary follows. $\qquad\square$

Remark 5.3.1 Note that if $\langle \psi_1 | \psi_2 \rangle = 0$, then $\sigma(p_1, p_2) = \pi/2$ and there are infinitely many horizontal planes spanned by ψ_1 and ψ_2. Indeed, a real plane spanned by $\psi_1' = e^{i\varphi_1} \psi_1$ and $\psi_2' = e^{i\varphi_1} \psi_2$ is still horizontal. Hence, there are infinitely many geodesics connecting p_1 and p_2 on $\mathcal{P}(\mathcal{H})$. Such points are called *conjugated*. Hence, any two orthogonal vectors in $S(\mathcal{H})$ give rise to conjugated points in $\mathcal{P}(\mathcal{H})$. Note that any two conjugated points divide each geodesic passing through them into two arcs of equal length. $\qquad\diamond$

5.3.2 Distance between pure quantum states

Let us consider two arbitrary points $p_1, p_2 \in \mathcal{P}(\mathcal{H})$. The Fubini–Study metric enables one to compute the distance between p_1 and p_2 as a length of the geodesic connecting these points, that is, the Fubini–Study length between p_1 and p_2 equals $\sigma(p_1, p_2)$. Note, however, that there are other ways to define a distance between pure quantum states represented by the state vectors ψ_1 and ψ_2. Denote by $\hat{P}_1$ and $\hat{P}_2$ the corresponding one-dimensional projectors, i.e.,

$$\hat{P}_1 = |\psi_1\rangle\langle\psi_1| \quad \text{and} \quad \hat{P}_2 = |\psi_2\rangle\langle\psi_2|,$$

and introduce the following distance functions:

$$D_{\mathrm{Tr}}(\hat{P}_1, \hat{P}_2) := \mathrm{Tr}\,|\hat{P}_1 - \hat{P}_2|\,, \tag{5.148}$$

and

$$D_{\mathrm{HS}}(\hat{P}_1, \hat{P}_2) := \sqrt{\mathrm{Tr}\,(\hat{P}_1 - \hat{P}_2)^2}\,, \tag{5.149}$$

where $|A| := \sqrt{A^*A}$ is the positive square root of A^*A. As we shall see in section 5.4.2 the distance D_{HS} corresponds to the Hilbert–Schmidt norm. The reader can easily show that

$$D_{\mathrm{Tr}}(\hat{P}_1, \hat{P}_2) = 2\sqrt{1 - |\langle\psi_1|\psi_2\rangle|^2}\,, \tag{5.150}$$

and

$$D_{\mathrm{HS}}(\hat{P}_1, \hat{P}_2) = \sqrt{2(1 - |\langle\psi_1|\psi_2\rangle|^2)}\,. \tag{5.151}$$

Another way to measure the distance between $\hat{P}_1$ and $\hat{P}_2$ uses the norm in the original Hilbert space $\mathcal{H}$ and it is usually called the Fubini–Study distance

$$
\begin{aligned}
D_{\mathrm{FS}}^2(\hat{P}_1, \hat{P}_2) \;&:=\; \inf_\varphi \|\psi_1 - e^{i\varphi}\psi_2\|^2 \\
&=\; \inf_\varphi \langle\psi_1 - e^{i\varphi}\psi_2|\psi_1 - e^{i\varphi}\psi_2\rangle\,.
\end{aligned}
\tag{5.152}
$$

Performing the scalar product the above formula leads to:

$$\inf_\varphi \langle\psi_1 - e^{i\varphi}\psi_2|\psi_1 - e^{i\varphi}\psi_2\rangle = 2(1 - \sup_\varphi \mathrm{Re}\,\langle\psi_1|e^{i\varphi}\psi_2\rangle) = 2(1 - |\langle\psi_1|\psi_2\rangle|)\,,$$

and hence,

$$D_{\mathrm{FS}}(\hat{P}_1, \hat{P}_2) = \sqrt{2(1 - |\langle\psi_1|\psi_2\rangle|)}\,. \tag{5.153}$$

This shows that all three distances (D_{Tr}, D_{HS}, D_{FS}) are closely related to the geodesic (or Fubini–Study) length $\sigma(\hat{P}_1, \hat{P}_2)$. It is therefore not surprising that they give rise to riemannian metrics closely related to the canonical Fubini–Study metric on $\mathcal{P}(\mathcal{H})$.

Finally, let us compute the explicit form of the (riemannian) Fubini–Study metric. Take any local section $\psi : \mathcal{P}(\mathcal{H}) \longrightarrow S(\mathcal{H})$ and define the corresponding metric tensor as follows

$$ds_{\mathrm{FS}}^2 =: g_{ij}(x)dx^i dx^j\,, \tag{5.154}$$

where

$$\cos\sqrt{ds_{\mathrm{FS}}^2} = |\langle\psi(x)|\psi(x+dx)\rangle|\,.$$

A Taylor expansion applied to $\psi(x + dx)$ gives

$$\psi(x + dx) = \psi(x) + \partial_i \psi(x)dx^i + \frac{1}{2}\partial_i \partial_j \psi(x)\, dx^i dx^j + \dots , \tag{5.155}$$

and hence one obtains

$$\langle \psi(x)|\psi(x + dx)\rangle = 1 + \langle \psi(x)|\partial_i \psi(x)\rangle dx^i + \frac{1}{2}\langle \psi(x)|\partial_i \partial_j \psi(x)\rangle dx^i dx^j + \dots . \tag{5.156}$$

Therefore, up to second order terms

$$|\langle \psi(x)|\psi(x + dx)\rangle|$$
$$= 1 + \frac{1}{2}\mathrm{Re}\left(\langle \psi(x)|\partial_i \partial_j \psi(x)\rangle + \langle \partial_i \psi(x)|\psi(x)\rangle\langle \psi(x)|\partial_j \psi(x)\rangle\right)dx^i dx^j .$$

Using

$$\mathrm{Re}\,\langle \psi|\partial_i \partial_j \psi\rangle = -\mathrm{Re}\,\langle \partial_i \psi|\partial_j \psi\rangle ,$$

and the following expansion of the cosine function:

$$\cos\sqrt{ds_{\mathrm{FS}}^2} = 1 - \frac{1}{2}\,ds_{\mathrm{FS}}^2 + \text{ higher order terms} ,$$

we obtain finally

$$g_{ij} = \mathrm{Re}\left(\langle \partial_i \psi|\partial_j \psi\rangle - \langle \partial_i \psi|\psi\rangle\langle \psi|\partial_j \psi\rangle\right) . \tag{5.157}$$

It is easy to show that g_{ij} is gauge invariant, i.e. it does not depend on the particular choice of the local section ψ, and hence defines a metric tensor on $\mathcal{P}(\mathcal{H})$. It is instructive to show that (5.157) reproduces the holomorphic-antiholomorphic components $g_{i\bar{j}}$ from the formula (5.42).

Example 5.3.1 (Two-level system — qubit) Consider a two-level quantum system living in $\mathcal{H} = \mathbb{C}^2$ — in quantum information theory this is called a *qubit*, see Nielsen and Chuang 2000. The corresponding quantum phase space

$$\mathbb{C}P^1 \cong S^2$$

may be parametrized by standard spherical angles (θ, φ). Take a local section

$$S^2 \ni x \longrightarrow \psi(x) \in S^3 = S(\mathbb{C}^2) ,$$

defined by

$$\psi(x) = \begin{vmatrix} \cos\left(\frac{\theta}{2}\right) e^{i\varphi} \\ \sin\left(\frac{\theta}{2}\right) \end{vmatrix} .$$

One can easily compute the following:

$$\langle \psi | \partial_\varphi \psi \rangle = i \cos^2 \left(\frac{\theta}{2} \right) , \quad \langle \psi | \partial_\theta \psi \rangle = 0 , \quad \langle \partial_\varphi \psi | \partial_\varphi \psi \rangle = \cos^2 \left(\frac{\theta}{2} \right) ,$$

$$\langle \partial_\varphi \psi | \partial_\theta \psi \rangle = -\frac{i}{4} \sin \theta , \quad \langle \partial_\theta \psi | \partial_\theta \psi \rangle = \frac{1}{4} .$$

Hence, the corresponding components of the Fubini–Study metric tensor (5.157) are given by

$$g_{\theta\theta} = \frac{1}{4} , \quad g_{\theta\varphi} = 0 , \quad g_{\varphi\varphi} = \frac{1}{4} \sin^2 \theta ,$$

i.e., the Fubini–Study metric agrees with the standard metric tensor on a sphere of radius $r = 1/2$. Moreover, any geodesic is a great circle with a length

$$L = 2\pi r = \pi ,$$

in perfect agreement with Proposition 5.3.1. Note that conjugated points correspond to antipodal points on S^2. $\diamond$

Let us note that the formula (5.157) for the Fubini–Study metric is a special case of the quantum metric tensor introduced in section 2.2.6. In particular, examples 2.2.4 and 5.3.1 define the same metric on S^2 — in the former case S^2 plays the role of the parameter manifold, and in the latter it serves as the space of quantum states. Note that, unlike the quantum metric tensor $g^{(n)}$ on the parameter manifold M, the Fubini–Study metric g on $\mathcal{P}(\mathcal{H})$ defines a proper riemannian metric.

5.3.3 Measurement process

Now, we are going to relate our previous geometrical considerations to the measurement process in quantum mechanics. The most important object studied in this context is the transition probability. For any state vector $\psi_0 \in S(\mathcal{H})$, one introduces a quantum mechanical probability distribution as a function on $S(\mathcal{H})$ defined by

$$S(\mathcal{H}) \ni \psi \longrightarrow |\langle \psi_0 | \psi \rangle|^2 \in \mathbb{R}_+ .$$

Note that $|\langle \psi_0 | \psi \rangle|$ does not depend upon the phases of ψ_0 and ψ and, hence, it enables one to define a function δ_{p_0} on the quantum phase space $\mathcal{P}(\mathcal{H})$, as follows:

$$\mathcal{P}(\mathcal{H}) \ni p \longrightarrow \delta_{p_0}(p) := |\langle \psi_0 | \psi \rangle|^2 , \tag{5.158}$$

where ψ_0 and ψ are arbitrary elements from the corresponding fibres, i.e., $\psi_0 \in \pi^{-1}(p_0)$ and $\psi \in \pi^{-1}(p)$. We may call δ_{p_0} a quantum mechanical probability distribution on the quantum phase space $\mathcal{P}(\mathcal{H})$. By identifying the points $p_0, p \in \mathcal{P}(\mathcal{H})$ with one-dimensional projectors $\hat{P}_0$ and $\hat{P}$, we have, equivalently,

$$\delta_{\hat{P}_0}(\hat{P}) = (\mathrm{Tr}\, \hat{P}_0 \hat{P}) . \tag{5.159}$$

Theorem 5.3.2 gives rise to the following simple interpretation of δ_{p_0}:

Corollary 5.3.2 *A quantum mechanical probability distribution on $\mathcal{P}(\mathcal{H})$ satisfies*

$$\delta_{p_0}(p) \ = \ \cos^2\left[\sigma(p_0, p)\right] ,$$

where $\sigma(p_0, p)$ is the minimal geodesic distance separating p_0 and p.

Suppose that we are dealing with a quantum observable $\hat{F}$. Assume for simplicity that $\hat{F}$ has a discrete, nondegenerate spectrum, i.e.,

$$\hat{F}\psi_k = f_k\psi_k , \tag{5.160}$$

with $\psi_k \in S(\mathcal{H})$. Denote by p_k the corresponding eigenstates in $\mathcal{P}(\mathcal{H})$:

$$p_k \ := \ \pi(\psi_k) ,$$

or, equivalently, define the one-dimensional projectors $\hat{P}_k$:

$$\hat{P}_k \ := \ |\psi_k\rangle\langle\psi_k| .$$

The spectral theorem implies the following spectral decomposition of $\hat{F}$:

$$\hat{F} = \sum_k f_k \hat{P}_k . \tag{5.161}$$

In the process of measurement any state $\hat{P}_0 \in \mathcal{P}(\mathcal{H})$ will collapse to one of the eigenstates $\hat{P}_k$ with probability equal to

$$\delta_{\hat{P}_0}(\hat{P}_k) \ = \ \mathrm{Tr}\,(\hat{P}_0\hat{P}_k) . \tag{5.162}$$

Corollary 5.3.2 implies, therefore, a suggestive picture of the measurement process on a true quantum phase space $\mathcal{P}(\mathcal{H})$: The probability of obtaining an eigenvalue f_k in measuring a quantum observable $\hat{F}$ is a monotonically decreasing function of the (minimal) separation of $\hat{P}_0$ and the corresponding eigenstate $\hat{P}_k$; the system is more likely to collapse to a nearby state than to a distant one.

Let $\hat{F}$ be a self-adjoint operator on $\mathcal{H}$ (a quantum observable) and let $f \in C^\infty(\mathcal{P}(\mathcal{H}))$ denote the corresponding observable function, i.e.,

$$f(\hat{P}) = \mathrm{Tr}(\hat{P}\hat{F}) .$$

Clearly,

$$f(\hat{P}_k) = f_k . \tag{5.163}$$

Recall that a hamiltonian vector field on $\mathcal{H}$, corresponding to $\hat{F}$, is given by

$$X_{\hat{F}}(\psi) = -\frac{i}{\hbar}\,\hat{F}\psi . \tag{5.164}$$

When calculated at ψ_k,

$$X_{\hat{F}}(\psi_k) = -\frac{if_k}{\hbar}\,\psi_k\,, \qquad (5.165)$$

and so is parallel to ψ_k, and, hence, is a vertical vector field. Therefore, the projected hamiltonian vector field X_f vanishes at all eigenstates $\hat{P}_k$, which means that a quantum observable function f has the eigenstates p_k as its critical points, i.e., $df(\hat{P}_k) = 0$.

At this point a natural question arises: Which functions on $\mathcal{P}(\mathcal{H})$ are quantum observable functions, i.e., correspond to some quantum observable on $\mathcal{H}$? It turns out (Ashtekar and Schilling 1998) that an observable function may be entirely characterized in terms of geometric structures on $\mathcal{P}(\mathcal{H})$, without reference to an underlying Hilbert space $\mathcal{H}$. One has the following

Proposition 5.3.3 *A function $f : \mathcal{P}(\mathcal{H}) \longrightarrow \mathbb{R}$ is a quantum observable function if and only if its hamiltonian vector field X_f is a Killing vector field of the corresponding Kähler metric g.*[8]

It turns out that in a similar manner one may deal also with the problem of observables with continuous spectra, and one may arrive at a complete formulation of quantum mechanics on a true quantum phase space, i.e., projective Hilbert space $\mathcal{P}(\mathcal{H})$. The interested reader is referred to Ashtekar and Schilling 1998 (see also Brody and Hughston 2001).

5.3.4 Pancharatnam phase

Consider a pair of vectors ψ and $\psi' = e^{i\alpha}\psi$ representing the same quantum state, i.e., such that

$$|\psi\rangle\langle\psi| = |\psi'\rangle\langle\psi'|\,.$$

It is clear that the relative phase between ψ and ψ' is α. However, when ψ and ψ' represent two different quantum states the definition of a relative phase is less obvious. Apparently no one had posed this problem until Pancharatnam (1956) came up with a physical interpretation of the relative phase between distinct polarization states of light.[9] It turns out that Pancharatnam's concept of relative phase has a quantal counterpart, with a surprisingly rich structure related to the geometry of the quantum phase space $\mathcal{P}(\mathcal{H})$.

Take two nonorthogonal vectors $\psi_1, \psi_2 \in S(\mathcal{H})$. We call the phase of their scalar product the *relative phase* or *phase difference* between ψ_1 and ψ_2, i.e.,

$$\langle\psi_1|\psi_2\rangle = re^{i\alpha_{12}} \implies \alpha_{12} := \text{phase difference between } \psi_1 \text{ and } \psi_2\,.$$

---

[8] X is a Killing vector field of a metric g if

$$\mathcal{L}_X\, g = 0\,,$$

where $\mathcal{L}_X$ denotes the Lie derivative with respect to X.

[9] We shall discuss Pancharatnam's idea in optics in more detail in section 6.1.4.

We say that ψ_1 and ψ_2 are *in phase* or *parallel* if

$$\langle \psi_1 | \psi_2 \rangle \quad \text{is real and positive .}$$

The above definition introduces the following relation between any two nonorthogonal vectors:

$$\psi \sim \phi \quad \Longleftrightarrow \quad \text{they are in phase .}$$

Clearly, ψ is in phase with itself, and if ψ is in phase with ϕ then ϕ is also in phase with ψ. However, the above relation is not transitive, if ψ_1 is in phase with ψ_2, and ψ_2 with ψ_3, then ψ_3 need not be in phase with ψ_1. Hence, the notion "to be in phase" does not define an equivalence relation.

Example 5.3.2 Consider the following three normalized vectors:

$$\psi_1 = \frac{1}{\sqrt{2}} \begin{pmatrix} 1 \\ 1 \end{pmatrix}, \quad \psi_2 = \begin{pmatrix} 1 \\ 0 \end{pmatrix}, \quad \psi_3 = \frac{1}{\sqrt{2}} \begin{pmatrix} 1 \\ i \end{pmatrix}.$$

Clearly:

- ψ_1 is in phase with ψ_2, since $\langle \psi_1 | \psi_2 \rangle = 1/\sqrt{2}$ is real and positive,

- ψ_2 is in phase with ψ_3, since $\langle \psi_2 | \psi_3 \rangle = 1/\sqrt{2}$ is real and positive,

- ψ_1 is not in phase with ψ_3 since $\langle \psi_1 | \psi_3 \rangle = -i/\sqrt{2}$ is not real. $\diamond$

One often calls the above rule of defining relative phases a *Pancharatnam connection*. In this way a principal $U(1)$-fibre bundle $S(\mathcal{H}) \longrightarrow P(\mathcal{H})$ is equipped with two connections:

1. a *natural connection* giving rise to a Aharonov–Anandan phase, and

2. a Pancharatnam connection defining relative phases between elements from $S(\mathcal{H})$.

What is the relation between these connections? The answer to this question is given by the following

Theorem 5.3.4 *Consider two points $p_1, p_2 \in P(\mathcal{H})$ and let γ be a shorter arc of the geodesic connecting p_1 and p_2. Moreover, let*

$$\tilde{\gamma} : \quad t \longrightarrow \psi(t) \in S(\mathcal{H}),$$

be a horizontal lift of γ with respect to the natural connection in a principal fibre bundle $S(\mathcal{H}) \longrightarrow P(\mathcal{H})$. Then any two points in $\tilde{\gamma}$ are in phase, i.e., a parallel transport of ψ keeps $\psi(t)$ in phase with $\psi(0)$.

Proof. Let C be a geodesic in $S(\mathcal{H})$ projecting to γ in $P(\mathcal{H})$. Any geodesic (a part of a great circle) on a unit sphere $S(\mathcal{H})$ is uniquely defined by a real plane in $\mathcal{H}$ spanned by two vectors ψ_1 and ψ_2. Keeping the notation of section 5.3.1, we know that the geodesic (more precisely, the shorter arc of the closed geodesic)

$$\psi(\vartheta) = \lambda_1(\vartheta)\psi_1 + \lambda_2(\vartheta)\psi_2 \, ,$$

is a horizontal lift of γ if and only if

$$\langle \psi_1 | \psi_2 \rangle \text{ is real and positive} \, ,$$

that is, ψ_1 and ψ_2 are in phase. Now, it is easy to show that

$$\langle \psi(\vartheta_1) | \psi(\vartheta_2) \rangle = \cos(\vartheta_1 - \vartheta_2) > 0 \, , \tag{5.166}$$

since $\vartheta_1, \vartheta_2 \in [0, \vartheta_0]$, and ϑ_0, defined in (5.144), satisfies $\vartheta_0 \in [0, \pi/2]$. Therefore, any two points belonging to the horizontal lift $\tilde{\gamma}$ are in phase. $\square$

Notice that we called a real plane spanned by two parallel vectors a parallel plane; this name is now fully justified. It should be stressed that only points belonging to the same arc of the geodesic connecting ψ_1 and ψ_2 are in phase. (Clearly, the geodesic passing through ψ_1 and ψ_2 is closed, i.e., there is a second arc from ψ_2 to ψ_1. Points belonging to different arcs are, in general, not in phase.)

Consider, now, two vectors $\psi_0, \psi_1 \in S(\mathcal{H})$ and let α be their relative phase. Denote by $p_0, p_1 \in \mathcal{P}(\mathcal{H})$ the corresponding projections, i.e., $p_0 = \pi(\psi_0)$ and $p_1 = \pi(\psi_1)$. The following theorem clarifies the geometric character of α:

Theorem 5.3.5 *Let γ be the shortest geodesic connecting p_0 and p_1. Then*

$$\alpha = - \int_{\tilde{\gamma}} \mathcal{A} \, , \tag{5.167}$$

where $\mathcal{A}$ is a connection one-form corresponding to the natural connection in $S(\mathcal{H}) \longrightarrow P(\mathcal{H})$, and $\tilde{\gamma}$ is an arbitrary lift of γ connecting ψ_0 and ψ_1.

Proof. Let A be a local connection form in $\mathcal{P}(\mathcal{H})$. We have, therefore,

$$\int_{\tilde{\gamma}} \mathcal{A} = \int_{\gamma} A \, . \tag{5.168}$$

Denote by $\tilde{\gamma}^0$ a horizontal lift of γ passing through ψ_0 (cf. Fig. 5.3), i.e., let

$$\tilde{\gamma} : \quad [0, 1] \ni t \longrightarrow \psi(t) \, ,$$

and

$$\tilde{\gamma}^0 : \quad [0, 1] \ni t \longrightarrow \phi(t) \, .$$

Clearly, both lifts are related by time-dependent phase factor, i.e.,

$$\phi(t) = e^{if(t)} \psi(t) \, , \tag{5.169}$$

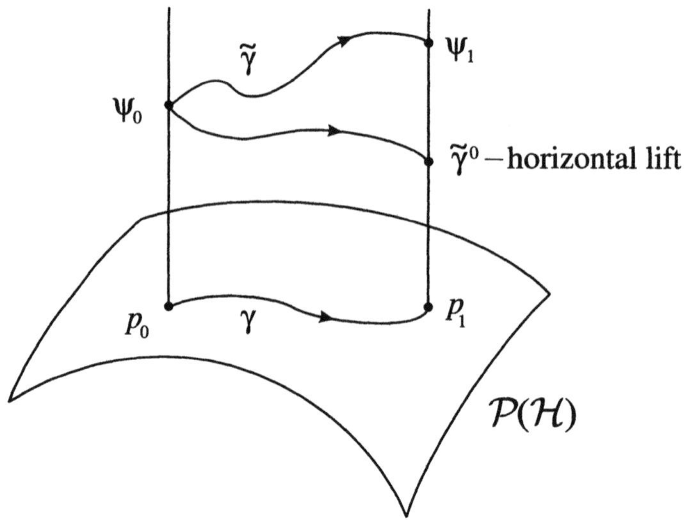

Figure 5.3: Lifts of γ passing through ψ_0: $\tilde{\gamma}^0$ — a horizontal lift; and $\tilde{\gamma}$ — an arbitrary lift connecting ψ_0 and ψ_1.

such that $f(0) = 0$. Now, by the very definition of a local connection form one has

$$A^0 = \langle \phi | id | \phi \rangle = A - df , \tag{5.170}$$

and hence

$$\int_\gamma A = \int_\gamma (A^0 + df) = f(1) , \tag{5.171}$$

since the integral of A^0 vanishes. Finally, the relative phase between ψ_0 and ψ_1 may be computed as follows:

$$e^{i\alpha} = \frac{\langle \psi_0 | \psi_1 \rangle}{|\langle \psi_0 | \psi_1 \rangle|} = \frac{\langle \phi(0) | e^{-if(1)} \phi(1) \rangle}{|\langle \psi_0 | \psi_1 \rangle|} = e^{-if(1)} , \tag{5.172}$$

since $\phi(0)$ and $\phi(1)$ are in phase. Thus, one concludes that

$$\alpha = - \int_\gamma A , \tag{5.173}$$

which completes the proof. □

Let us turn to physical applications. Consider a sequence of projections (filtering quantum measurements) that bring an initial state vector ψ_1 to itself after projections on the state vectors $\psi_2, \psi_3, \ldots, \psi_n$ and finally on ψ_1.[10] The resulting vector ψ_{final}

---

[10]Berry and Klein (1996) illustrated the theoretical treatment of the measurement process by experimenting with polarized light beams.

will differ by a phase factor from the initial one, i.e.,

$$\psi_{\text{final}} = e^{i\varphi}\psi_1 \, . \tag{5.174}$$

The relative phase φ is easy to compute. One obtains

$$e^{i\varphi} = \langle \psi_1|\psi_n \rangle \langle \psi_n|\psi_{n-1} \rangle \ldots \langle \psi_3|\psi_2 \rangle \langle \psi_2|\psi_1 \rangle \, . \tag{5.175}$$

Note that by defining the one-dimensional projectors

$$\hat{P}_k := |\psi_k \rangle \langle \psi_k| \, ,$$

we have, equivalently,

$$e^{i\varphi} = \text{Tr}\,(\hat{P}_n \cdot \hat{P}_{n-1} \cdot \ldots \cdot \hat{P}_2 \cdot \hat{P}_1) \, , \tag{5.176}$$

and hence φ corresponds to a geometric quantity defined entirely on $\mathcal{P}(\mathcal{H})$, i.e., without reference to an original Hilbert space $\mathcal{H}$. The phase shift φ is usually called the *Pancharatnam phase*, and has a beautiful geometric interpretation. Denote by $\gamma_{k+1,k}$ a geodesic connecting $\hat{P}_k$ and $\hat{P}_{k+1}$, and define a closed curve

$$C := \gamma_{1,n} * \gamma_{n,n-1} * \ldots * \gamma_{3,2} * \gamma_{2,1} \, .$$

Due to Theorem 5.3.5 we have that the

$$\text{Pancharatnam phase} \, = \, -\oint_C A \, = \, -\int_\Sigma F \, , \tag{5.177}$$

with $F = dA$ being a curvature two-form on $\mathcal{P}(\mathcal{H})$, and Σ any two-dimensional region such that $\partial\Sigma = C$.

Example 5.3.3 Consider a finite-dimensional Hilbert space $\mathcal{H} = \mathbb{C}^n$. The corresponding quantum phase space is the projective Hilbert space $\mathbb{C}P^n$. Due to (5.128), one has

$$\varphi = -2\int_\Sigma \omega \, , \tag{5.178}$$

with ω being a (Kähler) symplectic form in $\mathbb{C}P^n$. In particular, for a two-level system — a qubit — we have (cf. Example 5.2.1)

$$\varphi = -\frac{1}{2}\Omega(C) \, , \tag{5.179}$$

with $\Omega(C)$ being the solid angle subtended by C. $\qquad\qquad\diamond$

Remark 5.3.2 Samuel and Bhandari (Samuel and Bhandari 1988) showed that one can define a geometric phase for any (not necessarily closed) curve in $\mathcal{P}(\mathcal{H})$. Indeed, consider an open curve C_1 joining points p_1 and p_2 and define a closed curve

$$C := \gamma_{1,2} * C_1 \, ,$$

where $\gamma_{1,2}$ stands for geodesic connecting p_2 and p_1. One has, therefore,

$$\Delta\varphi_{AA} := \oint_C A = \int_{C_1} A + \int_{\gamma_{1,2}} A , \qquad (5.180)$$

where $\Delta\varphi_{AA}$ is the Aharonov–Anandan phase corresponding to the closed curve C. Hence,

$$\int_{C_1} A := \Delta\varphi_{AA} + \int_{\gamma_{1,2}} A \qquad (5.181)$$

uniquely defines a geometric object on $P(\mathcal{H})$ corresponding to the open path C_1. $\diamondsuit$

5.4   Geometric phase for mixed states

5.4.1   Mixed states in quantum mechanics

Up to now we have considered only pure quantum states, that is, elements from the projective Hilbert space $P(\mathcal{H})$. However, pure states form only a very limited class of quantum states. The most general state, the so-called *mixed state*, is represented by a *density operator* or *density matrix* in $\mathcal{H}$, that is, a hermitian trace class operator[11] ρ in $\mathcal{H}$ such that

$$\rho \geq 0 \quad \text{and} \quad \mathrm{Tr}\,\rho = 1 .$$

Let us denote by P the space of mixed quantum states. Note that a density operator ρ defines a pure state if and only if ρ is a projection of rank one, and hence

$$P(\mathcal{H}) = \left\{ \rho \in P \,\middle|\, \rho^2 = \rho \right\} .$$

It is well known that P is a convex space and that pure states define a set of extremal points for P. The quantum dynamics on the space of mixed states is described by the von Neumann equation:

$$i\hbar \frac{d}{dt} \rho = [\hat{H}, \rho] . \qquad (5.182)$$

If $\hat{F}$ is a quantum observable, then the expectation value of $\hat{F}$ in a mixed state ρ is defined by

$$\rho \longrightarrow \mathrm{Tr}\,(\hat{F}\rho) .$$

To discuss the structure of P in more detail, let us consider a finite-dimensional case which is of great importance for applications. Let us note that if $\mathcal{H} = \mathbb{C}^n$, then $i\rho \in$

---

[11] A linear, self-adjoint operator A in $\mathcal{H}$ is *trace class* if $\mathrm{Tr}\,AA^* < \infty$.

$u(n)$ for any $\rho \in \mathcal{P}$. Moreover, any density operator ρ may be uniquely written as follows:

$$\rho = \frac{1}{n} \mathbb{1}_n + i\alpha , \tag{5.183}$$

with $\alpha \in su(n)$. For each $\rho \in \mathcal{P}$ there exists a unitary matrix $U \in U(n)$, such that $\rho = UDU^*$, where D is a diagonal hermitian matrix. Let

$$\lambda_1 > \lambda_2 > \ldots > \lambda_m \geq 0 , \tag{5.184}$$

denote the eigenvalues of ρ, with multiplicities $(k_1, k_2, \ldots, k_m)$, i.e.,

$$k_1 + \ldots + k_m = n .$$

The normalization condition $\mathrm{Tr}\, \rho = \mathrm{Tr}\, D = 1$ implies that

$$\lambda_1 k_1 + \ldots + \lambda_m k_m = 1 .$$

One has, therefore,

$$D = \begin{pmatrix} \lambda_1 \mathbb{1}_{k_1} & & \\ & \ddots & \\ & & \lambda_m \mathbb{1}_{k_m} \end{pmatrix} . \tag{5.185}$$

Observe that all density matrices unitary equivalent to D lie on the orbit

$$\mathcal{O}_D = \left\{ UDU^* \,\middle|\, U \in U(n) \right\} ,$$

which is nothing but the coadjoint orbit for $U(n)$ passing through D (cf. section 1.2.2). This orbit is a quotient space

$$\mathcal{O}_D \cong U(n)/G_D ,$$

where G_D denotes an isotropy subgroup of D, i.e.,

$$G_D \cong \left\{ U \in U(n) \,\middle|\, UDU^* = D \right\} . \tag{5.186}$$

It is clear that

$$G_D \cong U(k_1) \times \ldots \times U(k_m) ,$$

and, hence, that

$$\mathcal{O}_D \cong \frac{U(n)}{U(k_1) \times \ldots \times U(k_m)} . \tag{5.187}$$

A manifold defined as a quotient space, as in the above formula, is called a *complex flag manifold* and is denoted by $P^{\mathbb{C}}_{k_1\ldots k_m}$. In particular, a flag manifold with $k_1 = k$ and $k_2 = n - k$, i.e.,

$$G^{\mathbb{C}}_{n,k} \cong \frac{U(n)}{U(k) \times U(n-k)} \, ,$$

is called a *complex Grassmann manifold*. It is a space of k-dimensional complex planes in $\mathbb{C}^n$. Clearly, if $k = 1$ it reproduces a complex projective space

$$G^{\mathbb{C}}_{n,1} \cong \mathbb{C}P^n \, .$$

It follows from (5.187) that

$$\dim_{\mathbb{R}} P^{\mathbb{C}}_{k_1\ldots k_m} = n^2 - (k_1^2 + \ldots + k_m^2) = 2 \sum_{1 \le i < j \le m} k_i k_j \, . \tag{5.188}$$

In particular,

$$\dim_{\mathbb{R}} \mathbb{C}P^n = 2(n-1) \, .$$

A sequence $(k_1, \ldots, k_m)$ is called an *orbit type*, and it uniquely determines all the geometric properties of the corresponding orbit, that is, any two orbits with the same orbit types are diffeomorphic. We have shown that the entire phase space of a quantum system is stratified into coadjoint orbits for the unitary group $U(n)$.

Example 5.4.1 (Qubit) Any density matrix in $\mathbb{C}^2$ may be written as follows:

$$\rho = \frac{1}{2}(\mathbb{1}_2 + \mathbf{x} \cdot \boldsymbol{\sigma}) \, ,$$

where $\mathbf{x} \in \mathbb{R}^3$, and $\boldsymbol{\sigma}$ stands for the vector of Pauli matrices. Hence, we have

$$\rho = \frac{1}{2} \begin{pmatrix} 1 + x_3 & x_1 - ix_2 \\ x_1 + ix_2 & 1 - x_3 \end{pmatrix} \, .$$

One easily finds two eigenvalues $\lambda_{\pm}$ of ρ. They are given by

$$\lambda_{\pm} = \frac{1 \pm |\mathbf{x}|}{2} \, .$$

The positivity of ρ requires $|\mathbf{x}| \le 1$. Therefore, the space of mixed states $\mathcal{P}$ may be represented as a unit ball in $\mathbb{R}^3$ — the so-called Bloch ball. Note that pure states correspond to points on a Bloch sphere — the boundary of the Bloch ball, i.e., $\mathbb{C}P^1 = \partial \mathcal{P}_2$ (cf. Fig. 5.4). These are one-dimensional projections with eigenvalues $\lambda_+ = 1$ and $\lambda_- = 0$. There are only two possible orbit types in this case:

1. $(1, 1)$: An orbit of this type is a two-dimensional sphere with radius $|\mathbf{x}| = \lambda_+ - \lambda_-$,

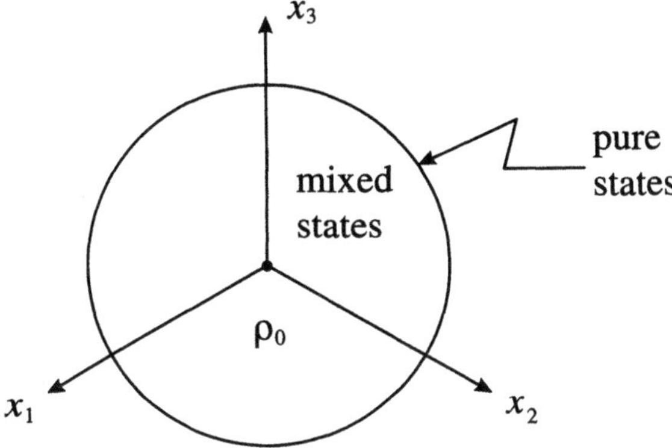

Figure 5.4: Space of mixed states $\mathcal{P}$ for a two-level system. Pure qubit states lie on the boundary $\partial\mathcal{P} = \mathbb{C}P^1$. The maximally mixed state ρ_0 lies in the center.

2. $(2,0)$: An orbit of this type consists of one element only: $\rho = \mathbb{1}_2/2$, which corresponds to the center of the unit ball. One calls this point a maximally mixed state. $\diamond$

For $n > 2$ the situation is much more complicated, e.g., for $n = 3$ we have three orbit types:

1. Six-dimensional $P^{\mathbb{C}}_{111}$,

2. The four-dimensional Grassmann manifold $G^{\mathbb{C}}_{3,1} \cong G^{\mathbb{C}}_{3,2}$ (e.g., an orbit of pure states $\mathbb{C}P^2$ is of this type),

3. $P^{\mathbb{C}}_3 = \{\frac{1}{3}\mathbb{1}_3\}$ — the maximally mixed state.

However, even in this case we do not know what these orbits "look like", i.e., the geometric characterization of these orbits, as subsets of the entire space of states, is not known.

5.4.2   Uhlmann's non-abelian geometric factor

Recall that the space of quantum pure states — projective Hilbert space $\mathcal{P}(\mathcal{H})$ — has an elegant interpretation as a $U(1)$-fibre bundle $S(\mathcal{H}) \longrightarrow \mathcal{P}(\mathcal{H})$. This fibre bundle, or, equivalently, Hopf fibration, is endowed with the canonical connection, whose holonomy is interpreted as the abelian Aharonov–Anandan geometric factor $e^{i\varphi_{AA}}$. Now, following Uhlmann (Uhlmann 1986, 1987, 1989, 1991), we are going to show that a similar construction may be performed in the case of mixed quantum states. The key idea of Uhlmann's approach to the mixed state geometric factor is to lift the system density operator ρ, acting on the Hilbert space $\mathcal{H}$, to an extended Hilbert space

$$\mathcal{H}^{\text{ext}} := \mathcal{H} \otimes \mathcal{H}' ,\tag{5.189}$$

where $\mathcal{H}'$ denotes another Hilbert space.[12] We shall take $\mathcal{H}' = \mathcal{H}$. Denote by $\mathcal{B}(\mathcal{H})$ a space of bounded operators in $\mathcal{H}$ and define the space of Hilbert–Schmidt operators, as follows:

$$\mathcal{H}^{\mathrm{HS}} := \left\{ W \in \mathcal{B}(\mathcal{H}) \,\middle|\, \mathrm{Tr}\,(WW^*) < \infty \right\} .$$

Such operators are endowed with the following scalar product:

$$\langle W_1, W_2 \rangle_{\mathrm{HS}} := \mathrm{Tr}\,(W_1^* W_2) . \tag{5.190}$$

A pair $(\mathcal{H}^{\mathrm{HS}}, \langle \, , \, \rangle_{\mathrm{HS}})$ defines a Hilbert space. Clearly, the so-called Hilbert–Schmidt norm is given by

$$\|W\|_{\mathrm{HS}}^2 = \langle W, W \rangle_{\mathrm{HS}} . \tag{5.191}$$

It is evident that if $\dim \mathcal{H} < \infty$, then

$$\mathcal{H}^{\mathrm{ext}} = \mathcal{H} \otimes \mathcal{H} \cong \mathcal{H}^{\mathrm{HS}} .$$

Now, let $S(\mathcal{H}^{\mathrm{HS}})$ denote a unit sphere in $\mathcal{H}^{\mathrm{HS}}$, i.e.,

$$S(\mathcal{H}^{\mathrm{HS}}) := \left\{ W \in \mathcal{H}^{\mathrm{HS}} \,\middle|\, \|W\|_{\mathrm{HS}} = 1 \right\} .$$

If ρ is a mixed state in $\mathcal{P}$ then we call an element $W \in S(\mathcal{H}^{\mathrm{HS}})$ a *purification* of ρ (one also says that W *purifies* ρ) if

$$\rho = WW^* . \tag{5.192}$$

A purification of ρ is by no means unique. If W purifies ρ then any element of the form

$$WV , \quad \text{with} \quad V \in U(\mathcal{H}) ,$$

does also.[13] This notion gives rise to a natural equivalence relation in $\mathcal{H}^{\mathrm{HS}}$, as follows:

$$W_1 \sim W_2 \iff W_2 = W_1 V , \quad \text{for some } V \in U(\mathcal{H}) ,$$

that is, two Hilbert–Schmidt operators W_1 and W_2 are equivalent if they purify the same mixed state ρ:

$$W_1 W_1^* = W_2 W_2^* . \tag{5.193}$$

This observation enables us to introduce a natural map

$$\pi : S(\mathcal{H}^{\mathrm{HS}}) \longrightarrow \mathcal{P} ,$$

---

[12] In quantum information theory (see, e.g., Nielsen and Chuang 2000) the procedure of extension, $\mathcal{H} \longrightarrow \mathcal{H}^{\mathrm{ext}}$, is known as attaching an ancilla living in $\mathcal{H}'$.

[13] $U(\mathcal{H})$ denotes the group of unitary operators on $\mathcal{H}$. Clearly, if $\mathcal{H} = \mathbb{C}^n$, then $U(\mathcal{H}) \equiv U(n)$.

defined by

$$\pi(W) := WW^* .\tag{5.194}$$

Note, however, that the above map does not define a fibre bundle projection. This is because the different "fibres" may be completely different (i.e., they may not be the same space). This deficiency may be easily cured. Let us define the following subspace of $\mathcal{H}^{\mathrm{HS}}$:

$$\widetilde{\mathcal{H}}^{\mathrm{HS}} := \left\{ W \in \mathcal{H}^{\mathrm{HS}} \,\middle|\, \mathrm{Ker}\, W = \{0\} \right\} \subset \mathcal{H}^{\mathrm{HS}} ,\tag{5.195}$$

where $\mathrm{Ker}\, W$ stands for a kernel of W. Then the image

$$\pi(S(\widetilde{\mathcal{H}}^{\mathrm{HS}})) = \left\{ \rho \in \mathcal{P} \,\middle|\, \rho > 0 \right\} =: \mathcal{P}^+\tag{5.196}$$

contains only strictly positive (or faithful) density operators. Note that, for instance, the set of pure states $\mathcal{P}(\mathcal{H})$ is excluded from $\mathcal{P}^+$. Nevertheless, it turns out that $\widetilde{\mathcal{H}}^{\mathrm{HS}}$ is dense in $\mathcal{H}^{\mathrm{HS}}$ in the $||\cdot||_{\mathrm{HS}}$-topology. This procedure leads to the well-defined principal $U(\mathcal{H})$-fibre bundle

$$\pi \;:\; S(\widetilde{\mathcal{H}}^{\mathrm{HS}}) \;\longrightarrow\; \mathcal{P}^+ .\tag{5.197}$$

The final step is to define a connection in $S(\widetilde{\mathcal{H}}^{\mathrm{HS}}) \longrightarrow \mathcal{P}^+$. It turns out that this bundle possesses a natural connection, in perfect analogy to the $U(1)$-bundle $S(\mathcal{H}) \longrightarrow \mathcal{P}(\mathcal{H})$. Let X be a tangent vector to a bundle space at a point W. Identifying each tangent space $T_W S(\widetilde{\mathcal{H}}^{\mathrm{HS}})$ with a subspace of $\mathcal{H}^{\mathrm{HS}}$, the formula (5.194) implies

$$(T_W \pi)(X) = WX^* + XW^* .\tag{5.198}$$

Hence, X is a vertical vector at W if

$$WX^* + XW^* = 0 .\tag{5.199}$$

The space of vertical vectors at the point W is isomorphic to the Lie algebra $u(\mathcal{H})$ of the unitary group $U(\mathcal{H})$. The Lie algebra $u(\mathcal{H})$ consists of antihermitian operators in $\mathcal{H}$. Note that any vertical vector at W is of the form $X = WS$, with $S \in u(\mathcal{H})$. Indeed,

$$WX^* + XW^* = W(WS)^* + WSW^* = WS^*W^* + WSW^* = 0 ,\tag{5.200}$$

since $S^* = -S$. A *natural connection* is defined as follows: A tangent vector X at W is horizontal if it is orthogonal to the fibre passing through W, i.e., if

$$\langle X, Y \rangle_{\mathrm{HS}} = 0 ,\tag{5.201}$$

for all vertical vectors Y at W. Hence, X is horizontal if and only if

$$\langle X, WS \rangle_{\mathrm{HS}} = \mathrm{Tr}(X^*WS) = 0 ,\tag{5.202}$$

for all $S \in u(\mathcal{H})$. However, if $\mathrm{Tr}\,(AS) = 0$ for all antihermitian operators S, then A must necessarily be Hermitian. Hence $(X^*W)^* = X^*W$, or, equivalently,

$$X^*W - W^*X = 0 . \tag{5.203}$$

In summary, the vertical and horizontal subspaces V_W and H_W are given by

$$V_W = \left\{ X \in T_W \widetilde{H}^{\mathrm{HS}} \,\middle|\, X^*W + W^*X = 0 \right\} , \tag{5.204}$$

and

$$H_W = \left\{ X \in T_W \widetilde{H}^{\mathrm{HS}} \,\middle|\, X^*W - W^*X = 0 \right\} . \tag{5.205}$$

Now, consider a curve

$$t \longrightarrow \rho(t) \in \mathcal{P}^+ ,$$

and let

$$t \longrightarrow W(t) \in S(\widetilde{H}^{\mathrm{HS}})$$

be a lift of $\rho(t)$. We call a lift (or equivalently a purification) $W(t)$ a horizontal lift (horizontal purification), if $\dot{W}$ is a horizontal vector, that is, if

$$\dot{W}^*W = W^*\dot{W} . \tag{5.206}$$

Clearly, the above condition generalizes the law of parallel transport in the $U(1)$-bundle $S(\mathcal{H}) \longrightarrow \mathcal{P}(\mathcal{H})$, that is,

$$\mathrm{Im}\,\langle \psi | \dot{\psi} \rangle = 0 ,$$

which is equivalent to

$$\langle \dot{\psi} | \psi \rangle = \langle \psi | \dot{\psi} \rangle .$$

Let us note that equation (5.206) may be solved by the following ansatz:

$$\dot{W} = GW , \tag{5.207}$$

with hermitian G. This leads to the following equation for the density matrix $\rho = WW^*$:

$$\dot{\rho} = [G, \rho]_+ = G\rho + \rho G . \tag{5.208}$$

Example 5.4.2 Following Uhlmann 1989, consider the law of parallel transport (5.206) in the case of a two-level quantum system. Any density matrix in $\mathbb{C}^2$ reads as follows:

$$\rho = \frac{1}{2}(\mathbb{1}_2 + \mathbf{x} \cdot \boldsymbol{\sigma}) , \quad \mathbf{x} \in \mathbb{R}^3 .$$

Since G is hermitian, it may be uniquely represented by

$$G = y_0 \, 1\!1_2 + \mathbf{y} \cdot \boldsymbol{\sigma} , \qquad y_0 \in \mathbb{R} , \; \mathbf{y} \in \mathbb{R}^3 .$$

One easily finds that

$$G\rho + \rho G = (y_0 + \mathbf{x} \cdot \mathbf{y}) \, 1\!1_2 + \mathbf{y} \cdot \boldsymbol{\sigma} ,$$

and, hence, equation (5.208) implies that

$$\dot{y}_0 = -\mathbf{x} \cdot \mathbf{y} , \quad \text{and} \quad \dot{\mathbf{x}} = 2(y_0 \mathbf{x} + \mathbf{y}) .$$

Now,

$$\dot{\mathbf{x}} \cdot \mathbf{x} = 2(y_0 x^2 + \mathbf{x} \cdot \mathbf{y}) = 2 y_0 (x^2 - 1) ,$$

from which it follows that

$$y_0 = \frac{1}{2} \frac{\dot{\mathbf{x}} \cdot \mathbf{x}}{x^2 - 1} = \frac{1}{4} \frac{d}{dt} (\ln \Delta) ,$$

where $\Delta = \frac{1}{4}(1 - x^2) = \det \rho$. Having determined y_0, one finds the following expression for $\mathbf{y}$:

$$\mathbf{y} = \frac{1}{2} \dot{\mathbf{x}} - y_0 \mathbf{x} = \frac{1}{2} \Delta^{\frac{1}{2}} \frac{d}{dt} \left(\mathbf{x} \Delta^{-\frac{1}{2}} \right) .$$

In particular, if $\rho(t) = U(t)\rho U^*(t)$, then $\Delta(t) = \text{const.}$ and hence $G = \frac{1}{2} \dot{\mathbf{x}} \cdot \boldsymbol{\sigma}$. In this case equation (5.207) may be rewritten as follows: If $W = w_0 \, 1\!1_2 + \mathbf{w} \cdot \boldsymbol{\sigma}$, then

$$\dot{w}_0 = \frac{1}{2} \dot{\mathbf{x}} \cdot \mathbf{w} , \quad \text{and} \quad \dot{\mathbf{w}} = \frac{1}{2} (i \dot{\mathbf{x}} \times \mathbf{w} + w_0 \dot{\mathbf{x}}) .$$

Note that the above system has the following first integral:

$$\det W = w_0^2 - \mathbf{w}^2 .$$

Thus $\det \rho$ is conserved during the unitary evolution of $\rho(t)$, whereas $\det W$ remains constant along the horizontal lift of $\rho(t)$, that is, during parallel transport of $W(t)$. $\diamond$

Now let us define the corresponding connection one-form taking values in $u(\mathcal{H})$. Following the abelian case $S(\mathcal{H}) \longrightarrow P(\mathcal{H})$ where one has

$$\mathcal{A} = \text{Im} \, \langle \psi | d\psi \rangle \propto \langle \psi | d\psi \rangle - \langle d\psi | \psi \rangle , \tag{5.209}$$

the obvious guess for $\mathcal{A}$ would be

$$\mathcal{A} \propto W^* dW - dW^* W .$$

Clearly, this form is antihermitian and vanishes on horizontal vectors. However, it does not transform properly under the gauge transformation

$$W \longrightarrow WU , \quad \text{with } U \in U(\mathcal{H}) .$$

Following Uhlmann (Uhlmann 1991) let us instead define the one-form $\mathcal{A}$ as follows:

$$W^* dW - dW^* W =: W^* W \cdot \mathcal{A} + \mathcal{A} \cdot W^* W . \tag{5.210}$$

Obviously, this $\mathcal{A}$ also vanishes on horizontal vectors and is antihermitian: $\mathcal{A}^* = -\mathcal{A}$. Moreover, it does obey the transformation law of a connection one-form. Indeed, let $W' = WU$. One has

$$dW' = (dW)U + WdU , \quad \text{and} \quad dW'^* = (dU^*)W^* + U^* dW^* , \tag{5.211}$$

and, hence,

$$\begin{aligned}
W'^* dW' &- dW'^* W' \\
&= U^* \Big(W^* dW - (dW^*)W \Big) U + \Big(U^* W^* W dU - (dU^*)W^* W U \Big) \\
&= U^* \Big(W^* W \cdot \mathcal{A} + \mathcal{A} \cdot W^* W \Big) U \\
&\quad + \Big(U^* W^* W dU - (dU^*)W^* W U \Big) .
\end{aligned} \tag{5.212}$$

Using the definition of $\mathcal{A}$, this expression can also be written as

$$W'^* W' \cdot \mathcal{A}' + \mathcal{A}' \cdot W'^* W' = U^* W^* W U \cdot \mathcal{A}' + \mathcal{A}' \cdot U^* W^* W U . \tag{5.213}$$

Then, since $dU = -U(dU)^* U$, by equating the r.h.s.'s of equations (5.210) and (5.211), one obtains

$$\begin{aligned}
U^* W^* W U &\cdot \mathcal{A}' + \mathcal{A}' \cdot U^* W^* W U \\
&= U^* W^* W (\mathcal{A} - U dU^*) U + U^* (\mathcal{A} - U dU^*) W^* W U .
\end{aligned} \tag{5.214}$$

Hence

$$\mathcal{A}' = U^* \mathcal{A} U - (dU^*)U = U^* \mathcal{A} U + U^* dU , \tag{5.215}$$

which is the transformation law for a (nonabelian) connection form (cf. section 1.3.4).

Having defined a connection it is clear how to define the corresponding geometric phase factor for mixed states; it is a holonomy of the natural connection $\mathcal{A}$, i.e., if C is a closed curve in $\mathcal{P}^+$, then

$$C \longrightarrow \Phi(C) = P \exp \left(\oint_C A \right) \tag{5.216}$$

is a (nonabelian) geometric phase factor corresponding to the cyclic evolution of a mixed state in $\mathcal{P}^+$. In the above formula, A denotes a local connection form on $\mathcal{P}^+$, i.e., a pull-back of $\mathcal{A}$ with respect to some local section of the bundle. Clearly, $\Phi(C) \in U(\mathcal{H})$ for any loop C in $\mathcal{P}^+$.

5.4.3 Distance between mixed states

Recall that the space of pure states $\mathcal{P}(\mathcal{H})$ is endowed with a Fubini–Study metric, which enables one to define a distance between states. Actually, as we have shown, the Fubini–Study distance is a measure of the transition probability between corresponding (pure) quantum states. One may now ask how to measure the distance between mixed states, and, in fact, there are several possible ways. The simplest way is to define the trace distance, according to

$$D_{\mathrm{Tr}}(\rho_1, \rho_2) := \mathrm{Tr}|\rho_1 - \rho_2| \,, \tag{5.217}$$

where we define $|A| := \sqrt{A^*A}$ to be the positive square root of A^*A.

Example 5.4.3 (Trace distance for qubit states) Consider the following two qubit states:

$$\rho_k = \frac{1}{2}(\mathbb{1}_2 + \mathbf{x}_k \cdot \boldsymbol{\sigma}) \,, \qquad k = 1, 2 \,. \tag{5.218}$$

One has

$$\begin{aligned}
D_{\mathrm{Tr}}(\rho_1, \rho_2) &= \mathrm{Tr}|\rho_1 - \rho_2| = \frac{1}{2}\mathrm{Tr}|(\mathbf{x}_1 - \mathbf{x}_2) \cdot \boldsymbol{\sigma}| \\
&= \frac{1}{2}\,\mathrm{Tr}\sqrt{[(\mathbf{x}_1 - \mathbf{x}_2) \cdot \boldsymbol{\sigma}]^2} = |\mathbf{x}_1 - \mathbf{x}_2| \,,
\end{aligned}$$

since $(\mathbf{x} \cdot \boldsymbol{\sigma})^2 = |\mathbf{x}|^2 \mathbb{1}_2$. Hence, the trace distance for qubit states reproduces the euclidean distance in the Bloch ball. $\diamond$

Another possibility is to use the Hilbert–Schmidt distance, defined by

$$D_{\mathrm{HS}}(\rho_1, \rho_2) = ||\rho_1 - \rho_2||_{\mathrm{HS}} = \sqrt{\mathrm{Tr}(\rho_1 - \rho_2)^2} \,. \tag{5.219}$$

Actually, for the qubit states (5.218) one finds

$$D_{\mathrm{HS}}(\rho_1, \rho_2) = \frac{1}{\sqrt{2}}\,|\mathbf{x}_1 - \mathbf{x}_2| \,. \tag{5.220}$$

Both the trace and Hilbert–Schmidt distances were often used in quantum optics. Another approach to measure the distance between quantum states is based on the idea of purification discussed in the previous section. One introduces the so-called Bures distance (Bures 1969) which measures the transition probability for a pair of density operators. The definition of the Bures distance is perfectly analogous to the Fubini–Study one. For any pair $\rho_1, \rho_2 \in \mathcal{P}$, one defines

$$\begin{aligned}
D_{\mathrm{B}}^2(\rho_1, \rho_2) &:= \inf_{W_1, W_2} ||W_1 - W_2||_{\mathrm{HS}}^2 \\
&= \inf_{W_1, W_2} \mathrm{Tr}\left[(W_1^* - W_2^*)(W_1 - W_2)\right] \tag{5.221} \\
&= 2 - \sup_{W_1, W_2} \mathrm{Tr}\,(W_1^* W_2 + W_2^* W_1) \,,
\end{aligned}$$

with $W_1 \in \pi^{-1}(\rho_1)$, and $W_2 \in \pi^{-1}(\rho_2)$.[14]

It is clear that the trace of $W_1^* W_2 + (W_1^* W_2)^*$ is maximized if $W_1^* W_2$ is a positive, and hence hermitian, operator, i.e.,

$$W_1^* W_2 = W_2^* W_1 > 0 . \tag{5.222}$$

This condition is called the *Uhlmann parallelity condition* (Uhlmann 1986). We call two Hilbert–Schmidt operators W_1 and W_2 *parallel* if the above condition holds. Let us choose $W_1 = \sqrt{\rho_1}$. Using polar decomposition W_2 may be represented as follows:

$$W_2 = \sqrt{\rho_2}\, U ,$$

with $U \in U(\mathcal{H})$. Now, the parallelity condition implies the following condition for the unitary operator U:

$$\sqrt{\rho_1}\sqrt{\rho_2}\, U = U^* \sqrt{\rho_2}\sqrt{\rho_1} ,$$

and hence

$$W_1^* W_2 + W_2^* W_1 = 2 W_1^* W_2 = 2\sqrt{W_1^* W_2 W_2^* W_1} = 2\sqrt{\rho_1^{1/2} \rho_2\, \rho_1^{1/2}} ,$$

which leads to the following formula for the Bures distance:

$$D_B^2(\rho_1, \rho_2) = 2(1 - F(\rho_1, \rho_2)), \tag{5.223}$$

where the quantity

$$F(\rho_1, \rho_2) := \mathrm{Tr}\sqrt{\rho_1^{1/2} \rho_2\, \rho_1^{1/2}} \tag{5.224}$$

is called the *fidelity* of states ρ_1 and ρ_2.[15]

Note, that if ρ_1 and ρ_2 are both one-dimensional projectors in $\mathcal{H}$, i.e., $\rho_1 = |\psi_1\rangle\langle\psi_1|$ and $\rho_2 = |\psi_2\rangle\langle\psi_2|$, then

$$\sqrt{\rho_1} = \rho_1 \quad \text{and} \quad \sqrt{\rho_2} = \rho_2 .$$

One has therefore

$$
\begin{aligned}
F(\rho_1, \rho_2) &= \mathrm{Tr}\sqrt{\rho_1^{1/2} \rho_2\, \rho_1^{1/2}} = \mathrm{Tr}\sqrt{|\psi_1\rangle\langle\psi_1|\psi_2\rangle\langle\psi_2|\psi_1\rangle\langle\psi_1|} \\
&= \mathrm{Tr}\sqrt{|\langle\psi_1|\psi_2\rangle|^2 |\psi_1\rangle\langle\psi_1|} = |\langle\psi_1|\psi_2\rangle|\, \mathrm{Tr}\sqrt{|\psi_1\rangle\langle\psi_1|} = |\langle\psi_1|\psi_2\rangle| ,
\end{aligned}
$$

and hence

$$D_B^2(\rho_1, \rho_2) = D_{FS}^2(\rho_1, \rho_2) , \tag{5.225}$$

---

[14]This is not the original definition of Bures (Bures 1969) but a special version proposed by Uhlmann (Uhlmann 1976, 1986). See also Alberti and Uhlmann 2000 for a recent review; Uhlmann 2000; and Alberti 2003.

[15]This notion was introduced by Jozsa (Jozsa 1994).

that is, when restricted to the space of pure states, i.e., projective Hilbert space, the Bures distance reproduces the Fubini–Study one. It turns out that the Bures distance, contrary e.g., to the trace distance, defines the riemannian metric on the space of density matrices $\mathcal{P}$ — we shall call it the *Bures metric*.

Example 5.4.4 (Bures metric for qubit states) Following Hübner (Hübner 1992, see also Hübner 1993 and Uhlmann 1996), we derive the Bures metric for qubit states. Consider two states ρ_1 and ρ_2, defined as in (5.218). Clearly, $\mathbf{x}_k$ belongs to the Bloch ball and hence $|\mathbf{x}_k| \leq 1$. Let us define

$$M := \sqrt{\rho_1}\, \rho_2 \sqrt{\rho_1} \,. \tag{5.226}$$

To compute the fidelity $F(\rho_1, \rho_2) = \mathrm{Tr}\sqrt{M}$, let us note that M is a positive 2×2 matrix and hence it has two non-negative eigenvalues m_1 and m_2. Thus

$$\mathrm{Tr}\sqrt{M} = \sqrt{m_1} + \sqrt{m_2} \,. \tag{5.227}$$

The above equation implies that

$$(\mathrm{Tr}\sqrt{M})^2 = m_1 + m_2 + 2\sqrt{m_1 m_2} = \mathrm{Tr}\, M + 2\sqrt{\det M} \,, \tag{5.228}$$

so to find the fidelity $F(\rho_1, \rho_2)$ one has to compute $\mathrm{Tr}\, M$ and $\det M$. One easily finds

$$
\begin{aligned}
\mathrm{Tr}\, M &= \mathrm{Tr}\,(\sqrt{\rho_1}\, \rho_2 \sqrt{\rho_1}) = \mathrm{Tr}\,(\rho_1 \rho_2) = \frac{1}{4}\mathrm{Tr}\left[(\mathbb{1}_2 + \mathbf{x}_1 \cdot \boldsymbol{\sigma})(\mathbb{1}_2 + \mathbf{x}_2 \cdot \boldsymbol{\sigma})\right] \\
&= \frac{1}{4}\mathrm{Tr}\left[(1 + \mathbf{x}_1 \cdot \mathbf{x}_2)\mathbb{1}_2\right] = \frac{1}{2}(1 + \mathbf{x}_1 \cdot \mathbf{x}_2) \,,
\end{aligned}
\tag{5.229}
$$

and

$$\det M = \det(\sqrt{\rho_1}\, \rho_2 \sqrt{\rho_1}) = (\det \rho_1)(\det \rho_2) = \frac{1}{16}(1 - |\mathbf{x}_1|^2)(1 - |\mathbf{x}_2|^2) \,, \tag{5.230}$$

where we have used

$$\det\left[\frac{1}{2}(\mathbb{1}_2 + \mathbf{x} \cdot \boldsymbol{\sigma})\right] = \frac{1}{4}(1 - |\mathbf{x}|^2) \,.$$

Hence, one has for the fidelity

$$
\begin{aligned}
F(\rho_1, \rho_2) &= \mathrm{Tr}\sqrt{M} = \left(\mathrm{Tr}\, M + 2\sqrt{\det M}\right)^{1/2} \\
&= \frac{1}{\sqrt{2}}\left(1 + \mathbf{x}_1 \cdot \mathbf{x}_2 + \sqrt{(1 - |\mathbf{x}_1|^2)(1 - |\mathbf{x}_2|^2)}\right)^{1/2} \,, \tag{5.231}
\end{aligned}
$$

from which one can easily find the corresponding formula for the Bures distance $D_{\mathrm{B}}^2(\rho_1, \rho_2) = 2(1 - F(\rho_1, \rho_2))$. In particular, by restricting the Bures distance to the Bloch sphere, i.e., pure states satisfying $|\mathbf{x}| = 1$, one obtains

$$D_{\mathrm{B}}^2(\rho_1, \rho_2) \equiv D_{\mathrm{FS}}^2(\rho_1, \rho_2) = 2\left(1 - \frac{1}{2}(1 + \mathbf{x}_1 \cdot \mathbf{x}_2)\right) \,, \tag{5.232}$$

and, hence, if $\mathbf{x}_1 \cdot \mathbf{x}_2 = \cos \Theta$, then the geodesic distance $\sigma(\mathbf{x}_1, \mathbf{x}_2)$ between $\mathbf{x}_1$ and $\mathbf{x}_2$ is given by

$$\cos[\sigma(\mathbf{x}_1, \mathbf{x}_2)] = \frac{1}{2}(1 + \cos \Theta) . \tag{5.233}$$

Now let us look for the metric tensor $ds_{\mathrm{B}}^2 = g_{ij}^{\mathrm{B}} dx^i \, dx^j$, defined by

$$D_{\mathrm{B}}(\rho, \rho + d\rho) = ds_{\mathrm{B}} .$$

Using $\rho = \frac{1}{2}(\mathbb{1}_2 + \mathbf{x} \cdot \boldsymbol{\sigma})$ and $d\rho = d\mathbf{x} \cdot \boldsymbol{\sigma}$ one finds

$$D_{\mathrm{B}}^2(\rho, \rho + d\rho) = 2 - \sqrt{2} \left(1 + |\mathbf{x}|^2 + \mathbf{x} \cdot d\mathbf{x} + \sqrt{(1 - |\mathbf{x}|^2)(1 - |\mathbf{x} + d\mathbf{x}|^2)} \right)^{1/2} ,$$
$$\tag{5.234}$$

and expanding to second order finally gives

$$\begin{aligned}
ds_{\mathrm{B}}^2 &= \frac{1}{4} \left((d\mathbf{x})^2 + \frac{(\mathbf{x} \cdot d\mathbf{x})^2}{1 - |\mathbf{x}|^2} \right) \\
&= \frac{1}{4} \left(dx_1^2 + dx_2^2 + dx_3^2 + \frac{(x_1 dx_1 + x_2 dx_2 + x_3 dx_3)^2}{1 - x_1^2 - x_2^2 - x_3^2} \right) .
\end{aligned} \tag{5.235}$$

Hence the Bures metric tensor reads

$$g_{ij}^{\mathrm{B}} = \frac{1}{4} \left(\delta_{ij} + \frac{x_i x_j}{1 - r^2} \right) , \tag{5.236}$$

with $r = |\mathbf{x}|$. Clearly, it defines a non-euclidean metric. Note, however, that introducing a new variable x_4 via

$$x_4^2 := 4 \det \rho = 1 - |\mathbf{x}|^2 ,$$

one obtains

$$ds_{\mathrm{B}}^2 = \frac{1}{4} \left(x_1^2 + x_2^2 + x_3^2 + x_4^2 \right) , \tag{5.237}$$

which shows that the Bloch ball can be isometrically embedded into a hemisphere of S^3, of radius $\frac{1}{2}$, defined by $x_4 \geq 0$ (cf. Fig. 5.5). Clearly, the point $(0, 0, 0, 1)$ corresponds to the maximally mixed state $\rho_0 = \frac{1}{2}\mathbb{1}_2$. $\diamond$

Now, following section 5.3.1, we may reproduce analogous results for the space of mixed states $\mathcal{P}$. Let

$$[0, 1] \ni t \longrightarrow \rho(t) \in \mathcal{P} ,$$

be a curve in $\mathcal{P}$ and consider one of its purifications, e.g.,

$$[0, 1] \ni t \longrightarrow W(t) \in \mathcal{H}^{\mathrm{HS}} .$$

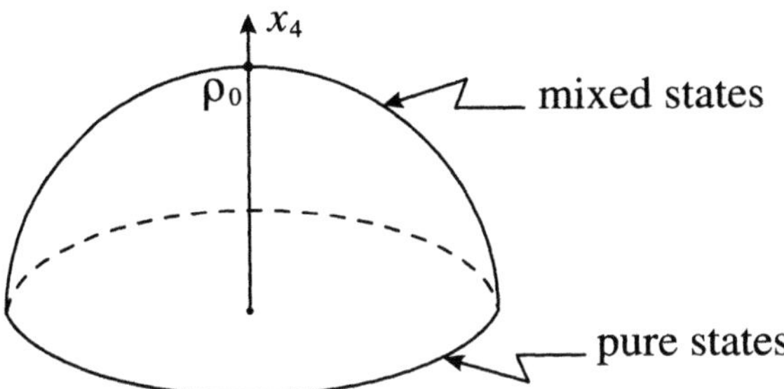

Figure 5.5: Hemisphere of S^3 representing qubit mixed states. Pure states define its boundary, i.e., the equatorial sphere S^2. The maximally mixed state ρ_0 is situated at the "north pole."

Clearly, a length of $W(t)$ is given by

$$L(W) = \int_0^1 \sqrt{\langle \dot{W}, \dot{W} \rangle_{\mathrm{HS}}}\, dt \ . \qquad (5.238)$$

As in the case of pure states, one can show that $L(W)$ is minimal if $W = W(t)$ is a horizontal purification, i.e., the condition (5.206) holds. We define the length of $\rho = \rho(t)$ as the length of its horizontal purification. Take two points $\rho_1, \rho_2 \in \mathcal{P}$ and consider a geodesic (with respect to Bures metric) connecting ρ_1 and ρ_2. Denote by $\sigma(\rho_1, \rho_2)$ the length of the shortest geodesic connecting ρ_1 and ρ_2. Using arguments similar to those used in the proof of Theorem 5.3.2, we arrive at the following

Proposition 5.4.1 *The length of the shortest geodesic connecting ρ_1 and ρ_2 in $\mathcal{P}$ is given by*

$$\cos\left[\sigma(\rho_1, \rho_2)\right] = F(\rho_1, \rho_2) \ .$$

Now, fix an arbitrary state ρ_0 and define the transition probability

$$\delta_{\rho_0} : \mathcal{P} \longrightarrow \mathbb{R}$$

by

$$\delta_{\rho_0}(\rho) = F(\rho_0, \rho) \ .$$

The above proposition gives us the geometric interpretation of the square of the fidelity $F^2(\rho, \sigma)$ as a transition probability between ρ and σ, by relating it to the geodesic distance between these states.

5.4.4   How to measure geometric phase

It is not completely clear what is a physical interpretation of the Uhlmann geometric phase for density matrices.[16] Recently, some authors (see Sjöqvist et al. 2000) proposed a different approach to geometric phases for mixed states. This new approach is nicely related to an experimental context of quantum interferometry. Recall, that for pure states one may define a relative (Pancharatnam) phase between state vectors $|A\rangle$ and $|B\rangle$, as follows:

$$\text{Pancharatnam phase } = \arg\langle A|B\rangle . \qquad (5.239)$$

The idea of relative phase is well suited for interferometric experiments. Shifting the phase of $|A\rangle$ by χ: $|A\rangle \longrightarrow e^{i\chi}|A\rangle$ results in the following change in the intensity I of $|A\rangle + |B\rangle$

$$I = \left| e^{i\chi}|A\rangle + |B\rangle \right|^2 = 2 + 2|\langle A|B\rangle| \cos[\chi - \arg\langle A|B\rangle] , \qquad (5.240)$$

which attains its maximum at the Pancharatnam relative phase $\varphi = \arg\langle A|B\rangle$.

This observation may be generalized to the case of mixed state. Suppose that the mixed state described by a density matrix ρ_0 undergoes unitary evolution, such that $\rho(t) = U(t)\rho_0 U^*(t)$. Let

$$\rho_0 = \sum_k w_k |k\rangle\langle k| , \qquad \rho(t) = \sum_k w_k |k(t)\rangle\langle k(t)| , \qquad (5.241)$$

be spectral decompositions of ρ_0 and $\rho(t)$, respectively. Clearly, $|k(t)\rangle = U(t)|k\rangle$. Now perform a phase shift $|k\rangle \longrightarrow e^{i\chi}|k\rangle$ and define the total interference profile

$$\begin{aligned}
I &= \sum_k I_k = \sum_k w_k \left| e^{i\chi}|k\rangle + |k(t)\rangle \right|^2 \\
&= 2 + 2\sum_k w_k |\langle k|k(t)\rangle| \cos[\chi - \arg\langle k|k(t)\rangle] ,
\end{aligned} \qquad (5.242)$$

where we have used $\sum_k w_k = 1$, that is, the total profile I is an incoherent average of the individual pure states' interference profiles I_k, defined in (5.240). Interestingly, the above formula may be rewritten in a perfect analogy to (5.240). Denote by φ_k the relative phase between $|k\rangle$ and $|k(t)\rangle$, i.e., $\varphi_k = \arg\langle k|k(t)\rangle = \arg\langle k|U(t)|k\rangle$, and define the so-called visibility $v_k := |\langle k|k(t)\rangle| = |\langle k|U(t)|k\rangle|$. One obtains

$$\begin{aligned}
I &= 2 + 2\sum_k w_k v_k \cos(\chi - \varphi_k) \\
&= 2 + 2\left[\cos\chi \left(\sum_k w_k v_k \cos\varphi_k \right) + \sin\chi \left(\sum_k w_k v_k \sin\varphi_k \right) \right] , \qquad (5.243)
\end{aligned}$$

---

[16]Some suggestions were proposed in Dąbrowski and Jadczyk 1989 and Dąbrowski and Grosse 1990.

and hence, defining

$$\varphi := \arg\left(\sum_k w_k v_k e^{i\varphi_k}\right),$$ (5.244)

and

$$v := \left|\sum_k w_k v_k e^{i\varphi_k}\right|,$$ (5.245)

the intensity formula may be rewritten as

$$I = 2 + 2v\cos(\chi - \varphi).$$ (5.246)

Let us observe that

$$\varphi = \arg\left[\mathrm{Tr}\,(U(t)\rho_0)\right], \qquad v = |\mathrm{Tr}\,(U(t)\rho_0)|,$$ (5.247)

that is

$$\mathrm{Tr}\,(U(t)\rho_0) = v e^{i\varphi}.$$ (5.248)

Following Sjöqvist et al. 2000, one says that $\rho(t)$ acquires a phase with respect to ρ_0 if $\arg\,[\mathrm{Tr}\,(U(t)\rho_0)]$ is nonvanishing. In this way one may define a parallel transport of a mixed state ρ: ρ is parallel transported if $\rho(t)$ is in phase with $\rho(t + dt)$. One has

$$\rho(t + dt) = U(t + dt)\rho_0 U^*(t + dt) = U(t + dt)U^*(t)\rho(t)U(t)U^*(t + dt),$$

and therefore the phase difference between $\rho(t)$ and $\rho(t + dt)$ is given by

$$d\varphi = \arg\left\{\mathrm{Tr}\left[\rho(t)U(t + dt)U^*(t)\right]\right\},$$ (5.249)

one may say that $\rho(t)$ and $\rho(t + dt)$ are in phase if

$$\mathrm{Tr}\left[\rho(t)U(t + dt)U^*(t)\right] \quad \text{is real and positive}.$$

Now, due to the normalization condition $\mathrm{Tr}\,\rho(t) = 1$ one has

$$\begin{aligned}
\mathrm{Tr}\left[\rho(t)U(t + dt)U^*(t)\right] &= \mathrm{Tr}\left[\rho(t)\left(U(t) + \dot{U}(t)dt\right)U^*(t)\right] \\
&= 1 + \mathrm{Tr}\left[\rho(t)\dot{U}(t)U^*(t)\right]dt,
\end{aligned}$$

and the hermiticity of $\rho(t)$ implies that the quantity $\mathrm{Tr}\,[\rho(t)\dot{U}(t)U^*(t)]$ is purely imaginary. Therefore, the parallel transport condition for the mixed state ρ undergoing the unitary evolution may be stated as follows:

$$\mathrm{Tr}\left[\rho(t)\dot{U}(t)U^*(t)\right] = 0.$$ (5.250)

On the space of density matrices the above condition may be rewritten as

$$\mathrm{Tr}\left[\rho\,(d_{\mathrm{U}}U)\,U^*\right] = 0\,,\tag{5.251}$$

where d_{U} denotes the exterior derivative on the space of density matrices. Note that $\rho(t)$ determines the unitary $N \times N$ matrix $U(t)$, up to N phase factors.[17] These phase factors may be uniquely fixed by the parallel condition

$$\langle k(t)|\dot{k}(t)\rangle = 0\,,\qquad k = 1,\dots,N\,.\tag{5.252}$$

The geometric phase introduced by Sjöqvist et al. (2000) fulfills two basic properties:

1. If ρ_0 is a pure state $\rho_0 = |\psi_0\rangle\langle\psi_0|$, then the formula (5.247) reproduces the relative Pancharatnam phase between $|\psi_0\rangle$ and $|\psi(t)\rangle = U(t)|\psi_0\rangle$;

2. If $\rho(t) = |\psi(t)\rangle\langle\psi(t)|$ then the parallel transport condition (5.250) reduces to corresponding condition for pure states $\langle\psi(t)|\dot{\psi}(t)\rangle = 0$.

The above result may be easily reproduced using the purification procedure (see the derivation of the Uhlmann phase). Actually, any mixed state ρ_0 can be obtained by tracing out some degrees of freedom of the larger system, which was in a pure state

$$|\Psi_0\rangle = \sum_k \sqrt{w_k}\,|k\rangle|k\rangle_{\mathrm{a}}\,,\tag{5.253}$$

where $|k\rangle_{\mathrm{a}}$ belong to the Hilbert space of the auxiliary system (ancilla). Then

$$\rho_0 = \mathrm{Tr}_{\mathrm{a}}|\Psi_0\rangle\langle\Psi_0|\,,\tag{5.254}$$

where Tr_{a} denotes the partial trace over the ancilla Hilbert space. Now, evolving $|\Psi_0\rangle$ according to a local unitary operation $\mathcal{U}(t) = U(t)\otimes\mathbb{1}_{\mathrm{a}}$, one finds

$$|\Psi(t)\rangle = \mathcal{U}(t)|\Psi_0\rangle = \sum_k \sqrt{w_k}\,|k(t)\rangle|k\rangle_{\mathrm{a}}\,,\tag{5.255}$$

and, hence,

$$\langle\Psi_0|\Psi(t)\rangle = \sum_k w_k\langle k|k(t)\rangle = \mathrm{Tr}\,(U(t)\rho_0)\,,\tag{5.256}$$

which shows that $\varphi = \arg[\mathrm{Tr}\,(U(t)\rho_0)]$ is the relative phase between the purifications $|\Psi_0\rangle$ and $|\Psi(t)\rangle$.

Example 5.4.5 (Qubit mixed states) Consider a mixed state of a qubit, i.e., a two-level quantum system (see Example 5.4.1). Any density matrix can be written as

$$\rho = \frac{1}{2}(\mathbb{1}_2 + r\,\mathbf{n}\cdot\boldsymbol{\sigma})\,,\tag{5.257}$$

---

[17] N is the dimension of the corresponding Hilbert space.

where $\mathbf{n} = \mathbf{r}/r$ and $r \leq 1$. Suppose that during a unitary evolution of $\rho(t)$, $\mathbf{n}(t)$ traces out a geodesically closed curve C on the Bloch sphere, i.e., $C = C_1 * C_2$ with C_1 an open curve and C_2 a geodesic segment, cf. Remark 5.3.2. Let Ω be the solid angle subtended by C. Recall that the two eigenvectors of ρ, namely,

$$\rho \, |\pm; \mathbf{n} \cdot \boldsymbol{\sigma}\rangle = \frac{1 \pm r}{2} \, |\pm; \mathbf{n} \cdot \boldsymbol{\sigma}\rangle \,,$$

acquire Pancharatnam phases $\varphi_{\pm} = \mp\frac{1}{2}\Omega$ and have identical visibilities $v_+ = v_- =: \eta$. Therefore, the Pancharatnam relative phase corresponding to the evolution of ρ is given by formula (5.244), as follows:

$$
\begin{aligned}
\varphi &= \arg\left(w_+ v_+ e^{-i\Omega/2} + w_- v_- e^{i\Omega/2} \right) \\
&= \arg\left\{ \frac{\eta}{2}\left[\left(e^{-i\Omega/2} + e^{i\Omega/2} \right) + r\left(e^{-i\Omega/2} - e^{i\Omega/2} \right) \right] \right\} \\
&= \arg\left[\eta\left(\cos\left(\frac{\Omega}{2}\right) - i\sin\left(\frac{\Omega}{2}\right) \right) \right] = -\arctan\left(r\tan\left(\frac{\Omega}{2}\right) \right) , \quad (5.258)
\end{aligned}
$$

where we have used $w_\pm = \frac{1}{2}(1 \pm r)$. Moreover, according to (5.245) one easily finds for the visibility

$$v = \eta\sqrt{\cos^2\left(\frac{\Omega}{2}\right) + r^2\sin^2\left(\frac{\Omega}{2}\right)}. \qquad (5.259)$$

Observe that, for a cyclic evolution, $\eta = 1$ but the visibility $v < 1$. One has $v = 1$ only for pure states, corresponding to $r = 1$. Clearly, for pure states the formula (5.258) reproduces the geometric phase $\Omega/2$. However, for mixed states the solid angle formula is no longer valid. $\qquad\qquad\qquad\qquad\qquad\qquad\qquad\qquad\qquad\diamond$

A natural question is, what is the relation between the mathematical formulation of Uhlmann and the more "experimental" approach of Sjöqvist et al? Note that if $W(t)$ defines a parallel purification in the Hilbert–Schmidt space $\mathcal{H}^{\mathrm{HS}}$, i.e., the condition (5.206) is satisfied, then one may define the Uhlmann phase

$$\varphi_{\mathrm{Uhlmann}} := \arg\langle W(t), W_0 \rangle_{\mathrm{HS}} = \arg\left[\mathrm{Tr}(W^*(t)W_0) \right]. \qquad (5.260)$$

Are $\varphi_{\mathrm{Uhlmann}}$ and the φ of Sjöqvist et al. the same for unitary evolution of ρ? It turns out that in general these two phases are different. This problem was studied by Slater and the interested reader is referred to Slater 2001 and 2002.

Further reading

Section 5.1. Complex manifolds are discussed in Kobayashi and Nomizu 1969; Chern 1967; Wells 1979, and Choquet-Bruhat et al. 1982. Many authors contributed to the geometrical formulation of quantum mechanics. The geometric approach was initiated by Tom Kibble (Kibble 1978, 1979). For recent contributions see Cirelli, Mania and Pizzocchero 1990; Ashtekar and Schilling 1998; and Brody and Hughston 2001.

Section 5.2. Layton, Huang and Chu (1990) derived a formula for the Aharonov–Anandan phase for an arbitrary spin in a magnetic field. For a three-level system, which is equivalent to a spin-one particle in a magnetic field, the detailed calculations were performed by Bouchiat and Gibbons (1988). For other examples see also Moore 1991. Other geometric aspects of Aharonov–Anandan geometric phases are studied in Moore 1991; Anandan 1992; Mukunda and Simon 1993a, 1993b; Pati 1991, 1995; and Chruściński 1995.

A detailed discussion of the classification of bundles in connection with the Aharonov–Anandan geometric phase may be found in Bohm, Boya, Mostafazadeh and Rudolph 1993; Mostafazadeh and Bohm 1993; Mostafazadeh 1996.

Section 5.3. For a recent review of the Pancharatnam phase in quantum mechanics see Sjöqvist 2002. It has been demonstrated by Wagh, Rakhecha, Fisher and Ioffe (1998), that the Pancharatnam relative phase for an internal spin degree of freedom may be tested in interferometry. Recently, Shi-Liang et al. (2000) derived a formula for a Pancharatnam phase for a quantum spin-half particle subjected to an arbitrary magnetic field.

Section 5.4. For more information about the structure of the space of quantum (mixed) states, the reader is referred, e.g., to Bloore 1976; Adelman, Corbett and Hurst 1993; Gibbons 1992; Chruściński 1990, 1991; and Petz and Sudar 1996.

The geometric phases for density matrices of three-level systems in an $SU(3)$ representation were studied in Arvind et al. 1997, Khanna et al. 1997 and Byrd 1999. A general discussion of geometric phases for n-level systems may be found in Boya et al. 1998. For a detailed description of the differential geometry corresponding to a three-level system we refer to Byrd 1998 and Ercolessi et al. 2001. For the interferometric approach to the mixed states geometric phase see also Bhandari et al. 2002 and Anandan et al. 2002. For other aspects of nonabelian phases see Herdegen 1989, and Chruściński 1994.

Recently, the structure of the space of quantum states has received considerable attention due to the investigation of quantum entanglement, see, e.g., Kuś and Życzkowski 2001 and the forthcoming monograph of Bengtsson and Życzkowski.

Problems

5.1. Show that a Hermitian metric h on a complex manifold M is invariant under a complex structure J, i.e., $h(Ju, Jv) = h(u, v)$, for any vector fields u and v on M.

5.2. Show that the general formula for the Fubini–Study metric (5.28) reproduces the Kähler metric (5.42) in the case of n-dimensional projective complex space $\mathbb{C}P^n$.

5.3. Show that the function $\mathcal{K} : \mathbb{C}P^n \to \mathbb{R}$, given by

$$\mathcal{K} := \ln \sqrt{1 + \overline{w}_k w^k} \, ,$$

defines a Kähler potential for the Fubini–Study metric (5.42).

5.4. Verify that the Kähler two-form K defined in (5.45) is closed.

5.5. Derive the Hamilton equation (5.67) on the projective Hilbert space $\mathcal{P}(\mathcal{H})$.

5.6. Show that on a real plane in $\mathcal{H}$ an arc of the great circle on $S(\mathcal{H})$ passing through $\psi_1, \psi_2 \in S(\mathcal{H})$, that is, $\psi(\vartheta) = \lambda_1(\vartheta)\,\psi_1 + \lambda_2(\vartheta)\,\psi_2$, satisfies

   (1) $\langle \partial_\vartheta \psi(\vartheta) | \partial_\vartheta \psi(\vartheta) \rangle = 0$,

   (2) $\langle \psi(\vartheta_1) | \psi(\vartheta_2) \rangle = \cos(\vartheta_1 - \vartheta_2)$.

5.7. Show that the Fubini–Study metric (5.157), rewritten in terms of (holomorphic–antiholomorphic) coordinates $(w_k, \overline{w}^k)$ on $\mathbb{C}P^n$, reproduces formula (5.42).

5.8. Prove Proposition 5.3.3.

5.9. Find all observable functions on $\mathbb{C}P^1$.

5.10. Show that the Uhlmann connection is antihermitian.

5.11. Find the Uhlmann connection in the principal $SU(2)$-fibre bundle over the space of faithful density matrices in $\mathbb{C}^2$ (Dittmann and Rudolph 1992). Show that it reproduces a canonical connection in the instanton bundle $S^7 \longrightarrow S^4 \cong \mathbb{H}P^1$.

5.12. Prove the following properties of the trace distance:

   (1) The trace distance is preserved under unitary transformations, i.e.,

$$D_{\mathrm{Tr}}(U\rho U^*, U\sigma U^*) = D_{\mathrm{Tr}}(\rho, \sigma) \, .$$

   (2) The trace distance is convex in its first argument, i.e.,

$$D_{\mathrm{Tr}}\left(\sum_k p_k \rho_k, \sigma \right) \le \sum_k p_k D_{\mathrm{Tr}}(\rho_k, \sigma) \, ,$$

where $p_k \ge 0$ and $\sum_k p_k = 1$. By symmetry it is also convex in the second argument.

5.13. Derive the Hilbert–Schmidt distance for the qubit states.

5.14. Show that the fidelity is symmetric, i.e., $F(\rho, \sigma) = F(\sigma, \rho)$.

5.15. Show that for qubit states

$$\rho = \frac{1}{2}(\mathbb{1}_2 + \mathbf{x} \cdot \boldsymbol{\sigma}), \qquad \sigma = \frac{1}{2}(\mathbb{1}_2 + \mathbf{y} \cdot \boldsymbol{\sigma}),$$

the fidelity is given by the following formula:

$$F(\rho, \sigma) = \frac{1}{2}\left(1 + \sum_{\alpha=1}^{4} x_\alpha y_\alpha\right),$$

where $x_4 = 2\sqrt{\det\rho}$ and $y_4 = 2\sqrt{\det\sigma}$.

5.16. Prove the following coordinate-free formula for the Bures metric for qubit states:

$$D_B^2(\rho, \rho + d\rho) = \frac{1}{2}\,\mathrm{Tr}\,(d\rho)^2 + (d\sqrt{\det\rho})^2 \ .$$

5.17. Derive the following formula for the Bures metric on the space of density matrices in $\mathbb{C}^n$:

$$ds_B^2 = D_B^2(\rho, \rho + d\rho) = \frac{1}{2}\sum_{\mu,\nu=1}^{n} \frac{|\langle\mu|d\rho|\nu\rangle|^2}{\lambda_\mu + \lambda_\nu} \ ,$$

where the λ's are the eigenvalues of ρ, i.e., such that $\rho|\mu\rangle = \lambda_\mu|\mu\rangle$.

6
Geometric Phases in Action

6.1 Optical manifestation of geometric phases

Both Berry's phase and Hannay's angles could have been discovered long before they were. In 1938 Russian physicist Sergei M. Rytov investigated the rotation of the polarization vector of light travelling along the coiled ray. Actually, as was shown by V.V. Vladimirskii in 1941, Rytov's observation finds an elegant interpretation in terms of geometric properties of a coiled ray. It turns out that rotation of polarization may be interpreted as a simple manifestation of the geometric phase. Actually, the similar conclusion was made by Bortolotti in 1926, however, both Bortolotti and Rytov–Vladimirskii papers were completely unknown to optical community.

Another optical manifestation of the geometric phase is due to the (then young) Indian scientist Pancharatnam. He showed in 1956 that there is a natural method of defining the relative phase between two light beams in different polarization states. It turns out that this relative phase has a purely geometric origin and is called now the *Pancharatnam phase*.

6.1.1 Spins and helicities

One can describe a circularly polarized electromagnetic wave as a set of photons of definite helicity. Recall that a helicity of a photon is the eigenvalue of the projection of the photon spin on its momentum, i.e.,

$$\Lambda = \frac{\mathbf{S} \cdot \mathbf{p}}{|\mathbf{p}|},$$

where $\mathbf{S}$ is the photon spin operator:

$$\mathbf{S} = (S_1, S_2, S_3), \quad \text{with} \quad (S_i)_{kl} = i\epsilon_{ikl}, \tag{6.1}$$

and $\mathbf{p}$ stands for the photon's momentum. The corresponding eigenvalue problem has the following form:

$$\Lambda \Psi_\pm = \pm \Psi_\pm. \tag{6.2}$$

Note that the helicity of a photon may be treated as one would the spin $\mathbf{S}$ of a massive particle. Consider an arbitrary smooth curve C in $\mathbb{R}^3$:

$$t \longrightarrow \mathbf{r}(t) \in \mathbb{R}^3.$$

Let $\mathbf{t}(t)$ be a unit tangent vector at $\mathbf{r}(t)$, i.e.,

$$\mathbf{t}(t) := \frac{\dot{\mathbf{r}}(t)}{|\dot{\mathbf{r}}(t)|}.$$

Clearly, $t \longrightarrow \mathbf{t}(t) \in S^2$ defines a new curve, call it $\tilde{C}$, on a unit sphere S^2. This sphere is usually called a *sphere of directions*. Any curve in $\mathbb{R}^3$ gives rise to a unique curve on a sphere of directions S^2, and the map

$$C \longrightarrow \tilde{C}$$

is called a *Gauss map*.

Suppose, now, that a particle carries a spin S, that is, the corresponding spin operator $\mathbf{S}$ lives in the Hilbert space $\mathcal{H} \cong \mathbb{C}^{2S+1}$. Let us describe the transport of a particle spin state $\psi \in \mathbb{C}^{2S+1}$ along a curve C such that the corresponding curve $\tilde{C}$ is closed, i.e., $\mathbf{t}(0) = \mathbf{t}(T)$ for some $T > 0$. Clearly, any two unit vectors $\mathbf{t}_1$ and $\mathbf{t}_2$ are related by an $SO(3)$-rotation; in fact, there are infinitely many rotations from $\mathbf{t}_1$ to $\mathbf{t}_2$. Let us choose the following one:

$$\mathbf{t}_2 = R(\boldsymbol{\theta}) \, \mathbf{t}_1, \tag{6.3}$$

a rotation around the $\boldsymbol{\theta}$-axis by an angle $|\boldsymbol{\theta}|$, with

$$\boldsymbol{\theta} := \mathbf{t}_1 \times \mathbf{t}_2. \tag{6.4}$$

Note that the rotation defined above is, in a sense, the simplest one: $\mathbf{t}_1$ is rotated into $\mathbf{t}_2$ along a geodesic — an arc of a great circle in a plane perpendicular to $\boldsymbol{\theta}$. The rotation $R(\boldsymbol{\theta})$ gives rise to a unitary operator in $\mathbb{C}^{2S+1}$, as follows:

$$U(\boldsymbol{\theta}) := \exp[-i R(\boldsymbol{\theta})]. \tag{6.5}$$

Note that if $\mathbf{t}_1$ and $\mathbf{t}_2$ satisfy (6.3) then the operators $\mathbf{t}_1 \cdot \mathbf{S}$ and $\mathbf{t}_2 \cdot \mathbf{S}$ — the projections of $\mathbf{S}$ onto $\mathbf{t}_1$ and $\mathbf{t}_2$, respectively — are related by

$$\mathbf{t}_2 \cdot \mathbf{S} = U(\boldsymbol{\theta}) \, (\mathbf{t}_1 \cdot \mathbf{S}) \, U(\boldsymbol{\theta})^{-1}. \tag{6.6}$$

Moreover, if ψ_1 is an eigenvector of $\mathbf{t}_1 \cdot \mathbf{S}$, i.e., such that

$$(\mathbf{t}_1 \cdot \mathbf{S})\,\psi_1 = m\psi_1 \ , \tag{6.7}$$

then

$$\psi_2 := U(\boldsymbol{\theta})\,\psi_1 \tag{6.8}$$

is an eigenvector of $\mathbf{t}_2 \cdot \mathbf{S}$ with the same eigenvalue m:

$$(\mathbf{t}_2 \cdot \mathbf{S})\,\psi_2 = m\psi_2 \ . \tag{6.9}$$

Suppose that the initial state $\psi(0) = |m\rangle$ is an eigenstate of $\mathbf{t}(0) \cdot \mathbf{S}$, i.e.,

$$\mathbf{t}(0) \cdot \mathbf{S}\,|m\rangle = m|m\rangle \ , \tag{6.10}$$

for some m. As we move on a sphere of directions along a curve $\widetilde{C}$ we perform a continuous sequence of rotations $\mathbf{t}(t) \longrightarrow \mathbf{t}(t + dt)$ from $t = 0$ to $t = T$. The product of all these rotations is a rotation around the $\mathbf{t}(0)$-axis. The state at time T is not changed by this rotation because it is represented by an eigenvector of $\mathbf{t}(0) \cdot \mathbf{S}$. Hence the product rotation just multiplies the initial state vector $|m\rangle$ by a phase factor, such that, for instance,

$$|\text{final}\rangle = U(\boldsymbol{\varphi}_m)|m\rangle \ , \tag{6.11}$$

with

$$\boldsymbol{\varphi}_m = \mathbf{t}(0)\,\varphi \ , \tag{6.12}$$

and thus

$$|\text{final}\rangle = e^{-im\varphi}|m\rangle \ . \tag{6.13}$$

How to find φ? Let $\mathbf{x}$ be an arbitrary vector tangent to S^2 at the point $\mathbf{t}(0)$ and define the transport of $\mathbf{x}$ along $\widetilde{C}$ as follows: $\mathbf{x}(t + dt)$ is obtained from $\mathbf{x}(t)$ by applying the same rotation that leads from $\mathbf{t}(t)$ to $\mathbf{t}(t + dt)$. More precisely,

$$\mathbf{x}(t) \longrightarrow \mathbf{x}(t + dt) = \mathbf{x}(t) + \dot{\mathbf{x}}(t)\,dt \ ,$$

where

$$\dot{\mathbf{x}}(t) = \boldsymbol{\Omega}(t) \times \mathbf{x}(t) \ , \tag{6.14}$$

and

$$\boldsymbol{\Omega}(t) := \mathbf{t}(t) \times \dot{\mathbf{t}}(t) \ . \tag{6.15}$$

Hence

$$\dot{\mathbf{x}} = (\mathbf{t} \cdot \mathbf{x})\,\dot{\mathbf{t}} - (\dot{\mathbf{t}} \cdot \mathbf{x})\,\mathbf{t} = -(\dot{\mathbf{t}} \cdot \mathbf{x})\,\mathbf{t} \ , \tag{6.16}$$

since $\mathbf{t} \cdot \mathbf{x} = 0$. Note, however, that (6.16) is the defining equation of a parallel transport on S^2, cf. formula (2.153). Therefore, the total angle of rotation after traversing a curve $\widetilde{C}$ on S^2 reads

$$\varphi = \Omega(\widetilde{C}) \,, \tag{6.17}$$

where $\Omega(\widetilde{C})$ stands for the solid angle subtended by $\widetilde{C}$ on the sphere of directions.

Formula (6.17), together with (6.13), generalizes (2.117). Note, that we did not use any specific Hamiltonian to perform the transport of spin states along C. The total phase shift appears more as a property of the spin states than of the Hamiltonian.[1] Note that all of the above arguments apply equally well to nonrelativistic particles or relativistic particles with either zero (photons) or nonzero mass.

6.1.2   Chiao–Tomita–Wu phase

Raymond Y. Chiao, Akira Tomita and Yong-Shi Wu[2] were the first to check the validity of the solid angle formula (6.17) in a simple optical experiment. Consider a linearly polarized electromagnetic wave propagating in the direction of a wave vector $\mathbf{k}(t)$ — e.g., by sending a light along an optical fibre. Suppose that the optical fibre is coiled such that $\mathbf{t}(0) = \mathbf{t}(T)$ for some $T > 0$, i.e., the initial and final directions of the fibre coincide. If the shape of the fiber is represented by a curve C, then under the Gauss map C is mapped onto a closed curve $\widetilde{C}$ on the sphere of directions.

Now, in the plane Σ perpendicular to $\mathbf{t}(0) = \mathbf{t}(T)$ let us introduce an orthonormal basis $(\boldsymbol{\epsilon}_1, \boldsymbol{\epsilon}_2)$. Suppose that $\boldsymbol{\epsilon}_1$ is the initial polarization vector, that is, $\boldsymbol{\epsilon}(0) = \boldsymbol{\epsilon}_1$. What is the final polarization $\boldsymbol{\epsilon}(T)$? Clearly, $\boldsymbol{\epsilon}(0)$ and $\boldsymbol{\epsilon}(T)$ differ by a $SO(2)$-rotation in the plane Σ (cf. Fig. 6.1), i.e.,

$$\boldsymbol{\epsilon}(T) = R(\varphi)\,\boldsymbol{\epsilon}(0) \,. \tag{6.18}$$

Introducing circular polarization vectors

$$\boldsymbol{\epsilon}_\pm = \frac{\boldsymbol{\epsilon}_1 \pm i\boldsymbol{\epsilon}_2}{\sqrt{2}} \,, \tag{6.19}$$

one has

$$\boldsymbol{\epsilon}(0) = \frac{\boldsymbol{\epsilon}_+ + \boldsymbol{\epsilon}_-}{\sqrt{2}} \,, \tag{6.20}$$

that is, the initial linear polarization is a superposition of right ($s = 1$) and left ($s = -1$) polarized waves. Now, the helicity eigenstates (with eigenvalues $s = \mp 1$) acquire

---

[1] This fundamental aspect was stressed by Jordan (Jordan 1987, 1988a, 1988b).

[2] Rotation of the polarization of light travelling along an optical fibre bent in a nonplanar curve was first observed in the laboratory by Neil Ross (Ross 1984) and then in a series of experiments by Chiao and Wu (Chiao and Wu 1986).

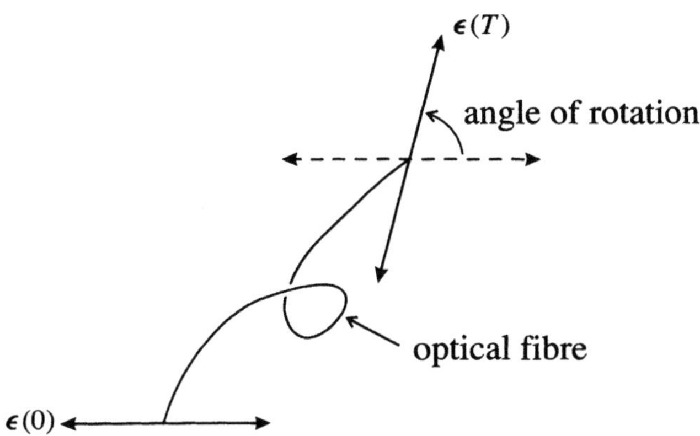

Figure 6.1: Rotation of the polarization vector.

a geometric phase $s\,\Omega(\widetilde{C})$ after passage through the optical fibre, where $\Omega(\widetilde{C})$ is the solid angle subtended by $\widetilde{C}$ on the sphere of directions. Hence, the final polarization is

$$
\begin{aligned}
\epsilon(T) &= \frac{1}{\sqrt{2}}\left(e^{-i\Omega(\widetilde{C})}\epsilon_+ + e^{i\Omega(\widetilde{C})}\epsilon_-\right) = \epsilon_1\cos\Omega(\widetilde{C}) + \epsilon_2\sin\Omega(\widetilde{C}) \\
&= R(\Omega(\widetilde{C}))\,\epsilon(0)\,.
\end{aligned}
\tag{6.21}
$$

Thus, the geometric phase that appears for circularly polarized photons corresponds to rotation of the linear polarization vector ϵ by the angle

$$
\varphi = \Omega(\widetilde{C})\,,
\tag{6.22}
$$

which proves our basic relation (6.17).

6.1.3   Rytov's law and Fermi–Walker transport

It turns out that the geometric law governing the transport of polarization vector ϵ along an optical fibre was initially observed in 1941, in a remarkable paper by Vladimirskii which was an extension of an earlier paper by Rytov (Rytov 1938). Vladimirskii showed that, along a light ray, the vectors $\mathbf{E}$ and $\mathbf{H}$ of the electric and magnetic field, respectively, perform a rotation with respect to a natural Frenet's triple $(\mathbf{t}, \mathbf{n}, \mathbf{b})$ (recall, that $\mathbf{t}$ stands for a vector tangent to, $\mathbf{n}$ for a vector normal to, and $\mathbf{b}$ for a vector binormal to the curved ray — cf. section 3.3.4). At each point of the curve, these vectors are related by the Serret–Frenet formulae (3.169):

$$
\begin{aligned}
\frac{d\mathbf{t}}{ds} &= \kappa\mathbf{n}\,, \\
\frac{d\mathbf{n}}{ds} &= -\kappa\mathbf{t} + \tau\mathbf{b}\,, \\
\frac{d\mathbf{b}}{ds} &= -\tau\mathbf{n}\,,
\end{aligned}
\tag{6.23}
$$

where s parametrizes the curve describing the light ray/optical fibre. Define now the following unit vectors:

$$\mathbf{e} = \frac{\mathbf{E}}{E} \quad \text{and} \quad \mathbf{h} = \frac{\mathbf{H}}{H} ,$$

which give rise to an orthonormal basis in the plane perpendicular to $\mathbf{t}$. Hence, a pair $(\mathbf{e}, \mathbf{h})$ is related to a pair $(\mathbf{n}, \mathbf{b})$ by an $SO(2)$ rotation, i.e.,

$$\begin{pmatrix} \mathbf{e} \\ \mathbf{h} \end{pmatrix} = \begin{pmatrix} \cos\phi & -\sin\phi \\ \sin\phi & \cos\phi \end{pmatrix} \begin{pmatrix} \mathbf{n} \\ \mathbf{b} \end{pmatrix} , \tag{6.24}$$

with $\phi = \phi(s)$. Let us study the transport of $\mathbf{e}$ and $\mathbf{h}$ along the curve describing the optical fibre. The condition $|\mathbf{e}| = 1$ implies that $\mathbf{e} \cdot \dot{\mathbf{e}} = 0$, and hence

$$\dot{\mathbf{e}} = \alpha \mathbf{t} + \beta (\mathbf{t} \times \mathbf{e}) . \tag{6.25}$$

We shall assume that the medium is not gyrotropic, i.e., that $\beta = 0$ (cf. Vinitskii et al. 1990). As a result we obtain the so-called *Rytov law*:

$$\dot{\mathbf{e}} = (\dot{\mathbf{e}} \cdot \mathbf{t}) \mathbf{t} = -(\mathbf{e} \cdot \dot{\mathbf{t}}) \mathbf{t} , \tag{6.26}$$

which is nothing but the law for a parallel transport of $\mathbf{e}$ along the curve

$$s \longrightarrow \mathbf{t}(s) \in S^2 ,$$

defined in (2.153). Let us note that, by the very definition of a parallel transport on S^2, vector $\mathbf{e}$ (and of course $\mathbf{h}$) does not rotate around $\mathbf{t}$. However, it does rotate with respect to the $(\mathbf{n}, \mathbf{b})$ basis. Using results from section 3.3.4, one finds that $\mathbf{e}$ and $\mathbf{h}$ rotate with respect to $(\mathbf{n}, \mathbf{b})$ with an angular velocity

$$\dot{\phi}(s) = \tau(s) . \tag{6.27}$$

Hence, after a time T one has

$$\begin{pmatrix} \mathbf{e}(T) \\ \mathbf{h}(T) \end{pmatrix} = \begin{pmatrix} \cos\Omega(\tilde{C}) & -\sin\Omega(\tilde{C}) \\ \sin\Omega(\tilde{C}) & \cos\Omega(\tilde{C}) \end{pmatrix} \begin{pmatrix} \mathbf{e}(0) \\ \mathbf{h}(0) \end{pmatrix} . \tag{6.28}$$

The law of transporting vectors $\mathbf{e}$ and $\mathbf{h}$ along the curve in $\mathbb{R}^3$ is related to another geometric concept well known in General Relativity and called *Fermi–Walker transport*. Consider a curve

$$s \longrightarrow \mathbf{x}(s) \in \mathbb{R}^n ,$$

and let

$$\mathbf{u} := \frac{d\mathbf{x}}{ds} , \quad \text{and} \quad \mathbf{a} := \frac{d\mathbf{u}}{ds} ,$$

denote the velocity and acceleration vectors, respectively. Let $\mathbf{Y} = \mathbf{Y}(s)$ be an arbitrary vector defined along the curve $\mathbf{x}(s)$. One defines the *Fermi–Walker derivative* of $\mathbf{Y}$ along the curve, as follows:

$$\frac{D\mathbf{Y}}{Ds} := \frac{d\mathbf{Y}}{ds} + (\mathbf{Y} \cdot \mathbf{a})\,\mathbf{u} - (\mathbf{Y} \cdot \mathbf{u})\,\mathbf{a} . \tag{6.29}$$

A vector $\mathbf{Y}$ is Fermi–Walker transported (or transported according to a Fermi–Walker rule) if

$$\frac{D\mathbf{Y}}{Ds} = 0 . \tag{6.30}$$

Note that both $\mathbf{e}$ and $\mathbf{h}$ are Fermi–Walker transported, since

$$\frac{D\mathbf{e}}{Ds} = \frac{D\mathbf{h}}{Ds} = 0 . \tag{6.31}$$

This property does not hold for $\mathbf{n}$ and $\mathbf{b}$. Rather, one easily finds that

$$\frac{D\mathbf{n}}{Ds} = \tau\,\mathbf{b} , \quad \text{and} \quad \frac{D\mathbf{b}}{Ds} = -\tau\,\mathbf{n} . \tag{6.32}$$

Remark 6.1.1 It turns out that the parallel transport law for the polarization of the electromagnetic wave was discovered by Bortolotti (1926). Bortolotti studied the evolution of a linearly polarized wave in a medium with varying index of refraction $n = n(\mathbf{r})$, and observed that this evolution is governed to be a parallel transport with respect to a metric connection whose components are determined by $\nabla \log n^2$.[3] The parallel transport of the polarization vector was later on independently discovered and studied by Luneburg (1964). $\diamond$

Remark 6.1.2 Some authors have raised the question of whether or not the effect of rotation of the polarization vector is quantum or classical. Chiao and Wu (1986) suggest that one ... *would rather think of this effect as a topological feature of classical Maxwell theory which originates at the quantum level, but survives the correspondence-principle limit ($\hbar \longrightarrow 0$) into the classical level*. However, as was already noted by Feynman, the *quantum equations* for photons are just the same as the classical Maxwell equations. Let us define two complex vector fields, that is, complex induction

$$\mathbf{F} = \mathbf{D} + i\mathbf{B} ,$$

and complex intensity

$$\mathbf{G} = \mathbf{E} + i\mathbf{H} .$$

Following Berry 1989a, we rewrite the Maxwell equations in an inhomogeneous medium, defined by the constitutive relations

$$\mathbf{D} = \epsilon(\mathbf{r})\mathbf{E} , \quad \text{and} \quad \mathbf{B} = \mu(\mathbf{r})\mathbf{H},$$

---

[3]The authors thank Prof. I. Białynicki-Birula for this remark.

as follows:

$$i\dot{\mathbf{F}} = \nabla \times \mathbf{G} .$$

Introducing the following spin-one representation of $su(2)$:

$$\mathbf{S} = \left(\begin{pmatrix} 0 & 0 & 0 \\ 0 & 0 & -i \\ 0 & i & 0 \end{pmatrix} , \begin{pmatrix} 0 & 0 & i \\ 0 & 0 & 0 \\ -i & 0 & 0 \end{pmatrix} , \begin{pmatrix} 0 & -i & 0 \\ i & 0 & 0 \\ 0 & 0 & 0 \end{pmatrix} \right) ,$$

we have

$$-i\hbar \nabla \times \mathbf{G} = \mathbf{p} \times \mathbf{G} = -i(\mathbf{p} \cdot \mathbf{S})\mathbf{G} ,$$

where $\mathbf{p} = -i\hbar \nabla$ stands for the quantum mechanical momentum operator. The Maxwell equations may therefore be rewritten formally as the following Schrödinger equation:

$$i\hbar \partial_t \Psi := \widehat{H}_{\text{Maxwell}} \Psi ,$$

where

$$\Psi(t, \mathbf{r}) = \left| \begin{matrix} \epsilon^{1/2}(\mathbf{r})\mathbf{E}(t, \mathbf{r}) + i\mu^{1/2}(\mathbf{r})\mathbf{H}(t, \mathbf{r}) \\ \epsilon^{1/2}(\mathbf{r})\mathbf{E}(t, \mathbf{r}) - i\mu^{1/2}(\mathbf{r})\mathbf{H}(t, \mathbf{r}) \end{matrix} \right| ,$$

and the "Maxwell Hamiltonian" is defined by

$$\widehat{H}_{\text{Maxwell}} := c \begin{pmatrix} \mathbf{\Pi} \cdot \mathbf{S} & i\hbar\boldsymbol{\xi} \cdot \mathbf{S} \\ -i\hbar\boldsymbol{\xi} \cdot \mathbf{S} & -\mathbf{\Pi} \cdot \mathbf{S} \end{pmatrix} .$$

In the above formula, c stands for the velocity of light and the vectors $\mathbf{\Pi}$ and $\boldsymbol{\xi}$ are defined in terms of the index of refraction:

$$n(\mathbf{r}) = \sqrt{\epsilon(\mathbf{r})\mu(\mathbf{r})} ,$$

as follows:

$$\mathbf{\Pi} = n^{1/2}(\mathbf{r}) \, \mathbf{p} \, n^{1/2}(\mathbf{r}) ,$$

and

$$\boldsymbol{\xi}(\mathbf{r}) = \frac{1}{4n(\mathbf{r})} \nabla \log \frac{\epsilon(\mathbf{r})}{\mu(\mathbf{r})} .$$

The reader can easily check that $\widehat{H}_{\text{Maxwell}}$ defines a hermitian operator. Hence, the Maxwell equations have, formally, a Schrödinger-like form. $\diamond$

6.1.4   Pancharatnam phase

In 1956, the young Indian physicist Pancharatnam, in the course of a study of polarization transformations of light waves propagating through optically anisotropic crystals, posed the following question: *how does one define a phase difference between two light waves that are in different polarization states?*[4] Pancharatnam concluded that the most reasonable definition would be the one that defines two waves to be in phase when their interference gives maximum intensity (Pancharatnam 1956).

Let us briefly recall the description of the polarization states for an electromagnetic wave (see, e.g., Born and Wolf 1959, Jackson 1999). A monochromatic plane wave travelling in a direction $\mathbf{k}$ is described by

$$\mathbf{E}(\mathbf{x}, t) = (E_+ \boldsymbol{\epsilon}_+ + E_- \boldsymbol{\epsilon}_-) \, e^{i\mathbf{k}\mathbf{x} - i\omega t} \, , \qquad (6.33)$$

where $\omega = k/c$ (with c the velocity of light), and $\boldsymbol{\epsilon}_\pm$ are defined in (6.19). Two complex numbers $E_\pm \in \mathbb{C}$ determine a polarization state of the wave:

- If $E_\pm \neq 0$, and E_+/E_- is a complex number, then the wave is *elliptically* polarized.

- If $E_\pm \neq 0$, and E_+/E_- is a real number, then the wave is *linearly* polarized.

- If $E_+ = 0$ or $E_- = 0$, then the wave is *circularly* polarized (one calls it right-handed polarization if $E_- = 0$ and left-handed if $E_+ = 0$).

Define a unit complex vector $\mathbf{d} \in \mathbb{C}^2$ by

$$\mathbf{d} = (d_+, d_-) \; := \; \frac{(E_+, E_-)}{|\mathbf{E}|} \, , \qquad (6.34)$$

or, using Dirac notation,

$$\mathbf{d} \longrightarrow |d\rangle = \begin{vmatrix} d_+ \\ d_- \end{vmatrix} \, .$$

One calls $\mathbf{d}$ a *polarization vector*. Due to the normalization condition $\langle d|d\rangle = 1$, we have $|d\rangle \in S^3 \subset \mathbb{C}^2$. Now, the state of polarization is uniquely determined by the vector

$$\mathbf{s} := \langle d|\boldsymbol{\sigma}|d\rangle \, . \qquad (6.35)$$

Clearly

$$\langle d|d\rangle = 1 \; \Longrightarrow \; |\mathbf{s}| = 1 \, ,$$

---

[4] We recommend to the reader a beautiful article by M.V. Berry called *Pancharatnam, virtuoso of the Poincaré sphere: an appreciation*, Berry 1994; see also Berry 1987a.

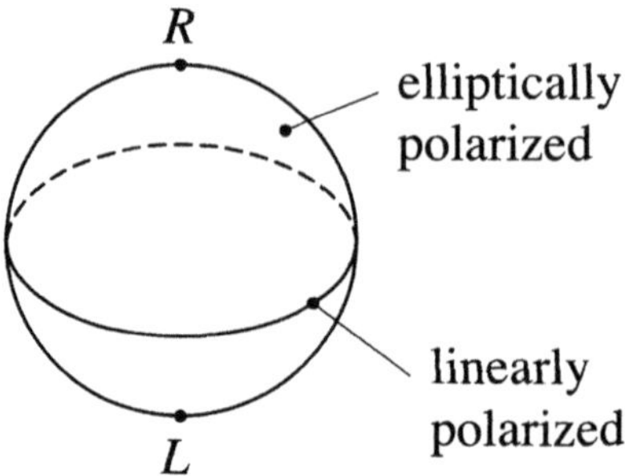

Figure 6.2: Poincaré sphere: north and south poles correspond to right- and left-handed circular polarizations, respectively, equatorial points represent linear polarizations, and the remaining points define elliptical polarizations.

that is, $\mathbf{s} \in S^2$. Note, that two polarization vectors $|d_1\rangle$ and $|d_2\rangle$ differing by a phase factor, i.e., such that

$$|d_2\rangle = e^{i\lambda}|d_1\rangle , \tag{6.36}$$

define the same polarization state:

$$\langle d_1|\boldsymbol{\sigma}|d_1\rangle = \langle d_2|\boldsymbol{\sigma}|d_2\rangle .$$

Hence, the

$$\text{space of polarization states} = S^3/U(1) = \mathbb{C}P^1 \cong S^2 ,$$

and the map

$$\text{polarization vector } \mathbf{d} \longrightarrow \text{polarization state } [\mathbf{d}] := \mathbf{s}$$

defines a Hopf bundle $S^3 \longrightarrow S^2$.

The above two-dimensional sphere, which serves as the space of polarization states, is called a *Poincaré sphere* (see Fig. 6.2). In summary, we have established that the space of possible polarization states for a monochromatic plane wave is isomorphic to the phase space of a two-level quantum system (more precisely, the space of pure states).

If $|A\rangle$ and $|B\rangle$ are the polarization vectors of two waves, then the intensity resulting from their interference is given by

$$I := ((\langle A| + \langle B|)(|A\rangle + |B\rangle)) = |A|^2 + |B|^2 + 2\mathrm{Re}\langle A|B\rangle . \tag{6.37}$$

Due to the Pancharatnam definition of the relative phase (cf. section 5.3.4), waves with polarizations $|A\rangle$ and $|B\rangle$ are in phase if and only if

$$\langle A|B\rangle \quad \text{is real and positive} .$$

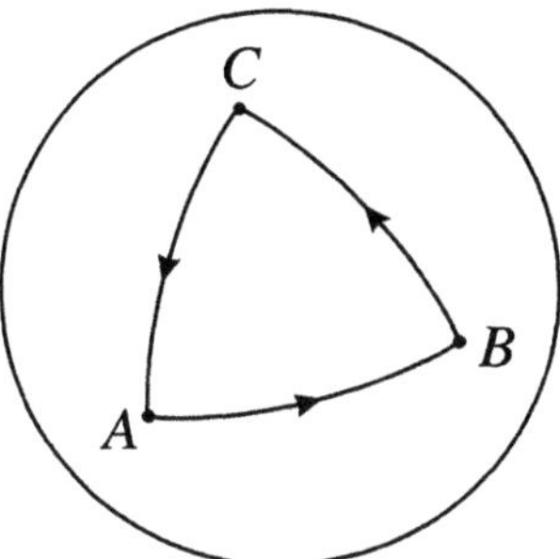

Figure 6.3: Geodesic triangle ABC on the Poincaré sphere.

Now, it is clear from section 5.3.4 that if $|B\rangle$ is in phase with $|A\rangle$, and $|C\rangle$ with $|B\rangle$, then $|C\rangle$ is, in general, not in phase with $|A\rangle$. Let $|A'\rangle$ be another polarization vector defining the same polarization state as $|A\rangle$, that is, $|A'\rangle = e^{i\varphi}|A\rangle$. Suppose that $|A'\rangle$ is in phase with $|C\rangle$. Then

$$\varphi = -\frac{1}{2}\,\Omega_{ABC}\,, \tag{6.38}$$

where Ω_{ABC} is the solid angle of the geodesic triangle ABC on the Poincaré sphere (see Fig. 6.3). This phase shift is a special example of the general phase shift formula (5.177), and is called a Pancharatnam phase.

6.2   Quantum mechanics as a gauge theory

It turns out that the geometric phase found by Berry is closely related to another effect observed by Yakir Aharonov and David Bohm in 1959 in Bristol. They showed that in quantum mechanics, in contrast to the classical case, the behavior of a charged particle in a region where there is no electromagnetic field can nevertheless be affected by a nonvanishing electromagnetic potential. For example, if the trajectory of a charged particle encloses a solenoid producing a magnetic flux, then the wave function of the particle acquires an additional phase factor, which is proportional to the flux magnitude and may be easily explained as Berry's geometric phase. Actually, this phase depends only upon the topology of the space — for instance, by excluding the solenoid the resulting space is no longer simply connected, and hence is topologically nontrivial. Therefore, some authors use the name *topological phase* to stress the topological origin of the Berry phase. However, in general, Berry's phase depends not only of the topology of the space, but also on the geometry of the closed curve in question. In the following section we derive the Aharonov–Bohm effect and the closely related Aharonov–Casher effect, using the gauge invariance of standard nonrelativistic quantum mechanics under an $U(1) \times SU(2)$ gauge group.

6.2.1 Classical particles in gauge theory

Recall that the classical dynamics of a charged particle interacting with an electromagnetic field, described by the field tensor

$$F_{\mu\nu} = \partial_\mu A_\nu - \partial_\nu A_\mu \, ,$$

is governed by the Lorentz equation:

$$\dot{p}_\mu = \frac{q}{c} F_{\mu\nu} u^\nu \, , \tag{6.39}$$

where u^μ denotes particle four-velocity, $p^\mu = mu^\mu$ stands for four-momentum, and q for the electric charge (space-time indices $\mu = 0, 1, 2, 3$ are raised and lowered by the Minkowski metric tensor $g_{\mu\nu} = \text{diag}[1, -1, -1, -1]$, see e.g. Jackson (1999)). In the non-abelian theory with a group G, the corresponding field strength is defined by

$$G^a_{\mu\nu} = \partial_\mu A^a_\nu - \partial_\nu A^b_\mu + f^a_{bc} A^b_\mu A^c_\nu \, ,$$

and the dynamics of a particle carrying a spin-like variable I^a is given by the Wong equations (Wong 1970, see Balachandran et al. 1983 for the review):

$$\dot{p}_\mu = G^a_{\mu\nu} u^\nu I_a \, , \tag{6.40}$$

and

$$\dot{I}^a = f^a_{bc} A^b_\mu u^\mu I^c \, , \tag{6.41}$$

where $I = \sum_a I^a \lambda_a$, and the λ's define a basis of the Lie algebra $\mathfrak{g}$ of G. Equivalently, one may write

$$\dot{p}_\mu = \text{Tr}(G_{\mu\nu} \cdot I) u^\nu \, , \quad \text{and} \quad \dot{I} = [A_\mu, I] u^\mu \, , \tag{6.42}$$

where I, A_μ and $G_{\mu\nu}$ are $\mathfrak{g}$-valued zero-, one- and two-forms, respectively. They transform under the adjoint representation of the internal symmetry group G, as follows:

$$\begin{aligned}
A_\mu &\longrightarrow U \cdot A_\mu \cdot U^{-1} + U \cdot \partial_\mu U^{-1} \, , \\
G_{\mu\nu} &\longrightarrow U \cdot G_{\mu\nu} \cdot U^{-1} \, , \\
I &\longrightarrow U \cdot I \cdot U^{-1} \, .
\end{aligned}$$

Note that in the region of space-time where the field strength $F_{\mu\nu}$ or $G_{\mu\nu}$ vanishes the classical particle is free, i.e., it satisfies

$$\dot{p}^\mu = 0 \, .$$

This is so even if the corresponding gauge potential A does not vanish. One usually concludes that the potential does not have physical meaning in classical physics. However, this is no longer true in quantum physics. The archetypal example showing that quantum objects feel the gauge potential even if the field strength vanishes is the celebrated Aharonov–Bohm effect (Aharonov and Bohm 1959). In the next section we show that the Aharonov–Bohm effect is a simple consequence of the $U(1)$-invariance of the standard Schrödinger theory.

It turns out that there exists another closely related quantum mechanical effect — the Aharonov–Casher effect (Aharonov and Casher 1984) — which is connected with the Pauli nonrelativistic theory of spin. This effect is implied by the nonabelian $SU(2)$ symmetry of Pauli theory. Both effects find elegant explanations as abelian and nonabelian geometric phases.

6.2.2 $U(1)$-invariance and the Aharonov–Bohm effect

In the presence of an external electromagnetic field, described by an electromagnetic potential A_μ, the standard nonrelativistic quantum mechanics of a charged particle is govern by the Schrödinger equation

$$i\hbar\dot{\psi} = H\psi ,$$

where the Hamiltonian is given by

$$H := \frac{-\hbar^2}{2m} \left(\nabla - \frac{iq}{\hbar c}\mathbf{A} \right)^2 - q\Phi . \tag{6.43}$$

In the above formula, q and m denote the charge and mass of the particle, respectively, and the four-potential is defined by $A_\mu := (\Phi, -\mathbf{A})$. This theory is invariant under the simultaneous gauge transformation of the electromagnetic four-potential,

$$\mathbf{A} \longrightarrow \mathbf{A} + \nabla\chi , \quad \text{and} \quad \Phi \longrightarrow \Phi - \dot{\chi} , \tag{6.44}$$

and the corresponding phase transformation of the wave function:

$$\psi \longrightarrow \exp\left(-\frac{iq}{\hbar c}\chi \right) \psi . \tag{6.45}$$

In this way, the standard nonrelativistic quantum mechanics of a charged particle defines a $U(1)$ gauge theory. Let $x^0 = ct$ and set $x^\mu := (x^0, \mathbf{x})$. Introducing a covariant derivative

$$D_\mu := \partial_\mu + ia_\mu , \tag{6.46}$$

with

$$a_0 = \frac{q}{\hbar c}\Phi , \quad \text{and} \quad a_k = -\frac{q}{\hbar c}A_k , \quad k = 1, 2, 3 , \tag{6.47}$$

the Schrödinger equation may be rewritten in the following manifestly gauge invariant form:

$$i\hbar c D_0\psi = -\frac{\hbar^2}{2m} \sum_{k=1}^{3} D_k D_k \psi . \tag{6.48}$$

A key effect demonstrating this $U(1)$ gauge invariance is the celebrated *Aharonov–Bohm effect* (Aharonov and Bohm 1959). Consider an infinitely long solenoid along the z-axis (its radius in the xy-plane equals a). Suppose that the solenoid carries a magnetic flux $\Phi_0 = \pi a^2 B_0$, where $\mathbf{B} = (0, 0, B_0)$ is the magnetic field inside. Clearly, there is no magnetic field outside. One may easily find the corresponding vector potential $\mathbf{A}$ such that $\mathbf{B} = \nabla \times \mathbf{A}$. In a suitable gauge one has

$$
\mathbf{A}(x, y, z) = \frac{B_0 a^2}{2} \cdot
\begin{cases}
\frac{1}{a^2} (-y, x, 0), & r < a \\[2mm]
\frac{1}{x^2+y^2} (-y, x, 0), & r > a
\end{cases}
\tag{6.49}
$$

Note that, unlike $\mathbf{B}$, the vector potential is continuous at $r = a$ (r measures the distance from the z-axis, i.e., $r^2 = x^2 + y^2$). If we take cylindrical coordinates (r, ϕ, z), then $A_r = A_z = 0$ and

$$
A_\phi =
\begin{cases}
\frac{r B_0}{2}, & r < a \\[2mm]
\frac{\Phi_0}{2\pi r}, & r > a
\end{cases}
\tag{6.50}
$$

The corresponding Hamiltonian in the region outside the solenoid is given by the following formula:

$$
H = \frac{1}{2m} \left(\mathbf{p} - \frac{q}{c} \mathbf{A} \right)^2 = \frac{1}{2m} \left[p_r^2 + \frac{1}{r^2} \left(p_\phi - \frac{q \Phi_0}{2\pi c} \right)^2 + p_z^2 \right].
\tag{6.51}
$$

The classical canonical transformation

$$
p_\phi \longrightarrow p_\phi - \frac{q \Phi_0}{2\pi c},
$$

completely eliminates the flux term from the Hamiltonian, so that the solenoid does not influence the motion of classical charges outside it.

However, in quantum theory the situation is quite different. There is likewise a unitary transformation

$$
\psi \longrightarrow \psi' = e^{-iq\Phi_0\phi/hc} \psi
$$

that eliminates Φ_0 from the Schrödinger equation. Note, however, that the flux is now encoded into the boundary condition: if ψ is single-valued, i.e.,

$$
\psi(2\pi) = \psi(0),
$$

then the transformed function ψ' is not; rather

$$
\psi'(2\pi) = e^{-iq\Phi_0/\hbar c} \psi'(0).
$$

The experimental configuration for this effect is shown in Fig. 6.4. An electron wave is split into two coherent waves. They pass on opposite sides of the solenoid and then

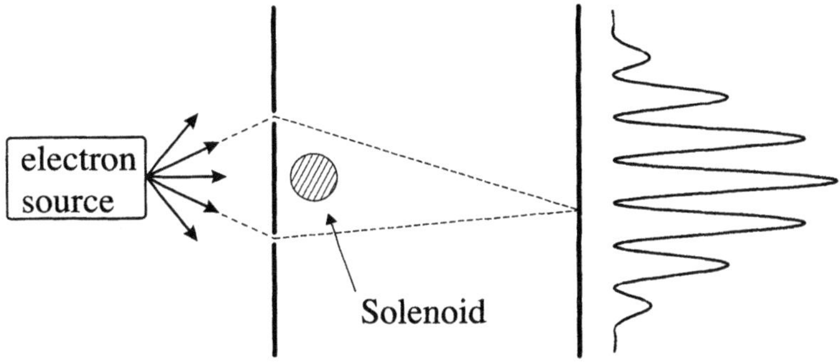

Figure 6.4: Aharonov–Bohm effect.

interfere. Although there are no magnetic fields outside the solenoid, i.e., in the region
in which the charged particles move, a relative phase shift between the two waves can
be observed as an interference pattern. The corresponding phase shift $\Delta\varphi$ is given by
the following formula:

$$\Delta\varphi = -\frac{q\,\Phi_0}{\hbar c} = -\frac{q}{\hbar c}\oint \mathbf{A}\cdot d\mathbf{l}\,, \tag{6.52}$$

where the integral is carried out along a closed curve formed by the union of the two
paths. Although the magnetic field vanishes everywhere outside the solenoid, the vector potential $\mathbf{A}$ cannot vanish there. This is because the loop integral of $\mathbf{A}$ around the
solenoid is equal to the magnetic flux

$$\Phi_0 = \int_\Sigma \mathbf{B}\cdot d\mathbf{S}\,, \tag{6.53}$$

through the solenoid. Note that the interference pattern is invariant under

$$\Phi_0 \longrightarrow \Phi_0 + n\frac{hc}{q}\,, \quad n \in \mathbb{Z}\,,$$

since

$$\Delta\varphi \longrightarrow \Delta\varphi - 2\pi n \cong \Delta\varphi\,.$$

It is clear that the Aharonov–Bohm phase shift $\Delta\varphi$ may be interpreted as a geometric
Berry phase that a charged particle accumulates by circling around a solenoid carrying
a nonzero magnetic flux.[5]

Remark 6.2.1 Consider the limiting case in which the magnetic field vanishes away
from a single line (a so-called *flux line*), e.g., the z-axis, and suppose that the motion of

---

[5]This was already observed by Berry (1984).

the charged particles is restricted to the xy-plane, i.e., the allowed region for particle motion is a punctured plane $M = \mathbb{R}^2 - \{0\}$. The magnetic field in the xy-plane is given by

$$B_z(\mathbf{x}) = \Phi_0\,\delta^2(\mathbf{x}) \, . \tag{6.54}$$

This field may be derived from the following vector potential:

$$\mathbf{A} = \frac{\Phi_0}{2\pi}\,\frac{(-y, x)}{r^2} \, , \tag{6.55}$$

with $r^2 = x^2 + y^2$ (we have already considered a field of this type in Example 1.1.9). In the above formula $\mathbf{A}$ denotes a two-dimensional vector field on M. Recall, that the fundamental group of a punctured plane M is:

$$\pi_1(M) \cong \mathbb{Z} \, .$$

Consider any closed curve C on M and let n be its winding number (cf. Example 1.4.3). It is therefore evident that

$$\oint_C \mathbf{A} \cdot d\mathbf{l} = n\Phi_0 \, . \tag{6.56}$$

This shows that the Aharonov–Bohm phase shift is an example of a *topological phase*, i.e., the shift $\Delta\varphi$ does not depend on the particular geometry of the closed curve C but only on its winding number n, which is a purely topological notion.

It turns out that the Aharonov–Bohm effect explains the possibility of *fractional* (or so-called θ–) *statistics* of anyons in two-dimensional systems, i.e., particles carrying electric charge q and magnetic flux Φ. We refer the reader to Wilczek 1982a, 1982b, 1990, Fröhlich and Studer 1993, and Morandi 1992 for more details. $\diamond$

6.2.3 $SU(2)$-invariance and the Aharonov–Casher effect

As is well known, spin is a purely quantum concept. A particle with a spin $\mathbf{S}$ carries a magnetic moment

$$\mu_{\text{spin}} = \frac{g\mu}{\hbar}\,\mathbf{S} \, ,$$

where g denotes the giromagnetic ratio ($g = 2$ for electron), and $\mu = e\hbar/2m_e c$ is the Bohr magneton (m_e stands for the electron mass). Clearly, the spin operator $\mathbf{S}$ is given by

$$\mathbf{S} := \frac{\hbar}{2}\,\mathbf{L}^{(s)} \, ,$$

where $\mathbf{L}^{(s)} = (L_1^{(s)}, L_2^{(s)}, L_3^{(s)})$ are hermitian generators of $su(2)$ in the spin-s representation.

Now, a magnetic moment may also be considered from the classical point of view. In particular, one has the following classical equation of motion for $\mathbf{S}$, in a slowly varying external electromagnetic field $(\mathbf{E}, \mathbf{B})$:[6]

$$\frac{d\mathbf{S}}{dt} = \frac{e}{2m_e c} \mathbf{S} \times \left[g\mathbf{B} - (g-1)\frac{\mathbf{v}}{c} \times \mathbf{E} \right] + O((v/c)^2) \,. \tag{6.57}$$

The terms proportional to g describe the precession of the spin (or magnetic moment) in the magnetic field (in the rest frame of the spin). The remaining term describes a purely kinematical effect known as *Thomas precession*. The above equation may be rewritten as

$$\frac{dS_k}{dt} = \frac{e}{2m_e c^2} \sum_{i,j=1}^{3} \epsilon_{kij} S_i \left[g B_j v^0 + (g-1)\epsilon_{ijm} E_j v_m \right] = \sum_{i,j=1}^{3} \sum_{\mu=0}^{3} \epsilon_{kij} S_i b^j{}_\mu v^\mu \,, \tag{6.58}$$

where $v^\mu := (c, \mathbf{v})$, and

$$b^a{}_0 = \frac{e}{2m_e c^2} g B_a \,, \tag{6.59}$$

$$b^a{}_l = \frac{e}{2m_e c^2} (g-1) \sum_{m=1}^{3} \epsilon_{alm} E_m \,, \qquad l = 1,2,3 \,, \tag{6.60}$$

for $a = 1, 2, 3$. As usual S_k, E_k and B_k ($k = 1, 2, 3$) are components of $\mathbf{S}$, $\mathbf{E}$ and $\mathbf{B}$, respectively. In this way, one obtains the following gauge potential for an $SU(2)$ gauge theory:

$$b_\mu(x) := i \sum_{a=1}^{3} b^a{}_\mu(x) L_a^{(s)} \,, \qquad \mu = 0, 1, 2, 3 \,. \tag{6.61}$$

Now, to define the quantum theory let us proceed as in the $U(1)$ case. Define a covariant derivative

$$D_\mu = \partial_\mu + i a_\mu + b_\mu \,, \tag{6.62}$$

and replace the free Schrödinger equation by

$$i\hbar c D_0 \psi^{(s)} = -\frac{\hbar^2}{2m} \sum_{k=1}^{3} D_k \cdot D_k \psi^{(s)} \,, \tag{6.63}$$

with $\psi^{(s)}$ being a $(2s+1)$-component complex spinor. Rewriting this equation as

$$i\hbar \partial_0 \psi^{(s)} = H_{\text{spin}} \psi^{(s)} \,, \tag{6.64}$$

---

[6]This equation, found by Thomas in 1927, is a nonrelativistic version of the relativistic equation derived in 1959 by Bergmann, Michel and Telegdi — see Jackson 1999 for more details.

one easily finds the following expression for the spin Hamiltonian[7]:

$$H_{\text{spin}} := q\Phi\, \mathbb{1}_{2s+1} - \boldsymbol{\mu}_{\text{spin}} \cdot \mathbf{B} + \frac{1}{2m}\, \Pi^2$$
$$- \frac{1}{2mc}\Big[\Pi \cdot \left(\boldsymbol{\mu}_{\text{spin}} \times \mathbf{E}\right) + \left(\boldsymbol{\mu}_{\text{spin}} \times \mathbf{E}\right) \cdot \Pi\Big], \tag{6.65}$$

where Π is defined as follows:

$$\Pi := \mathbf{p} - \frac{q}{c}\mathbf{A}. \tag{6.66}$$

In the above formulae q and m stands for the particle charge and mass, respectively. In this way we have shown that the standard nonrelativistic theory of spinning particles is manifestly $U(1) \times SU(2)$-invariant. The spin $SU(2)$ gauge transformations are defined as follows:

$$\begin{cases} b_\mu & \longrightarrow \quad U \cdot b_\mu \cdot U^{-1} + U \cdot \partial_\mu U^{-1} \\[2mm] \psi^{(s)} & \longrightarrow \quad U\psi^{(s)} \end{cases}, \tag{6.67}$$

with $U \in SU(2)$. Interestingly, apart from the standard Pauli Hamiltonian

$$H_{\text{Pauli}} := q\Phi\, \mathbb{1}_{2s+1} - \boldsymbol{\mu}_{\text{spin}} \cdot \mathbf{B} + \frac{1}{2m}\, \Pi^2,$$

we obtain by the requirement of $SU(2)$ symmetry an additional term, known as a spin-orbit interaction.

Consider now a system of quantum neutral particles with spin $s = 1/2$, and hence carrying a magnetic moment $\boldsymbol{\mu}_{\text{spin}}$, (e.g. neutrons) moving in a xy-plane in $\mathbb{R}^3$. Following Aharonov and Casher (Aharonov and Casher 1984), we study the influence of a static, external electric field on the dynamics of such particles. Consider the static electric field $\mathbf{E}$ produced by a uniformly charged wire placed along the z-axis, with constant charge Q per unit length (cf. Fig. 6.5), that is,

$$\mathbf{E}(x, y) = \frac{Q}{2\pi r^2}\, (x, y). \tag{6.68}$$

Using the formula for $\mathbf{E}$ one obtains the following expressions for the x and y components of b^3_μ:

$$\mathbf{b}(x, y) := (b^3_1, b^3_2) = \frac{\Lambda}{4\pi}\, \frac{(y, -x)}{r^2} \cdot \sigma_3, \tag{6.69}$$

where

$$\Lambda = (g - 1)\frac{eQ}{m_e c^2}.$$

---

[7] Actually, we have omitted a term of order $O(b^2)$, which is neglected in the nonrelativistic theory.

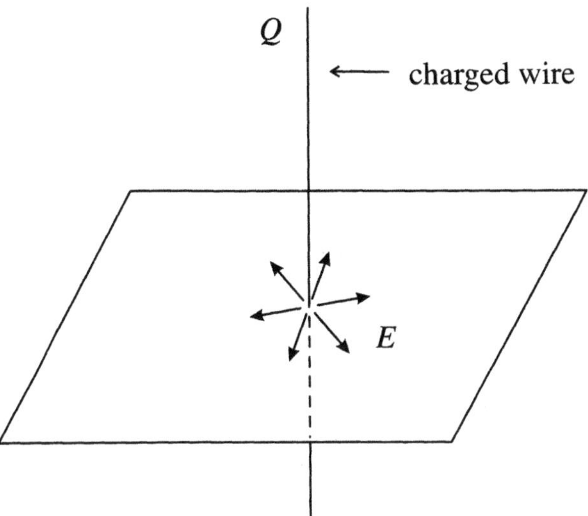

Figure 6.5: A charged wire along the z direction.

One easily finds that for a two-dimensional system confined to the x,y-plane, the only components of the $su(2)$-valued curvature

$$G^a_{\mu\nu} := \partial_\mu b^a_{\ \nu} - \partial_\nu b^a_{\ \mu} - 2\sum_{b,c=1}^{3} \epsilon_{abc} b^b_{\ \mu} b^c_{\ \nu} \tag{6.70}$$

read

$$G^3_{12}(\mathbf{x}) = -G^3_{21}(\mathbf{x}) = -\frac{\Lambda}{2\pi}\,\delta^2(\mathbf{x})\ . \tag{6.71}$$

Note the direct correspondence between (6.55) and (6.69), and between (6.54) and (6.71). The gauge field $G^3_{\mu\nu}$ vanishes outside the charged wire, that is, in the region in which the particle moves. In classical physics it means that in this region a particle is free. Moreover, if $\mathbf{S}(0)$ is perpendicular to the plane, then $\mathbf{S}(t) = \mathbf{S}(0)$. Hence, both particle momentum and spin are classically conserved. However, the scattering of quantum particles at the wire depends upon its charge density Q, via the corresponding holonomy element

$$U(C) = \mathrm{P}\exp\left(i\oint_C \mathbf{b}\cdot d\mathbf{l}\right) = \exp\left(i\Lambda\,\sigma_3\right) = \begin{pmatrix} e^{i\Lambda} & 0 \\ 0 & e^{-i\Lambda} \end{pmatrix}, \tag{6.72}$$

where C is any closed curve enclosing the wire. Clearly, the Aharonov–Casher phase factor $U(C)$ has a purely topological nature and, like the Aharonov–Bohm one, depends on the winding number of the closed curve C.

6.3   Phases in molecular physics

The manifestation of geometric phase in molecular physics was observed long before Berry's celebrated paper (Berry 1984). In 1958 Longuet-Higgens et al. drew attention to the fact that in some molecular systems the electron wave function acquires an additional phase factor, whose existence could not be explained on the grounds of the standard Born–Oppenheimer approximation. This additional factor was found to appear when the coordinates of nuclei interchanged cyclically around a point where the energy spectrum was degenerate. In 1979 chemists C.A.S. Mead and D. Truhlar demonstrated that the nature of this effect closely resembles the Aharonov–Bohm effect, and suggested that is be called the *molecular Aharonov–Bohm effect.*

In the present section we briefly present two natural ways in which the geometric phase enters the game: one way is via accidental degeneracies of the spectrum, and the second way is via the improved Born–Oppenheimer approximation.

6.3.1   Degeneracies

As we have already noted, degeneracies of the spectrum play a crucial role in determining the geometric Berry's phase. Recall, for example, our basic system of a spin particle in a magnetic field (see section 2.2.5), where Berry's curvature $F^{(n)}$ corresponds to the field of a magnetic pole placed at the degeneracy point $\mathbf{B} = 0$. Consider an arbitrary quantum system parametrized by some parameters $\mathbf{R} = (R_1, \ldots, R_n) \in M$. Suppose that for some point $\mathbf{R}^*$ two energy eigenvectors $\psi_1(\mathbf{R})$ and $\psi_2(\mathbf{R})$ are degenerate, with energy $E^* := E(\mathbf{R}^*)$, i.e.,

$$H(\mathbf{R})\psi_i(\mathbf{R}) = E_i(\mathbf{R})\psi_i(\mathbf{R}) , \qquad i = 1, 2 , \tag{6.73}$$

and

$$E_1(\mathbf{R}^*) = E_2(\mathbf{R}^*) = E^* .$$

Near a degeneracy point $\mathbf{R}^*$ the system may be considered as a two-level system, and its Hamiltonian described by 2×2 $\mathbf{R}$-dependent matrix:

$$\begin{pmatrix} H_{11}(\mathbf{R}) & H_{12}(\mathbf{R}) \\ H_{12}^*(\mathbf{R}) & H_{22}(\mathbf{R}) \end{pmatrix} .$$

In looking for eigenvalues, one solves the following secular equation:

$$E^2 - E(H_{11} + H_{22}) + H_{11}H_{22} - |H_{12}|^2 = 0 .$$

In order to have a degenerate eigenvalue, the discriminant of the secular equation has to vanish, i.e.,

$$(H_{11} - H_{22})^2 + 4|H_{12}|^2 = 0 ,$$

which means that there are three conditions upon the matrix elements H_{ij}, namely

$$H_{11} = H_{22}, \quad \mathrm{Re}\, H_{12} = 0, \quad \mathrm{Im}\, H_{12} = 0. \tag{6.74}$$

The above system of equations define an $(n-3)$-dimensional submanifold of M (with n being the dimension of M). If $n = 3$, then degeneracies define isolated points in M.

In particular, for a system with time-reversal symmetry, the Hamiltonian H is real and $\mathrm{Im}\, H_{12} \equiv 0$. Hence, the subspace of degenerate points defines an $(n-2)$-dimensional submanifold of M. This fact was noted by von Neumann and Wigner (1929). Note that if we introduce the three parameters

$$X := \frac{1}{2}(H_{11} - H_{22}), \quad Y := \frac{1}{4}\,\mathrm{Re}\, H_{12}, \quad Z := \frac{1}{4}\,\mathrm{Im}\, H_{12}, \tag{6.75}$$

the formula for energy eigenvalues implies that near degeneracy, i.e., near a point $(X, Y, Z) = 0$, the level surface of the eigenvalues of H, considered as a function of (X, Y, Z), forms a double cone with its apex at the degeneracy point. That is,

$$E_{1,2}(X, Y) = E^* \pm \sqrt{X^2 + Y^2 + Z^2}, \tag{6.76}$$

where E^* denotes the energy of the crossing, i.e., $E^* = (H_{11} + H_{22})/2$. The apex of this cone is called a *diabolical point*.[8]

To see how the geometric phase is related to the degeneracies of the spectrum, let us consider a simple example studied by Herzberg and Longuet-Higgins (1963). Suppose that two electronic levels are degenerate at a degeneracy point $\mathbf{r}^*$. Hence, near $\mathbf{r}^*$ the Hamiltonian may be truncated to a 2×2 matrix acting only on the near-degenerate states. Up to an unimportant identity operator $\mathbb{1}_2$, any real symmetric 2×2 matrix may be written as follows:

$$H(\mathbf{r}) = x\sigma_3 + y\sigma_1 = \begin{pmatrix} x & y \\ y & -x \end{pmatrix} = r \begin{pmatrix} \cos\phi & \sin\phi \\ \sin\phi & -\cos\phi \end{pmatrix}, \tag{6.77}$$

where $r = \sqrt{x^2 + y^2}$. The eigenvalues are $E_\pm = \pm r$, and the corresponding eigenvectors $\psi_\pm$ are given by

$$\psi_+(\mathbf{r}) = \begin{vmatrix} \cos\left(\frac{\phi}{2}\right) \\ \sin\left(\frac{\phi}{2}\right) \end{vmatrix}, \quad \psi_-(\mathbf{r}) = \begin{vmatrix} -\sin\left(\frac{\phi}{2}\right) \\ \cos\left(\frac{\phi}{2}\right) \end{vmatrix}. \tag{6.78}$$

Now, let us trace a circuit on the xy-plane with the center at $(0, 0)$, and radius r. Note that

$$\psi_+(r, 0) = \begin{vmatrix} 1 \\ 0 \end{vmatrix}, \quad \psi_-(r, 0) = \begin{vmatrix} 0 \\ 1 \end{vmatrix}, \tag{6.79}$$

---

[8] For a discussion of the importance of diabolical points in molecular physics, see Berry 1985b.

whereas,

$$\psi_+(r, 2\pi) = \begin{vmatrix} -1 \\ 0 \end{vmatrix}, \quad \psi_-(r, 2\pi) = \begin{vmatrix} 0 \\ -1 \end{vmatrix}. \tag{6.80}$$

This means that the maps

$$\mathbb{R}^2 - \{0\} \ni \mathbf{r} \longrightarrow \psi_\pm(\mathbf{r}),$$

are not single-valued. In undergoing one circuit around the degeneracy point $\mathbf{r}^* = 0$, one obtains

$$\psi_\pm(r, 2\pi) = e^{i\pi}\,\psi_\pm(r, 0) = -\psi_\pm(r, 0). \tag{6.81}$$

Note, however, that performing a simple gauge transformation:

$$\chi_\pm(r, \phi) = e^{i\phi}\psi_\pm(r, \phi), \tag{6.82}$$

we arrive at single-valued eigenvectors $\chi_\pm(\mathbf{r})$, i.e., such that

$$\chi_\pm(r, 2\pi) = \chi_\pm(r, 0). \tag{6.83}$$

What is the basic difference between the single-valued $\chi_\pm$ and the multi-valued $\psi_\pm$? Let us note that $\chi_\pm$, contrary to $\psi_\pm$, give rise to a nontrivial *gauge potential* $A^{(\pm)}$. One easily finds that

$$\langle \psi_\pm | d | \psi_\pm \rangle = 0, \tag{6.84}$$

whereas

$$i \langle \chi_\pm | d | \chi_\pm \rangle = A_\phi^{(\pm)} d\phi + A_r^{(\pm)} dr, \tag{6.85}$$

with

$$A_r^{(\pm)} = 0 \quad \text{and} \quad A_\phi^{(\pm)} = -\frac{1}{2}. \tag{6.86}$$

This is an example of a situation that frequently arises in the theory of geometric phases in molecular physics: One has to choose between the vanishing of a gauge potential and the single-valuedness of the electronic eigenvectors. In general, it is not possible to have both.

Let C be a closed curve on $M := \mathbb{R}^2 - \{0\}$ enclosing the degeneracy point $\mathbf{r}^* = 0$. Any such curve defines an element of the fundamental group of M, which is isomorphic to the set of integers, i.e.,

$$C \longrightarrow [C] \in \pi_1(M) \cong \mathbb{Z}.$$

Evidently, if C_1 and C_2 are homotopically equivalent, then

$$\oint_{C_1} A^{(\pm)} = \oint_{C_2} A^{(\pm)}.$$

Moreover, if $n \in \mathbb{Z}$ denotes the winding number of C (cf. Example 1.4.3), then it is easy to see that

$$\oint_C A^{(\pm)} = (-1)^n \pi .$$ (6.87)

In the context of molecular physics the winding number $n \in \mathbb{Z}$ is called the *Longuet-Higgins charge*. It is the analog of the magnetic number (or magnetic charge) of a Dirac monopole, and the instanton number (or instanton charge) in the $SU(2)$ gauge theory.

Finally, let us note that it is easy to construct a principal fibre bundle such that the corresponding holonomy element reproduces a Longuet-Higgins phase shift $\pm\pi$. Note that the radial coordinate r does not play any role, and hence we may reduce the parameter space M to a one-dimensional sphere S^1. Both spaces have the same fundamental group, i.e.,

$$\pi_1(M) = \pi_1(S^1) \cong \mathbb{Z} .$$

The corresponding fibre bundle is a principal $\mathbb{Z}_2$-bundle over S^1, constructed in perfect analogy to a Möbius strip (cf. Example 1.3.1). In the next section, we shall study more complicated examples of molecular systems giving rise to nonabelian geometric phases.

6.3.2 Time-reversal invariant fermionic system

Another interesting system displaying a nonabelian geometric phase factor is the time-reversal invariant fermionic system. In section 2.2.5, we studied a spin system in a magnetic field defined by the following Hamiltonian:

$$H(\mathbf{B}) = \mathbf{B} \cdot \mathbf{J} ,$$ (6.88)

where $\mathbf{J}$ denotes a spin vector. The spectrum of $H(\mathbf{B})$ is given by

$$\mathrm{Spectrum} = \left\{ m|\mathbf{B}| \;\Big|\; m = J, -J+1, \ldots, J \right\},$$

and hence $H(\mathbf{B})$ is nondegenerate away from a point $\mathbf{B} = 0$, i.e., on a punctured three-dimensional space $M = \mathbb{R}^3 - \{0\}$. It is convenient to "normalize" the spectrum of $H(\mathbf{B})$ by restricting $\mathbf{B}$ to a unit sphere S^2, i.e., $|\mathbf{B}| = 1$. Denote by $P_m(\mathbf{B})$ the spectral projection on the mth eigenspace of $H(\mathbf{B})$. The corresponding mth spectral complex line bundle over S^2

$$\pi_m : \mathcal{P}_m \longrightarrow S^2 ,$$

is defined by

$$\pi_m^{-1}(\mathbf{B}) := \left\{ \psi \in \mathbb{C}^{2J+1} \;\Big|\; P_m(\mathbf{B})\psi = \psi \right\} .$$ (6.89)

It is an associated bundle to a principal $U(1)$ Hopf bundle $S^3 \longrightarrow S^2$. Hence, a spectral bundle is uniquely characterized by the corresponding Chern number. We have already shown that (see formula (2.118))

$$\text{Chern number of } m\text{th spectral bundle} = -2m \,.$$

Moreover, if C is a closed curve on S^2 then the corresponding Berry adiabatic phase is (cf. (2.117))

$$\text{Berry's phase} = -m\, \Omega(C) \,, \tag{6.90}$$

where $\Omega(C)$ is the solid angle subtended by C.

Let us turn to a system displaying time reversal invariance. For the importance of time reversal in molecular physics see, e.g., the review article Mead 1992. Time reversal in quantum mechanics is implemented by an anti-unitary operator Θ, such that

$$\Theta^2 = \begin{cases} \mathbb{1} & \text{for bosons} \\ -\mathbb{1} & \text{for fermions} \end{cases} \,.$$

The distinction between bosons and fermions comes from the fact that the angular momentum $\mathbf{J}$ is odd under time reversal, i.e.,

$$\Theta \mathbf{J} = -\mathbf{J}\, \Theta \,. \tag{6.91}$$

Such an operator can be represented by

$$\Theta \psi := U \overline{\psi} \,, \tag{6.92}$$

where U is a unitary operator acting in the Hilbert space $\mathcal{H}$ and $\overline{\psi}$ denotes the complex conjugate of ψ. In the usual representation, where J_1 and J_3 are real and J_2 is imaginary, the unitary operator U represents a rotation by π around the y-axis, i.e.,

$$U = \exp\left(-i\pi J_2\right) \,. \tag{6.93}$$

Hence,

$$\Theta^2 = U^2 = \exp\left(-2i\pi J_2\right) \,, \tag{6.94}$$

defines a 2π-rotation around the y-axis, and therefore $\Theta^2 = \mathbb{1}$ for bosons, and $-\mathbb{1}$ for fermions. Note that if $\mathcal{H}$ is finite-dimensional, then $\Theta^2 = -\mathbb{1}$ implies that $\dim_{\mathbb{C}} \mathcal{H}$ is even. Indeed, for any $\psi \in \mathcal{H}$ one has

$$\langle \psi | \Theta \psi \rangle = \langle \Theta^2 \psi | \Theta \psi \rangle = -\langle \psi | \Theta \psi \rangle \,, \tag{6.95}$$

where we have used the anti-unitarity of Θ, i.e., $\langle \phi | \psi \rangle = \langle \Theta \psi | \Theta \phi \rangle$. Thus, ψ and $\Theta \psi$ are orthogonal, which proves the even-dimensionality of $\mathcal{H}$. An immediate consequence of the above observation is (so-called) *Kramer's degeneracy*, i.e., if H is a hermitian operator commuting with Θ, then each eigenvalue of H has degeneracy of even degree. Obviously, the dipole Hamiltonian (6.88) is odd under time reversal, i.e.,

$$\Theta H(\mathbf{B}) = -H(\mathbf{B})\Theta . \tag{6.96}$$

Mead (see Mead 1992) proposed to study a time-reversal invariant quadrupole system described by the following Hamiltonian:

$$H(Q) := \sum_{k,l=1}^{3} Q_{kl} J_k J_l , \tag{6.97}$$

where Q_{kl} are the components of a real 3×3 symmetric and traceless matrix — a quadrupole matrix. For any quadrupole Q one has

$$\Theta H(Q) = H(Q)\Theta . \tag{6.98}$$

Due to Kramer's degeneracy, any quadrupole Hamiltonian $H(Q)$ has always degenerate eigenvalues. Recall that a hermitian operator over $\mathbb{R}$ ($\mathbb{C}$, or $\mathbb{H}$) is *simple* if it has no degenerate eigenvalues. Hence, the time-reversal invariant $H(Q)$ is never simple over $\mathbb{C}$. Interestingly, one may prove (see Avron, Sadun, Segert and Simon 1989) that for $Q \neq 0$, the time-reversal invariant $H(Q)$ with odd J is simple over $\mathbb{H}$. The space of quadrupole matrices is a five-dimensional real vector space endowed with the following scalar product:

$$(Q_1, Q_2) := \frac{3}{2} \operatorname{Tr}(Q_1 Q_2) . \tag{6.99}$$

A convenient orthonormal basis is given by

$$Q_0 = \frac{1}{3} \begin{pmatrix} -1 & 0 & 0 \\ 0 & -1 & 0 \\ 0 & 0 & 2 \end{pmatrix}, \quad Q_1 = \frac{1}{\sqrt{3}} \begin{pmatrix} 0 & 0 & 1 \\ 0 & 0 & 0 \\ 1 & 0 & 0 \end{pmatrix}, \quad Q_2 = \frac{1}{\sqrt{3}} \begin{pmatrix} 0 & 0 & 0 \\ 0 & 0 & 1 \\ 0 & 1 & 0 \end{pmatrix},$$

$$Q_3 = \frac{1}{\sqrt{3}} \begin{pmatrix} 1 & 0 & 0 \\ 0 & -1 & 0 \\ 0 & 0 & 0 \end{pmatrix}, \quad Q_4 = \frac{1}{\sqrt{3}} \begin{pmatrix} 0 & 1 & 0 \\ 1 & 0 & 0 \\ 0 & 0 & 0 \end{pmatrix}.$$

The simplest fermionic system corresponds to $J = 1/2$. Note, however, that in this case $H(Q) = 0$ for any Q. To see this, let us observe that for $J = 1/2$, we have $J_k = \frac{1}{2}\sigma_k$ and, hence, it is easy to show that

$$\sum_{k,l=1}^{3} (Q_\alpha)_{kl} J_k J_l = 0 , \tag{6.100}$$

for any $\alpha = 0, 1, 2, 3, 4$. Let us therefore investigate fermionic systems with $J = 3/2$. It turns out that, for quadrupole Hamiltonians $H(Q)$, the case $J = 3/2$ plays a role similar to that of the case $J = 1/2$ for dipole Hamiltonians $H(\mathbf{B})$.

Recall that for a dipole Hamiltonian $H(\mathbf{B})$ with $J = 1/2$, there is a direct correspondence between unit vectors $\mathbf{B}$ and traceless hermitian operators in $\mathbb{C}^2$, with spectrum $\{-1, +1\}$. This means that if $H : \mathbb{C}^2 \longrightarrow \mathbb{C}^2$ is a traceless hermitian operator and $\mathrm{Spec}_{\mathbb{C}}(H) = \{-1, +1\}$, then $H = H(\mathbf{B})$ for some unit vector $\mathbf{B}$. A similar correspondence holds for quadrupole Hamiltonians $H(Q)$ with $J = 3/2$, that is, there is a direct correspondence between unit quadrupoles Q and traceless quaternionic hermitian operators in $\mathbb{H}^2$, with spectrum $\{-1, +1\}$. That is, if $H : \mathbb{H}^2 \longrightarrow \mathbb{H}^2$ is a traceless quaternionic Hermitian operator, and $\mathrm{Spec}_{\mathbb{H}}(H) = \{-1, +1\}$, then $H = H(Q)$ for some unit quadrupole Q.

Any two unit vectors $\mathbf{B}_1, \mathbf{B}_2 \in S^2$ are related by an $SO(3)$ rotation R, i.e., $\mathbf{B}_2 = R\,\mathbf{B}_1$. The corresponding dipole Hamiltonians $H(\mathbf{B}_1)$ and $H(\mathbf{B}_2)$ are unitary related, as follows:

$$H(\mathbf{B}_2) = U(R)H(\mathbf{B}_1)U(R)^{-1}\,, \tag{6.101}$$

where

$$SO(3) \ni R \longrightarrow U(R) \in SU(2)\,,$$

denotes a unitary representation of $SO(3)$ in $\mathbb{C}^2$. More precisely, it is a representation of the universal (two-fold) cover of $SO(3)$, that is, $\mathrm{Spin}(3) \cong SU(2)$. Therefore, it is only a projective representation of $SO(3)$. A similar picture is true for quadrupole Hamiltonians $H(Q)$ with $J = 3/2$. The set of unit quadrupoles defines a unit sphere

$$S^4 = \{ Q \mid (Q, Q) = 1 \} \subset \mathbb{R}^5\,.$$

Any two $Q_1, Q_2 \in S^4$ are related by an $SO(5)$ rotation R, i.e., $Q_2 = R\,Q_1$. It is well known that the universal (two-fold) cover of $SO(5)$ is $\mathrm{Spin}(5) \cong Sp(2)$, and the unitary representation of $\mathrm{Spin}(5)$ (or equivalently, the projective representation of $SO(5)$)

$$\mathrm{Spin}(5) \ni R \longrightarrow U(R) \in Sp(2)\,,$$

is realized by

$$H(Q_2) = U(R)H(Q_1)U(R)^{-1}\,. \tag{6.102}$$

Corollary 6.3.1 *For $J = 3/2$, any two quadrupole Hamiltonians $H(Q_1)$ and $H(Q_2)$, with $Q_1, Q_2 \in S^4$, are unitary related.*

We stress that this result holds only for $J = 3/2$.

Finally, let $P_\pm(Q)$ denote the spectral projections onto the positive and negative eigenspaces of $H(Q)$ (i.e., those corresponding to eigenvalues ± 1) with $Q \in S^4$:

$$P_\pm(Q) := \frac{1}{2}\,(\mathbb{1} \pm H(Q))\,. \tag{6.103}$$

To see that $P_\pm(Q)$ are projections in $\mathbb{H}^2$, observe that

$$[H(Q)]^2 = (Q, Q)\,\mathbb{1} = \mathbb{1} , \qquad (6.104)$$

for any unit quadrupole Q, and hence

$$[P_\pm(Q)]^2 = \frac{1}{4}(\mathbb{1} \pm 2H(Q) + \mathbb{1}) = P_\pm(Q) . \qquad (6.105)$$

Actually, for $J = 3/2$, the quadrupole operators $H(Q)$ form a Clifford algebra

$$H(Q_1)H(Q_2) + H(Q_2)H(Q_1) = 2(Q_1, Q_2) , \qquad (6.106)$$

(see Avron, Sadun, Segert and Simon 1989). Defining

$$T_\alpha := H(Q_\alpha) , \quad \alpha = 0, 1, 2, 3, 4 , \qquad (6.107)$$

one finds, for $J = 3/2$,

$$T_0 = \begin{pmatrix} 1 & 0 \\ 0 & -1 \end{pmatrix} , \quad T_1 = \begin{pmatrix} 0 & -\hat{\jmath} \\ \hat{\jmath} & 0 \end{pmatrix} , \quad T_2 = \begin{pmatrix} 0 & \hat{k} \\ -\hat{k} & 0 \end{pmatrix} ,$$

$$T_3 = \begin{pmatrix} 0 & 1 \\ 1 & 0 \end{pmatrix} , \quad T_4 = \begin{pmatrix} 0 & -\hat{\imath} \\ \hat{\imath} & 0 \end{pmatrix} ,$$

where $\hat{\imath}, \hat{\jmath}, \hat{k}$ denote basic quaternions (cf. Appendix B). The formula (6.106) may be equivalently rewritten as follows:

$$T_\alpha T_\beta + T_\beta T_\alpha = 2\delta_{\alpha\beta} . \qquad (6.108)$$

Denote by $\mathcal{P}_\pm$ the corresponding spectral bundles, i.e.,

$$\pi_\pm : \mathcal{P}_\pm \longrightarrow S^4 ,$$

with

$$\pi_\pm^{-1}(Q) := \left\{ \psi \in S^7 \ \middle|\ P_\pm(Q)\psi = \psi \right\} ,$$

where S^7 denotes the space of unit vectors in $\mathbb{H}^2$. Hence, spectral bundles are precisely Hopf bundles

$$S^7 \longrightarrow S^4$$

described in section 1.4.5. These bundles are entirely characterized by the corresponding Chern number. Following section 1.4.5 the reader may show that the

$$\text{Chern number of the } \mathcal{P}_\pm \text{ spectral bundle} = \pm 1 . \qquad (6.109)$$

Therefore, the above spectral bundles correspond to instantons (cf. section 1.4.3) with instanton numbers

$$k = \pm 1 \, .$$

Finally, the nonabelian geometric phase factor is defined as follows:

$$U(C) := \mathrm{P} \exp \left(\oint_C A \right) \in SU(2) \, , \tag{6.110}$$

where A denotes an $su(2)$-valued gauge potential (instanton) and C is any closed curve in the sphere S^4 of unit quadrupoles.

6.3.3   Born–Oppenheimer approximation

The Born–Oppenheimer approach (Born and Oppenheimer 1927, Born and Huang 1954) is one of the basic tools of molecular physics (see, e.g., Bohm 1993a, Schiff 1968). Roughly speaking one divides the dynamics of a complicated physical system (a molecule) into two parts: *fast motion*, which is described by *fast variables* $\mathbf{r}$ and $\mathbf{p}$ (the positions and momenta of the electrons); and *slow motion*, which is described by *slow variables* $\mathbf{R}$ and $\mathbf{P}$ (the variables of the nuclei). In the standard Born–Oppenheimer approach, one considers the slow variables as slowly changing classical parameters. However, in the full quantum theory $\mathbf{R}$ and $\mathbf{P}$ become quantum operators $\hat{\mathbf{R}}$ and $\hat{\mathbf{P}}$. The Hilbert space $\mathcal{H}$ of the composed system is a tensor product of the space for slow motion and the space for fast motion, i.e.,

$$\mathcal{H} = \mathcal{H}^{\mathrm{slow}} \otimes \mathcal{H}^{\mathrm{fast}} \, .$$

The spaces $\mathcal{H}^{\mathrm{slow}}$ and $\mathcal{H}^{\mathrm{fast}}$ are, in general, infinite-dimensional but often one obtains a good approximation if one restricts oneself to a finite-dimensional subspace of $\mathcal{H}^{\mathrm{slow}}$.

Consider a molecule with a Hamiltonian

$$\hat{H} = \frac{\hat{\mathbf{P}}^2}{2M} + \frac{\hat{\mathbf{p}}^2}{2m} + V(\mathbf{R}, \mathbf{r}) \, . \tag{6.111}$$

Since the light electrons instantaneously follow the motion of the heavy nuclei, the slow variables can also be understood as being the variables of the molecule as a whole, i.e., collective variables. In particular, for a diatomic molecule, $\mathbf{R}$ will be the vector along the internuclear axis and $\mathbf{P}$ its conjugate momentum. The potential $V(\mathbf{R}, \mathbf{r})$ is, in general, a highly complicated function of the nuclear and electronic positions $\mathbf{R}$ and $\mathbf{r}$, and possibly some other operators, like, e.g., spin. One can divide the molecular Hamiltonian into two parts, as follows:

$$\hat{H} = \frac{\hat{\mathbf{P}}^2}{2M} + \hat{h}(\mathbf{R}) \, , \tag{6.112}$$

where

$$\hat{h}(\mathbf{R}) = \frac{\hat{\mathbf{p}}^2}{2m} + V(\mathbf{R}, \mathbf{r}) \, . \tag{6.113}$$

Let us take as a basis in $\mathcal{H}^{\text{slow}}$, the generalized eigenvectors of $\hat{\mathbf{R}}$, such that

$$\hat{\mathbf{R}}|\mathbf{R}\rangle = \mathbf{R}|\mathbf{R}\rangle \, , \tag{6.114}$$

together with

$$\langle \mathbf{R}|\mathbf{R}'\rangle = \delta^{(3)}(\mathbf{R} - \mathbf{R}') \, .$$

Similarly, we take $|\mathbf{r}\rangle$ to be generalized eigenvectors of $\hat{\mathbf{r}}$:

$$\hat{\mathbf{r}}|\mathbf{r}\rangle = \mathbf{r}|\mathbf{r}\rangle \, , \tag{6.115}$$

satisfying the normalization condition

$$\langle \mathbf{r}|\mathbf{r}'\rangle = \delta^{(3)}(\mathbf{r} - \mathbf{r}') \, .$$

Clearly, we may define a generalized basis in $\mathcal{H}$ as follows:

$$|\mathbf{R}, \mathbf{r}\rangle := |\mathbf{R}\rangle \otimes |\mathbf{r}\rangle \, .$$

Now, for each eigenvalue $\mathbf{R} \in \mathbb{R}^3$, we define a basis $|n(\mathbf{R})\rangle$ in $\mathcal{H}^{\text{fast}}$ such that

$$|\mathbf{R}, n\rangle := |\mathbf{R}\rangle \otimes |n(\mathbf{R})\rangle$$

satisfies

$$\hat{h}(\mathbf{R})|\mathbf{R}, n\rangle = \epsilon_n(\mathbf{R})|\mathbf{R}, n\rangle, \tag{6.116}$$

together with $\langle n(\mathbf{R})|m(\mathbf{R})\rangle = \delta_{nm}$. Consider, now, the eigenvalue problem for the total Hamiltonian, i.e.,

$$\hat{H}|\Psi^E\rangle = E|\Psi^E\rangle \, . \tag{6.117}$$

Using the notation

$$\Psi^E(\mathbf{R}, \mathbf{r}) := \langle \mathbf{R}, \mathbf{r}|\Psi^E\rangle \, , \quad \text{and} \quad \Psi_n^E(\mathbf{R}) := \langle \mathbf{R}, n|\Psi^E\rangle \, ,$$

one obtains

$$\begin{aligned}
\psi^E(\mathbf{R}, \mathbf{r}) &= \sum_n \int d\mathbf{R}' \, \langle \mathbf{R}, \mathbf{r}|\mathbf{R}', n\rangle \langle \mathbf{R}', n|\Psi^E\rangle \tag{6.118}\\
&= \sum_n \int d\mathbf{R}' \, \langle \mathbf{R}|\mathbf{R}'\rangle \langle \mathbf{r}|n(\mathbf{R}')\rangle \Psi_n^E(\mathbf{R}') = \sum_n \phi_n(\mathbf{R}, \mathbf{r})\Psi_n^E(\mathbf{R}) \, ,
\end{aligned}$$

where

$$\phi_n(\mathbf{R}, \mathbf{r}) := \langle \mathbf{r} | n(\mathbf{R}) \rangle .$$

Now, in the adiabatic Born–Oppenheimer approximation one neglects mixing between different electronic levels, so we can take for the entire wavefunction $\psi^E(\mathbf{R}, \mathbf{r})$ just one term in the sum (6.118), corresponding to the nth electronic level. The eigenvalue problem (6.117) implies that

$$\left[-\frac{\hbar^2}{2M} \nabla_\mathbf{R}^2 + \hat{h}(\mathbf{R}, \mathbf{r}) \right] \phi_n(\mathbf{R}, \mathbf{r}) \Psi_n^E(\mathbf{R}) = E \phi_n(\mathbf{R}, \mathbf{r}) \Psi_n^E(\mathbf{R}) , \qquad (6.119)$$

where the electronic Hamiltonian reads

$$\hat{h}(\mathbf{R}, \mathbf{r}) = -\frac{\hbar^2}{2m} \nabla_\mathbf{r} + V(\mathbf{R}, \mathbf{r}) .$$

Now, by the very definition of $\phi_n(\mathbf{R}, \mathbf{r})$ one has

$$\hat{h}(\mathbf{R}, \mathbf{r}) \phi_n(\mathbf{R}, \mathbf{r}) = \epsilon_n(\mathbf{R}) \phi_n(\mathbf{R}, \mathbf{r}) .$$

Hence, multiplying both sides of (6.119) by $\phi_n^*(\mathbf{R}, \mathbf{r})$, and integrating over $\mathbf{r}$, one finds

$$\left[\frac{-\hbar^2}{2M} \mathbf{D}^2 + \epsilon_n(\mathbf{R}) \right] \psi_n^E(\mathbf{R}) = E \psi_n^E(\mathbf{R}) , \qquad (6.120)$$

where we have introduced a covariant derivative

$$\mathbf{D} := \nabla_\mathbf{R} - i\mathbf{A}^{(n)}(\mathbf{R}) , \qquad (6.121)$$

and the gauge potential $\mathbf{A}^{(n)}$ is given by

$$\mathbf{A}^{(n)}(\mathbf{R}) := i \langle n(\mathbf{R}) | \nabla_\mathbf{R} | n(\mathbf{R}) \rangle = i \int d\mathbf{r}\, \phi_n^*(\mathbf{R}, \mathbf{r}) \nabla_\mathbf{R} \phi_n(\mathbf{R}, \mathbf{r}) . \qquad (6.122)$$

The gauge potential $\mathbf{A}^{(n)}$ is called by chemical physicists a *Mead potential*. It was neglected in the conventional Born–Oppenheimer approximation. Mead and Truhlar (1979) called the result of the modification of the conventional Born-Oppenheimer approximation a *molecular Aharonov–Bohm effect*. Obviously, if the nth electronic level is N-times degenerate, i.e.,

$$\hat{h}(\mathbf{R}) | n_a(\mathbf{R}) \rangle = \epsilon_n(\mathbf{R}) | n_a(\mathbf{R}) \rangle , \quad a = 1, \dots, N , \qquad (6.123)$$

then instead of (6.120) we obtain

$$\sum_{c=1}^{N} \left[\frac{-\hbar^2}{2M} \sum_{b=1}^{N} \mathbf{D}_{ab} \mathbf{D}_{bc} + \epsilon_n(\mathbf{R}) \delta_{ac} \right] \psi_{nc}^E(\mathbf{R}) = E \psi_{na}^E(\mathbf{R}) , \qquad (6.124)$$

with

$$\mathbf{D}_{ab} := \delta_{ab}\,\nabla_{\mathbf{R}} - i\mathbf{A}_{ab}^{(n)}\,, \tag{6.125}$$

and the nonabelian gauge potential $\mathbf{A}^{(n)}$ is given by

$$\mathbf{A}_{ab}^{(n)} := i\langle n_b(\mathbf{R})|\nabla_{\mathbf{R}}|n_a(\mathbf{R})\rangle\,. \tag{6.126}$$

Clearly, the above formula reproduces the Wilczek–Zee gauge potential (cf. formula (2.178)). Hence, the Born–Oppenheimer treatment of molecular systems gives rise to a natural manifestation of adiabatic geometric phases. We stress that in the conventional approach to the Born-Oppenheimer approximation, the above-derived gauge structure is completely absent; "slow parameters" $\mathbf{R}$ are kept fixed and one considers the quantum dynamics of the "fast parameters" $\mathbf{r}$. The improved approach, which takes into account the back reaction[9] of the fast degrees of freedom on the slow ones, gives rise to the appearance of the gauge structure (i.e., the slow degrees of freedom move in the external gauge potential induced by the fast degrees of freedom) and leads in a natural way to a nonabelian geometric phase.

6.3.4   Diatomic molecule

As an application of the general scheme of the Born-Oppenheimer approximation, let us consider a diatomic molecule (Bohm 1993a,b). Such a molecule has an axial symmetry about the internuclear axis $\mathbf{R} \in \mathbb{R}^3$. The electronic states are classified according to the eigenvalues of

$$\Omega := \sqrt{\text{eigenvalue of } \mathbf{J}\cdot\mathbf{R}/R}\,,$$

where

$$\mathbf{J} := \mathbf{S}_{\text{elec}} + \mathbf{L}_{\text{elec}}\,,$$

is the total electronic angular momentum, i.e., the sum of the spin $\mathbf{S}_{\text{elec}}$ and the orbital angular momentum $\mathbf{L}_{\text{elec}}$. Consider the simplest situation of when the molecule is in the so-called Σ state, i.e., $\mathbf{L}_{\text{elec}} = 0$. Clearly, Ω can take values from the set $\{0, \frac{1}{2}, 1, \frac{3}{2}, 2, \dots\}$, and for for a given energy state this value is fixed. Since the molecular Hamiltonian is parity invariant, any energy eigenstate is doubly degenerate, i.e., if $\Omega = m$, then one has two eigenvectors in $\mathcal{H}^{\text{fast}}$, namely

$$|m(\mathbf{R})\rangle \quad \text{and} \quad |-m(\mathbf{R})\rangle\,,$$

such that

$$H(\mathbf{R})|\pm m(\mathbf{R})\rangle = E_m(\mathbf{R})|\pm m(\mathbf{R})\rangle\,.$$

---

[9]This back reaction will be further studied in section 6.3.5.

Therefore, the Born–Oppenheimer approximation leads to the $SU(2)$ gauge theory with gauge potential

$$A_{ab}^{(m)}(\mathbf{R}) = i \langle b(\mathbf{R})|d_{\mathbf{R}}|a(\mathbf{R}) \rangle , \qquad (6.127)$$

where $a, b = \pm m$. To find the gauge potential A and its curvature F, we shall follow the procedure developed in section 2.3.3. It is clear that in our discussion only the direction of $\mathbf{R}$ is important, and hence let us assume that $R = |\mathbf{R}| = 1$, and parametrize the corresponding parameter space — a two-dimensional sphere — using the standard spherical angles (θ, φ). Following (2.203), we define

$$|m(\theta, \varphi) \rangle := U(\theta, \varphi)|m(\mathbf{e}_3) \rangle , \qquad (6.128)$$

where $U(\theta, \varphi)$ is as introduced in (2.86). The corresponding connection form is given by the following 2×2 matrix:

$$A^{(m)}(\theta, \varphi) = A_\theta^{(m)}(\theta, \varphi) \, d\theta + A_\varphi^{(m)}(\theta, \varphi) \, d\varphi . \qquad (6.129)$$

One finds, for $|m| \neq 1/2$ (cf. formulae (2.213)–(2.214)),

$$A_\theta^{(m)} = 0 , \qquad (6.130)$$

$$A_\varphi^{(m)} = -m(1 - \cos\theta)\,\sigma_3 , \qquad (6.131)$$

and, for $m = 1/2$ (cf. formulae (2.215)–(2.216)),

$$A_\theta^{(1/2)} = \frac{\kappa}{2}(-\cos\varphi\sigma_2 - \sin\varphi\sigma_1) = -\frac{\kappa}{2} \begin{pmatrix} 0 & ie^{i\varphi} \\ -ie^{-i\varphi} & 0 \end{pmatrix} , \qquad (6.132)$$

$$A_\varphi^{(1/2)} = \frac{1}{2}\Big[(1 - \cos\theta)\sigma_3 + \kappa \sin\theta(-\cos\varphi\sigma_1 + \sin\varphi\sigma_2)\Big]$$

$$= \frac{1}{2} \begin{pmatrix} (1 - \cos\theta) & -\kappa \sin\theta e^{i\varphi} \\ -\kappa \sin\theta e^{-i\varphi} & -(1 - \cos\theta) \end{pmatrix} , \qquad (6.133)$$

where the parameter κ is defined by

$$\langle a(\mathbf{e}_3)|J_k|b(\mathbf{e}_3) \rangle = \frac{\kappa}{2}(\sigma_k)_{ab} , \qquad k = 1, 2 . \qquad (6.134)$$

Recall that $\langle a(\mathbf{e}_3)|J_3|b(\mathbf{e}_3) \rangle = \hbar a\delta_{ab}$. Clearly, the above formulae reproduce those from section 2.3.3 for $\kappa = 1/2$. Now, the corresponding curvature two-form $F^{(m)}$ is given by

$$F^{(m)} = F_{\theta\varphi}^{(m)} \, d\theta \wedge d\varphi ,$$

with

$$F_{\theta\varphi}^{(m)} = \partial_\theta A_\varphi^{(m)} - \partial_\varphi A_\theta^{(m)} - i[A_\theta^{(m)}, A_\varphi^{(m)}] .$$

One finds, for $|m| \neq 1/2$,

$$F_{\theta\varphi}^{(m)} = -m \sin\theta\, \sigma_3 , \qquad (6.135)$$

and, for $m = 1/2$,

$$F_{\theta\varphi}^{(1/2)} = -\frac{1}{2}(1 - \kappa^2) \sin\theta\, \sigma_3 . \qquad (6.136)$$

The reader should recognize the field of a magnetic pole (cf. section 1.4.2), or, more precisely, a pair of monopoles with the following magnetic charges:

$$g = \begin{cases} \pm m , & \text{for } |m| \neq \frac{1}{2} \\[2ex] \pm\frac{1}{2}(1 - \kappa^2) , & \text{for } |m| = \frac{1}{2} \end{cases} .$$

Standard vector notation gives the following formulae for the $so(3)$-valued vector field $\mathbf{F}^{(m)}$:

$$\mathbf{F}^{(m)} = -m\, \sigma_3\, \frac{\mathbf{R}}{R^3} , \qquad |m| \neq \frac{1}{2} , \qquad (6.137)$$

and

$$\mathbf{F}^{(1/2)} = -\frac{1}{2}(1 - \kappa^2)\, \sigma_3\, \frac{\mathbf{R}}{R^3} . \qquad (6.138)$$

Thus, if C is a closed curve in the two-sphere $R = 1$, then the doublet $|m\,\rangle$ and $|-m\,\rangle$ acquires a geometric Berry phase, as follows:

$$\left| \begin{matrix} m \\ -m \end{matrix} \right\rangle \longrightarrow \exp\left[-im\Omega(C)\,\sigma_3\right] \left| \begin{matrix} m \\ -m \end{matrix} \right\rangle , \qquad (6.139)$$

for $m \neq 1/2$, and

$$\left| \begin{matrix} 1/2 \\ -1/2 \end{matrix} \right\rangle \longrightarrow \exp\left[-\frac{i}{2}(1 - \kappa^2)\Omega(C)\,\sigma_3\right] \left| \begin{matrix} 1/2 \\ -1/2 \end{matrix} \right\rangle , \qquad (6.140)$$

for $m = 1/2$. As usual, $\Omega(C)$ stands for the solid angle subtended by C on the $\mathbf{R}$-sphere.

6.3.5 Quantum geometric forces

The Born–Oppenheimer approximation shows that the dynamics of the slow subsystem (describing the nuclei) should include additional reaction forces that depend upon the geometry of the fast subsystem in the space of slow variables. Using notation from section 6.3.3, let us define an effective Hamiltonian

$$\hat{H}_{\text{eff}}^{(n)}(\mathbf{R}, \mathbf{P}) := \langle n(\mathbf{R})|\hat{H}|n(\mathbf{R})\rangle = \int d\mathbf{r}\, \phi_n^*(\mathbf{R}, \mathbf{r})\, \hat{H}(\mathbf{R}, \mathbf{P}, \mathbf{r}, \mathbf{p})\, \phi_n(\mathbf{R}, \mathbf{r}) , \qquad (6.141)$$

which governs the dynamics of slow motion. If

$$\hat{H}(\mathbf{R}, \mathbf{P}, \mathbf{r}, \mathbf{p}) = \frac{1}{2} \sum_{ij} Q^{ij} \hat{P}_i \hat{P}_j + \hat{h}(\mathbf{R}, \mathbf{r}, \mathbf{p}) ,$$

where Q^{ij} is an inverse mass tensor (in section 6.3.3 we have $Q^{ij} = \delta_{ij}/M$), then, using the standard position representation for the momentum, i.e.,

$$\hat{P}_i = -i\hbar \frac{\partial}{\partial R^i} ,$$

it is easy to show that

$$\hat{H}_{\text{eff}}^{(n)}(\mathbf{R}, \mathbf{P}) = \frac{1}{2} \sum_{ij} Q^{ij} (\hat{P}_i - A_i^{(n)}(\mathbf{R}))(\hat{P}_j - A_j^{(n)}(\mathbf{R})) + \Phi^{(n)}(\mathbf{R}) + \epsilon_n(\mathbf{R}) ,$$

$$(6.142)$$

where the vector gauge potential is given by

$$A_i^{(n)}(\mathbf{R}) = i\langle n(\mathbf{R}) | \partial_i n(\mathbf{R}) \rangle ,$$

and the scalar potential by

$$\Phi^{(n)}(\mathbf{R}) = \frac{\hbar^2}{2} \sum_{ij} Q^{ij} g_{ij}^{(n)}(\mathbf{R}) . \qquad (6.143)$$

In the above formula, $g_{ij}^{(n)}$ denotes the quantum metric tensor, introduced in section 2.2.6, i.e.,

$$g_{ij}^{(n)} = \text{Re} \left(\langle \partial_i n | (\mathbb{1} - |n\rangle\langle n)) | \partial_j n \rangle \right) .$$

The effective Hamiltonian gives rise to the following reaction forces:

The Born–Oppenheimer force: $-\partial_i \epsilon_n(\mathbf{R})$;

The magnetic gauge force: $B_{ij}^{(n)}(\mathbf{R}) = \hbar \left(\partial_i A_j^{(n)}(\mathbf{R}) - \partial_j A_i^{(n)}(\mathbf{R}) \right)$;

The electric gauge force: $-\partial_i \Phi^{(n)}(\mathbf{R})$.

The classical Hamilton equations of motion of the slow variables, i.e.,

$$\dot{R}^i = \frac{\partial H_{\text{eff}}^{(n)}}{\partial P_i} , \quad \text{and} \quad \dot{P}_i = -\frac{\partial H_{\text{eff}}^{(n)}}{\partial R^i} ,$$

imply that

$$\ddot{R}^i = \sum_j Q^{ij} \left(\dot{R}^m B_{mj}(\mathbf{R}) - \partial_i (\epsilon_n(\mathbf{R}) + \Phi^{(n)}(\mathbf{R})) \right) , \qquad (6.144)$$

which is the Newton equation of motion in the "magnetic field" represented by a two-form $B = \frac{1}{2} B_{ij} dR^i \wedge dR^j$ and "electric field" represented by $\partial_i (\epsilon + \Phi) dR^i$. Hence, in the improved adiabatic (Born–Oppenheimer) approximation, the slow variable $\mathbf{R}$ is no longer frozen, but is governed by (6.144).

Example 6.3.1 Consider a composite system in which the spin $s = \frac{1}{2}$ of one (light) particle is coupled to the spatial coordinates $\mathbf{R}$ of a second, otherwise free (heavy), particle (Berry 1989b). The corresponding Hamiltonian reads as follows:

$$\hat{H}(\mathbf{R}, \mathbf{P}, \text{spin}) = \frac{\hat{\mathbf{P}}^2}{2M} + F(\hat{\mathbf{R}} \cdot \boldsymbol{\sigma}) , \tag{6.145}$$

where the function F defines an interaction between the light and heavy subsystems. One has, for the eigenvectors,

$$\mathbf{R} \cdot \boldsymbol{\sigma} \, |n(\mathbf{R})\rangle = nR \, |n(\mathbf{R})\rangle , \qquad n = \pm\frac{1}{2} .$$

Moreover, it is easy to show that the corresponding quantum geometric tensor $T_{ij}^{(n)}$ is given by

$$T_{ij}^{(n)} = \frac{1}{2R^2} \left\{ \frac{1}{2} \left(\delta_{ij} - \frac{R_i R_j}{R^2} \right) - in \, \epsilon_{ijk} \frac{R^k}{R} \right\} . \tag{6.146}$$

Clearly,

$$F_{ij}^{(n)} = -2\text{Im} \, T_{ij}^{(n)} = n \, \epsilon_{ijk} \frac{R^k}{R^3} , \tag{6.147}$$

and

$$g_{ij}^{(n)} = \text{Re} \, T_{ij}^{(n)} = \frac{1}{4R^2} \left(\delta_{ij} - \frac{R_i R_j}{R^2} \right) . \tag{6.148}$$

Note that the above formula, rewritten in spherical coordinates, reproduces (2.147). Now, one finds the following expression for $\Phi^{(n)}$:

$$\Phi^{(n)}(\mathbf{R}) = \frac{\hbar^2}{4MR^2} , \tag{6.149}$$

and hence the corresponding equation of motion for the slow variables reads

$$M\ddot{\mathbf{R}} = \frac{n\hbar}{2} \, \dot{\mathbf{R}} \times \frac{\mathbf{R}}{R^3} + \frac{\hbar^2}{2M} \frac{\mathbf{R}}{R^4} - nF'(\mathbf{R} \cdot \boldsymbol{\sigma}) \frac{\mathbf{R}}{R} . \tag{6.150}$$

Clearly, this equation for the heavy (mass M) particle involves the Lorentz force from the magnetic monopole of strength $n/2$, and the inverse cube gauge electric force, which repels the particle from the degeneracy point $\mathbf{R} = 0$.   ◇

6.4   The quantum Hall effect

The quantum Hall effect (QHE) is one of the most remarkable condensed matter phenomena discovered in the second half of the 20th century. In the following section we shall discuss only the integral quantum Hall effect, discovered in 1980 by Klaus von Klitzing. Soon after, in 1982, Tsui, Störmer and Gossard observed the fractional quantum Hall effect. For a review we refer the reader to the books of Prange and Girvin (1987); Morandi (1988a); Stone (1992); Das Sarma and Pinczuk (1997); and Janßen et al. (1994).

6.4.1   Preliminaries

Let us briefly recall basic facts about the classical Hall effect, discovered by Edwin Hall in 1879. Consider a homogeneous conducting plate, with dimensions (L_x, L_y) placed in a strong magnetic field $\mathbf{B} = B\mathbf{e}_z$. Imposing an electric current (I_x, I_y) gives rise to a change in voltage across the plate, given by

$$
\begin{aligned}
U_x &= RI_x + R_H I_y \,, && (6.151) \\
U_y &= -R_H I_x + \tilde{R} I_y \,, && (6.152)
\end{aligned}
$$

where $R, \tilde{R}$ are longitudinal resistances and R_H is the so-called Hall resistance. Usually, in performing an experiment, one has $I_y = 0$ and hence the above relations reduce to Ohm's law,

$$
U_x = RI_x \,, \tag{6.153}
$$

and the *classical Hall effect*:

$$
U_y = -R_H I_x \,. \tag{6.154}
$$

Therefore, knowing the longitudinal resistance R and measuring U_x and U_y, one finds, for the the Hall resistance,

$$
R_H = -R\frac{U_y}{U_x} \,. \tag{6.155}
$$

Let as assume that the electric current is produced by the constant electric field $\mathbf{E} = (E_x, E_y)$, with

$$
E_x = \frac{U_x}{L_x} \,, \qquad E_y = \frac{U_y}{L_y} \,. \tag{6.156}
$$

Defining the current density $\mathbf{j} = (j_x, j_y)$ by

$$
j_x = \frac{I_x}{L_y} \,, \qquad j_y = \frac{I_y}{L_x} \,, \tag{6.157}
$$

one introduces the resistivity tensor $\rho_{\mu\nu}$ according to the following relation:

$$E_\mu = \sum_\nu \rho_{\mu\nu} j_\nu , \qquad (\mu, \nu = x, y) . \tag{6.158}$$

One finds that

$$\rho_{xx} = \rho_{yy} = R\frac{L_x}{L_y} = \tilde{R}\frac{L_y}{L_x} , \tag{6.159}$$

and

$$\rho_{xy} = -\rho_{yx} = R_H . \tag{6.160}$$

The quantities ρ_{xx} and ρ_{xy} are called the longitudinal and Hall resistivities, respectively.[10] The corresponding conductivity tensor $\sigma_{\mu\nu}$ is the inverse of $\rho_{\mu\nu}$, and hence

$$\sigma_{xx} = \sigma_{yy} = \frac{\rho_{xx}}{\rho_{xx}^2 + \rho_{xy}^2} , \tag{6.161}$$

$$\sigma_{yx} = -\sigma_{xy} = \frac{\rho_{xy}}{\rho_{xx}^2 + \rho_{xy}^2} . \tag{6.162}$$

Let us observe that when $\rho_{xy} \neq 0$, the conductivity σ_{xx} vanishes if and only if the resistivity ρ_{xx} vanishes.

The goal of physical experiments and theoretical investigations is to determine σ_{xx} and σ_{yx} as functions of electron concentration $n := N/L_x L_y$, magnetic field B, and temperature T. In the Drude theory of the electrical conductivity of a metal, an electron is accelerated by the electric field for an average time τ — the relaxation or mean free time — before being scattered by impurities, lattice imperfections, and phonons into a state which has zero average velocity. The average drift velocity of the electron is

$$\mathbf{v} = -\frac{e\tau}{m}\mathbf{E} , \tag{6.163}$$

where m denotes the electron mass. The current density is thus

$$\mathbf{j} = -en\mathbf{v} = \sigma_0\mathbf{E} , \tag{6.164}$$

where

$$\sigma_0 = \frac{ne^2\tau}{m} , \tag{6.165}$$

and n stands for electron concentration. In the presence of a steady magnetic field $\mathbf{B}$, formula (6.163) is replaced by

$$\mathbf{v} = -\frac{e\tau}{m}\left(\mathbf{E} + \frac{\mathbf{v}}{c} \times \mathbf{B}\right) , \tag{6.166}$$

---

[10]In two-dimensional systems, resistivity and resistance are measured in the same dimensions.

that is, the vectors $\mathbf{j}$ and $\mathbf{E}$ are no longer parallel. Ohm's law (6.158) therefore gives

$$\begin{pmatrix} E_x \\ E_y \end{pmatrix} = \frac{1}{\sigma_0} \begin{pmatrix} 1 & \tau\omega_c \\ -\tau\omega_c & 1 \end{pmatrix} \begin{pmatrix} j_x \\ j_y \end{pmatrix} , \tag{6.167}$$

where $\omega_c = eB/mc$ is the cyclotron frequency. This implies that

$$\rho_{xx} = \frac{1}{\sigma_0} , \qquad \rho_{xy} = \frac{\tau\omega_c}{\sigma_0} . \tag{6.168}$$

Now, in the Drude model the Hall resistivity ρ_{xy} is independent of the friction and is given by

$$\rho_{xy} = \frac{1}{\sigma_H} , \tag{6.169}$$

and hence one finds for the Hall conductance

$$\sigma_H = \frac{en}{B} . \tag{6.170}$$

Let us note that, in two dimensions, the physical dimension of the conductivity is $(\text{charge})^2/\text{action}$. Thus its atomic unit is e^2/h, where h is the Planck constant. Hence

$$\sigma_H = \nu \frac{e^2}{h} , \tag{6.171}$$

where ν is the so-called filling factor, defined by

$$\nu = 2\pi \ell_B^2 n , \tag{6.172}$$

and $\ell_B = \sqrt{\hbar/eB}$ is so-called magnetic length. Note that

$$\nu = \frac{n}{B} \frac{h}{e} = \frac{n}{B} \Phi_0 , \tag{6.173}$$

where $\Phi_0 = h/e$ denotes the elementary quantum of magnetic flux. In this way one finds the following microscopic interpretation of the filling factor:

$$\nu = \frac{\text{number of electrons}}{\text{number of flux quanta}} . \tag{6.174}$$

As revealed by the experiments of von Klitzing, the behavior of a real two-dimensional system is dramatically different from that of an ideal two-dimensional electron gas. Indeed, the Hall conductivity σ_H is not at all linear, but was found to be a step function

$$\sigma_H = \ell \frac{e^2}{h} , \qquad \ell = 1, 2, \dots , \tag{6.175}$$

with plateaus of an unexpected precision of order 10^{-8}, cf. Fig. 6.6. It turns out that this quantization is universal and independent of all microscopic details such as the type of semiconductor material, the purity of the sample, the precise value of the magnetic field, and so forth. Note that the measurement of e^2/h is equivalent to a measurement of the fine structure constant, which is of fundamental importance in physics.

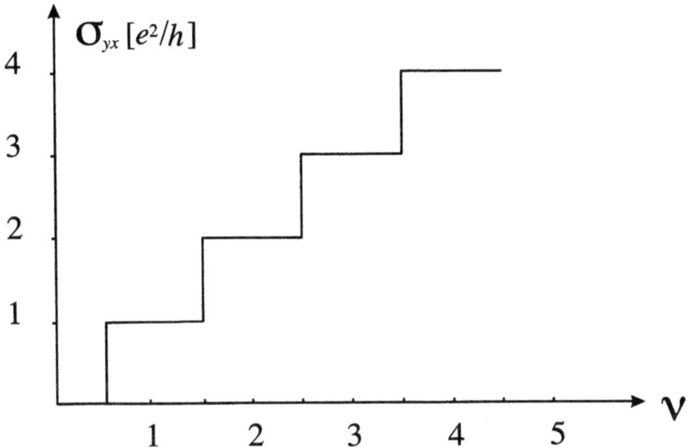

Figure 6.6: A schematic view of the integer quantum Hall effect.

6.4.2   Quantum dynamics in a magnetic field

We shall now briefly describe the quantum dynamics of electrons in a constant magnetic field $\mathbf{B} = B\mathbf{e}_z$. The Hamiltonian for electrons in the presence of the magnetic field is given by

$$\hat{H} = \frac{\hat{\Pi}^2}{2m} ,$$

(6.176)

where the kinetic momentum $\hat{\Pi}$ reads

$$\hat{\Pi} = \hat{\mathbf{p}} + \frac{e}{c}\mathbf{A} = -i\hbar\nabla + \frac{e}{c}\mathbf{A} ,$$

(6.177)

and $\mathbf{A}$ is a vector potential for $\mathbf{B}$, i.e., $\mathbf{B} = \nabla \times \mathbf{A}$. The reader can easily check the following commutation relation:

$$[\hat{\Pi}_x, \hat{\Pi}_y] = -i\hbar m\omega_c .$$

(6.178)

Now, introducing

$$\hat{a} := \frac{1}{\sqrt{2m\hbar\omega_c}} (\hat{\Pi}_x - i\hat{\Pi}_y) ,$$

(6.179)

one finds

$$[\hat{a}, \hat{a}^*] = 1 ,$$

(6.180)

and the quantum Hamiltonian takes the following form:

$$\hat{H} = \hbar\omega_c \left(\hat{a}^*\hat{a} + \frac{1}{2} \right) ,$$ (6.181)

and hence corresponds to a harmonic oscillator of frequency ω_c. The corresponding energy eigenvalues

$$\epsilon_n = \hbar\omega_c \left(n + \frac{1}{2} \right)$$ (6.182)

are called Landau levels. To find the corresponding eigenfunctions one needs to fix the gauge of $\mathbf{A}$. In particular, using the so-called Landau gauge,

$$\mathbf{A} = (0, Bx, 0) ,$$ (6.183)

and by writing the wave function $\psi(x, y)$ in the form

$$\psi_k(x, y) = e^{iky} \varphi_k(x) ,$$ (6.184)

one obtains

$$H_k \varphi_k(x) = \epsilon_k \varphi_k(x) ,$$ (6.185)

with

$$H_k = -\frac{\hbar^2}{2m} \frac{d^2}{dx^2} + \frac{1}{2} m\omega_c^2 \left(x + k\ell_B^2 \right)^2 ,$$ (6.186)

which corresponds to the Hamiltonian of a displaced oscillator. Hence, the corresponding (unnormalized) eigenfunctions ψ_{nk} are given by

$$\psi_{nk}(x, y) = \frac{1}{\sqrt{L_y}} e^{iky} e^{-(x+k\ell_B^2)^2/2\ell_B^2} H_n(x + k\ell_B^2) ,$$ (6.187)

where H_n is the nth Hermite polynomial. Each Landau level is highly degenerate, since the energy does not depend upon k. To count the number of states corresponding to a single Landau level, one assumes periodic boundary conditions in the y direction. Clearly, the use of the Landau gauge does not allow for periodic boundary conditions in the x direction. Note, however, that ψ_{nk} rapidly vanishes for x away from $-k\ell_B^2$. Let us suppose that the left-hand edge of the rectangular sample is at $x = -L_x$, and that the right-hand edge is at $x = 0$. Then the values of the wave vector k for which the basis state is substantially inside the sample, run from $k = 0$ to $k = L_x/\ell_B^2$. Thus, the total number of states in each Landau level is given by

$$\frac{L_y}{2\pi} \int_0^{L_x/\ell_B^2} dk = \frac{BL_x L_y}{\Phi_0} ,$$ (6.188)

and hence is equal to the number of flux quanta penetrating the sample. Therefore, using (6.174) we obtain another interpretation of the filling factor v, i.e.,

$$v = \frac{\text{number of electrons}}{\text{number of states per Landau level}} .$$ (6.189)

Clearly, $v = \ell \in \mathbb{Z}$ if the first ℓ Landau levels are totally occupied.

6.4.3 Fibre bundle approach to the QHE

To show that σ_{xy} is quantized we shall use the topological approach to the QHE developed by Thouless et al. (1982), Niu et al. (1985), and Avron and Seiler (1985) (see also Thouless 1997).

As a starting point let us consider a two-dimensional rectangular system in a perpendicular magnetic field with periodic boundary conditions for the wave functions in both directions, i.e., such that

$$\psi(x + L_x, y) = \psi(x, y), \qquad \psi(x, y + L_y) = \psi(x, y). \tag{6.190}$$

Imposing these conditions implies identifying the opposite edges of the rectangle, which is in turn topologically equivalent to considering a two-dimensional torus. Now, to apply the standard methods of condensed matter physics, we consider an infinite lattice with lattice constants L_x and L_y, respectively, that is, we repeat the original system infinitely many times in both the x and y directions. The Hamiltonian of the system is invariant under the discrete two-parameter group of so-called magnetic translations

$$T_{\mathbf{a}}\psi(\mathbf{x}) := \exp\left(-i\frac{e}{2\hbar}\mathbf{B} \cdot (\mathbf{a} \times \mathbf{x})\right) \psi(\mathbf{x} + \mathbf{a}), \tag{6.191}$$

where $\mathbf{a} = (n_x L_x, n_y L_y)$, $n_x, n_y = 0, 1, 2, \ldots$ is the lattice vector. Magnetic translations commute with the Hamiltonian but not among themselves. In fact, it is easy to show that

$$T_{\mathbf{a}} \cdot T_{\mathbf{b}} = \exp\left(2\pi i \frac{\Phi}{\Phi_0}\right) T_{\mathbf{b}} \cdot T_{\mathbf{a}}, \tag{6.192}$$

where $\Phi = BL_x L_y$ is the magnetic flux through the unit cell of our lattice (i.e., it is the flux through the sample). Clearly, for discrete values of B, such that

$$\Phi = m\Phi_0, \qquad m = 0, 1, 2, \ldots \tag{6.193}$$

(and thus corresponding to rational values of the filling factor $\nu = N/m$), the magnetic translations do commute, and hence we may apply the Bloch theorem well known from the quantum mechanics of periodic systems. The energy eigenvalues ϵ_α and eigenfunctions ψ_α, defined by

$$\hat{H}\psi_\alpha = \epsilon_\alpha \psi_\alpha, \tag{6.194}$$

are labeled by $\alpha = (n, \mathbf{k})$, where n stands for the band index and $\mathbf{k}$ is the Bloch wave vector. The Bloch theorem implies that

$$\psi_\alpha(\mathbf{x}) = e^{i\mathbf{k}\cdot\mathbf{x}} u_\alpha(\mathbf{x}), \tag{6.195}$$

where u_α is strictly periodic, i.e.,

$$u_\alpha(\mathbf{x}) = u_\alpha(\mathbf{x} + \mathbf{a}), \tag{6.196}$$

with **a** being any lattice vector. Moreover,

$$\epsilon_{n,\mathbf{k}+\mathbf{g}} = \epsilon_{n,\mathbf{k}} \, , \tag{6.197}$$

and

$$u_{n,\mathbf{k}+\mathbf{g}}(\mathbf{x}) = e^{-i\mathbf{g}\cdot\mathbf{x}} \, u_{n,\mathbf{k}} \, , \tag{6.198}$$

where $\mathbf{g} = (g_x, g_y)$ is a vector of the dual lattice, i.e.,

$$\mathbf{g} = 2\pi \left(\frac{n_x}{L_x}, \frac{n_y}{L_y} \right) \, . \tag{6.199}$$

These properties enable us to restrict the Bloch wave vector $\mathbf{k}$ to the first Brillouin zone, defined by

$$-\frac{\pi}{L_x} \le k_x < \frac{\pi}{L_x} \, , \qquad -\frac{\pi}{L_y} \le k_y < \frac{\pi}{L_y} \, . \tag{6.200}$$

Now, let us recall that the unit cell L_x, L_y is topologically equivalent to the two-dimensional torus. Therefore, it is convenient to parametrize the first Brillouin zone by the following two angles:

$$\varphi_x = k_x L_x \, , \qquad \varphi_y = k_y L_y \, . \tag{6.201}$$

Clearly, $-\pi \le \varphi_x, \varphi_y < \pi$. This construction gives rise to the following $U(1)$-fibre bundle over the two-dimensional torus:

$$u_{n,\varphi} \ \longrightarrow \ \varphi = (\varphi_x, \varphi_y) \in \text{Torus} \, . \tag{6.202}$$

Define $U(\mathbf{k}) := e^{i\mathbf{k}\cdot\mathbf{x}}$, and

$$\hat{A}(\mathbf{k}) := U^{-1}(\mathbf{k}) \hat{A} U(\mathbf{k}) \, , \tag{6.203}$$

for an arbitrary operator $\hat{A}$. With this notation one finds

$$\hat{H}(\mathbf{k}) u_\alpha = \epsilon_\alpha u_\alpha \, . \tag{6.204}$$

Now let us introduce the velocity operator $\hat{\mathbf{v}} = (\hat{v}_x, \hat{v}_y)$, as follows:

$$\hat{v}_x = \frac{1}{\hbar} \frac{\partial \hat{H}(\mathbf{k})}{\partial k_x} = \frac{L_x}{\hbar} \frac{\partial \hat{H}(\varphi)}{\partial \varphi_x} \, , \tag{6.205}$$

and analogously $\hat{v}_y$. The celebrated fluctuation-dissipation Kubo formula (Kubo 1966) relates the conductivity of a material to its current–current correlation function in the following way:

$$\sigma_{\mu\nu} = \frac{1}{L_x L_y} \sum_{\epsilon_n < \epsilon_F} \int_{\text{Torus}} \sigma_{\mu\nu}^{(n)}(\varphi) \, d\varphi_x \wedge d\varphi_y \, , \tag{6.206}$$

where ϵ_F denotes the Fermi level,

$$\sigma_{\mu\nu}^{(n)}(\boldsymbol{\varphi}) = i\frac{e^2\hbar}{(2\pi)^2}\sum_{m\neq n}\frac{(\hat{v}_\nu(\boldsymbol{\varphi}))_{nm}(\hat{v}_\mu(\boldsymbol{\varphi}))_{mn} - (\hat{v}_\mu(\boldsymbol{\varphi}))_{nm}(\hat{v}_\nu(\boldsymbol{\varphi}))_{mn}}{(\epsilon_n(\boldsymbol{\varphi}) - \epsilon_m(\boldsymbol{\varphi}))^2}\,, \qquad (6.207)$$

and the matrix elements $\hat{v}_{mn} := \langle u_m|\hat{v}|u_n\rangle$.[11] To simplify the formula for $\sigma_{\mu\nu}^{(n)}$, recall that

$$(\hat{v}_x(\boldsymbol{\varphi}))_{nm} = \frac{L_x}{\hbar}\left\langle u_{n,\varphi}\left|\frac{\partial\hat{H}(\boldsymbol{\varphi})}{\partial\varphi_x}\right|u_{m,\varphi}\right\rangle\,. \qquad (6.208)$$

Moreover, using

$$\left\langle u_{n,\varphi}\left|\frac{\partial u_{m,\varphi}}{\partial\varphi_x}\right.\right\rangle = \frac{1}{\epsilon_n(\boldsymbol{\varphi}) - \epsilon_m(\boldsymbol{\varphi})}\left\langle u_{n,\varphi}\left|\frac{\partial\hat{H}(\boldsymbol{\varphi})}{\partial\varphi_x}\right|u_{m,\varphi}\right\rangle\,, \qquad (6.209)$$

one finds

$$\sigma_{yx} = \frac{e^2}{h}\frac{i}{2\pi}\int_{\text{Torus}}\left(\left\langle\frac{\partial u_{n,\varphi}}{\partial\varphi_x}\left|\frac{\partial u_{n,\varphi}}{\partial\varphi_y}\right.\right\rangle - \left\langle\frac{\partial u_{n,\varphi}}{\partial\varphi_y}\left|\frac{\partial u_{n,\varphi}}{\partial\varphi_x}\right.\right\rangle\right)d\varphi_x\wedge d\varphi_y\,. \qquad (6.210)$$

The above formula has the same form as the corresponding formula for the Berry phase as the surface integral of the Berry curvature $F^{(n)}$, cf. (2.62). In geometric terms the Hall conductivity

$$\sigma_H = \frac{e^2}{h}\int_{\text{Torus}}\frac{iF^{(n)}}{2\pi}\,, \qquad (6.211)$$

where $F^{(n)}$ is the curvature two-form on the $U(1)$-bundle over the two-dimensional torus. Recalling that

$$\int_{\text{Torus}}\frac{iF^{(n)}}{2\pi} = \ell \in \mathbb{Z}\quad\text{(Chern number)}\,, \qquad (6.212)$$

we have proved the quantization of σ_H.

6.5 Spin, statistics and the geometric phase

In this section, following Berry and Robbins (1997, 2000) (see also the recent article by Harrison and Robbins 2003), we are going to show that the standard relation between spin and statistics in the nonrelativistic quantum mechanics of identical particles may

---

[11] Actually, formula (6.207) is valid only for a temperature $T = 0$. If $T > 0$ there is an additional T-dependent factor — see the literature on QHE.

$$\text{a)} \qquad\qquad\qquad\qquad \text{b)} \qquad\qquad\qquad\qquad \text{c)}$$

a)		b)		c)	
m_1	m_2	m_1	m_2	m_2	m_1
•	•	•	•	•	•
1	2	2	1	1	2

Figure 6.7: Three states of the two-particle system: a) corresponds to $|M(\mathbf{r})\rangle$; b) to $|M(-\mathbf{r})\rangle$; and c) to $|\overline{M}(\mathbf{r})\rangle$.

be interpreted as another manifestation of geometric phase. (This time it has a purely topological origin and hence it is an example of topological phase.)

We restrict ourselves to the two-particle case (for generalization to many particles we refer the reader to the original article by Berry and Robbins). Here we show that the celebrated Pauli sign $(-1)^{2S}$ appears as a geometric phase factor corresponding to the parallel transport of states of spins in a certain enlarged Hilbert space (which contains the original spin Hilbert space as a subspace). This parallel transport is realized via a unitary operation called *exchange rotation*. Each exchange rotation produces the corresponding Pauli sign. It turns out that the proper mathematical language to deal with this object is the Stiefel–Whitney class of the corresponding vector (or rather two-spin) bundle over the configuration space of two identical particles. As we shall see, this space is neither orientable nor simply connected.

6.5.1 The transported spin basis

Consider two identical particles with spin S in $\mathbb{R}^3$. If $\mathbf{r}_1$ and $\mathbf{r}_2$ denote their positions, then the wave function of the composite two-particle system depends on the relative position $\mathbf{r} := \mathbf{r}_1 - \mathbf{r}_2$. Clearly, exchange of particles leads to

$$\mathbf{r}_1 \longleftrightarrow \mathbf{r}_2, \qquad \text{or, equivalently,} \quad \mathbf{r} \longleftrightarrow -\mathbf{r},$$

together with the exchange of spins. Each particle carries $(2S + 1)$ spin states which are usually described, one for each particle, by

$$|S, m_1\rangle, \qquad \text{and} \qquad |S, m_2\rangle,$$

where $m_1, m_2 = -S, -S + 1, \ldots, S - 1, S$ represents the z-component of the spin of the corresponding particle. We shall denote by $M := \{m_1, m_2\}$ the spin state of the composite system. Then the exchange of spins corresponds to (cf. Fig. 6.7)

$$M = \{m_1, m_2\} \longleftrightarrow \overline{M} = \{m_2, m_1\}. \tag{6.213}$$

Now, apart from the fixed basis $|M\rangle$ let as define an $\mathbf{r}$-dependent (transported or co-moving) basis $|M(\mathbf{r})\rangle$ which represents spins in a way that depends on the relative position of the particles. The transported spin basis is defined by

$$|M(\mathbf{r})\rangle = U(\mathbf{r})|M\rangle, \tag{6.214}$$

where $U(\mathbf{r})$ is a unitary operator. It should be stressed that the above formula does not represent a simple $\mathbf{r}$-dependent change of basis in the corresponding Hilbert space of spins $\mathcal{H}_{\text{spins}} \cong \mathbb{C}^{(2S+1)^2}$; rather, the vectors $|M(\mathbf{r})\rangle$ live in an augmented d-dimensional space with $d > (2S+1)^2$. Thus $U(\mathbf{r})$ operates in an enlarged space and it is clear that neither $|M(\mathbf{r})\rangle$ nor $|M\rangle$ span this space (recall that $(2S+1)^2$ vectors $|M\rangle$ span the space of spins $\mathcal{H}_{\text{spins}}$). We require the transported basis $|M(\mathbf{r})\rangle$ to satisfy the following properties:

- It is smooth for all $\mathbf{r} \neq 0$.

- $|M(\mathbf{r})\rangle$ is single-valued.

- $|M(\mathbf{r})\rangle$ is parallel transported, i.e., the following equation is satisfied:

$$A_{M,M'} = i\langle M'(\mathbf{r})|\nabla M(\mathbf{r})\rangle = 0 , \tag{6.215}$$

  for arbitrary values of M and M'.

Clearly, the above formula is an analog of the Born–Fock gauge conditions (cf. (2.16) and (2.187)). By introducing the transported basis $|M(\mathbf{r})\rangle$, we may incorporate the indistinguishability of particles by identifying $\mathbf{r}$ and $-\mathbf{r}$. Clearly, $|M(-\mathbf{r})\rangle$ and $|\overline{M}(\mathbf{r})\rangle$ define the same spin state, and hence they may differ by an $\mathbf{r}$-dependent phase factor only, as follows:

$$|M(-\mathbf{r})\rangle = e^{i\phi(\mathbf{r})}|\overline{M}(\mathbf{r})\rangle . \tag{6.216}$$

Now, since the basis is single-valued, a double exchange would lead to

$$|M(\mathbf{r})\rangle = |M(-(-\mathbf{r}))\rangle = e^{i\phi(-\mathbf{r})}|\overline{M}(-\mathbf{r})\rangle = e^{i[\phi(-\mathbf{r})+\phi(\mathbf{r})]}|M(\mathbf{r})\rangle ,$$

and hence $\phi(\mathbf{r}) = \mu(\mathbf{r}) + \pi K$, where $\mu(\mathbf{r})$ is an odd function, i.e., $\mu(-\mathbf{r}) = -\mu(\mathbf{r})$, and K is an integer. Therefore, the exchange rule (6.216) may be rewritten in the form

$$|M(-\mathbf{r})\rangle = (-1)^K e^{i\mu(\mathbf{r})}|\overline{M}(\mathbf{r})\rangle . \tag{6.217}$$

Now, the parallel transport condition (6.215) implies

$$A_{M,M'}(-\mathbf{r}) = A_{\overline{M},\overline{M}'}(\mathbf{r}) - \nabla\mu(\mathbf{r})\,\delta_{M,M'} , \tag{6.218}$$

which shows that μ is constant. However, μ is an odd function, and hence it vanishes. Thus the exchange rule reduces to a sign exchange, i.e.,

$$|M(-\mathbf{r})\rangle = (-1)^K |\overline{M}(\mathbf{r})\rangle . \tag{6.219}$$

Having defined a parallel transported basis we may represent any spin state $|\Psi(\mathbf{r})\rangle$ of the two-particle system as follows:

$$|\Psi(\mathbf{r})\rangle = \sum_M \psi_M(\mathbf{r})|M(\mathbf{r})\rangle , \tag{6.220}$$

where $\psi_M(\mathbf{r})$ is a $(2S + 1)^2$-dimensional vector. Note that $|M(\mathbf{r})\rangle$ do not span the augmented d-dimensional space, the physical spin states span live in a $(2S + 1)^2$-dimensional subspace unitary related to $\mathcal{H}_{\text{spins}}$ via relation (6.214). Hence, $|\Psi(\mathbf{r})\rangle$ as a physical state is represented entirely in terms of $|M(\mathbf{r})\rangle$. Now, single-valuedness of the wave function requires

$$|\Psi(\mathbf{r})\rangle = |\Psi(-\mathbf{r})\rangle \,, \tag{6.221}$$

and, hence, the sign exchange rule (6.219) gives

$$|\Psi(-\mathbf{r})\rangle = \sum_M \psi_M(-\mathbf{r})(-1)^K |\overline{M}(\mathbf{r})\rangle = \sum_M \psi_{\overline{M}}(-\mathbf{r})(-1)^K |M(\mathbf{r})\rangle \,. \tag{6.222}$$

Thus

$$\psi_{\overline{M}}(-\mathbf{r}) = (-1)^K \psi_M(\mathbf{r}) \,. \tag{6.223}$$

Obviously, this resembles the usual spin-statistics relation. Recall that in the standard relation we have

$$\psi_{\overline{M},\text{fixed}}(-\mathbf{r}) = (-1)^{2S} \psi_{M,\text{fixed}}(\mathbf{r}) \,, \tag{6.224}$$

where $\psi_{M,\text{fixed}}(\mathbf{r})$ are defined with respect to the fixed basis $|M\rangle$, that is,

$$|\Psi_{\text{fixed}}(\mathbf{r})\rangle = \sum_M \psi_{M,\text{fixed}}(\mathbf{r}) |M\rangle \,. \tag{6.225}$$

What is the relation between $\psi_M(\mathbf{r})$ and $\psi_{M,\text{fixed}}(\mathbf{r})$? Note that $|\Psi_{\text{fixed}}\rangle$ and $|\Psi\rangle$ live in different spaces.

Proposition 6.5.1 *Both $\psi_{M,\text{fixed}}(\mathbf{r})$ and $\psi_M(\mathbf{r})$ satisfy the same Schrödinger equation and hence they are equal.*

Proof. We have to compare the action of the momentum operator $\mathbf{P}$ and the spin operators $\mathbf{S} = (\mathbf{S}_1, \mathbf{S}_2)$ on $\psi_{M,\text{fixed}}(\mathbf{r})$ and $\psi_M(\mathbf{r})$. The momentum operator has, in the fixed basis, the usual form, i.e.,

$$\mathbf{P}_{\text{fixed}} = -i\hbar\nabla \,,$$

and therefore, in the transported basis, it is defined by

$$\mathbf{P}(\mathbf{r}) = U(\mathbf{r})\mathbf{P}_{\text{fixed}}U^*(\mathbf{r}) \,.$$

Clearly, in the fixed basis, one has

$$\langle M|\mathbf{P}_{\text{fixed}}|\Psi_{\text{fixed}}(\mathbf{r})\rangle = -i\hbar\nabla\psi_{M,\text{fixed}}(\mathbf{r}) \,.$$

Now, due to

$$\langle M(\mathbf{r})|\mathbf{P}(\mathbf{r})|\Psi(\mathbf{r})\rangle = \langle M|U^*(\mathbf{r})U(\mathbf{r})\mathbf{P}_{\text{fixed}}U^*(\mathbf{r})|\Psi(\mathbf{r})\rangle \,,$$

and

$$U^*(\mathbf{r})|\Psi(\mathbf{r})\rangle = \sum_{M'} \psi_{M'}(\mathbf{r})\, U^*(\mathbf{r})|M'(\mathbf{r})\rangle = \sum_{M'} \psi_{M'}(\mathbf{r})\,|M'\rangle\,,$$

one obtains

$$\langle M(\mathbf{r})|\mathbf{P}(\mathbf{r})|\Psi(\mathbf{r})\rangle = \sum_{M'}\langle M|M'\rangle \mathbf{P}_{\text{fixed}}\psi_{M'}(\mathbf{r}) = -i\hbar\nabla\psi_M(\mathbf{r})\,. \tag{6.226}$$

Similarly, the action of $\mathbf{S}_{\text{fixed}}$, in the fixed basis, is given by

$$\langle M|\mathbf{S}_{\text{fixed}}|\Psi_{\text{fixed}}(\mathbf{r})\rangle = \sum_{M'} \psi_{M',\text{fixed}}(\mathbf{r})\,\langle M|\mathbf{S}_{\text{fixed}}|M'\rangle\,, \tag{6.227}$$

whereas

$$\begin{aligned}
\langle M(\mathbf{r})|\mathbf{S}(\mathbf{r})|\Psi(\mathbf{r})\rangle &= \langle M|U^*(\mathbf{r})U(\mathbf{r})\mathbf{S}_{\text{fixed}}U^*(\mathbf{r})|\Psi(\mathbf{r})\rangle \\
&= \sum_{M'} \psi_{M'}(\mathbf{r})\,\langle M|\mathbf{S}_{\text{fixed}}|M'\rangle\,.
\end{aligned} \tag{6.228}$$

This shows that the transported ($\psi_M(\mathbf{r})$) and fixed ($\psi_{M,\text{fixed}}(\mathbf{r})$) quantities satisfy the same Schrödinger equation, and hence they are the same functions, i.e.,

$$\psi_M(\mathbf{r}) = \psi_{M,\text{fixed}}(\mathbf{r})\,, \tag{6.229}$$

for all $\mathbf{r}$. $\square$

The above proposition immediately implies that

$$|\Psi(\mathbf{r})\rangle = U(\mathbf{r})|\Psi_{\text{fixed}}(\mathbf{r})\rangle\,. \tag{6.230}$$

Note that, unlike $|\Psi(\mathbf{r})\rangle$, which does not change under exchange of particles, $|\Psi_{\text{fixed}}(\mathbf{r})\rangle$ may change.

6.5.2 Schwinger representation

It is well known that one may represent the Lie algebra $su(2)$ in terms of the annihilation and creation operators of the harmonic oscillator — the so-called Schwinger representation (Schwinger 1965). For a single spin, two independent oscillators are required, that is, we have two sets a, a^* and b, b^* together with the standard commutation relations:

$$[a, a^*] = 1\,, \quad \text{and} \quad [b, b^*] = 1\,, \tag{6.231}$$

with all other commutators involving a's and b's vanishing (since the oscillators are independent). Let us construct a spin vector $\mathbf{S} = (S_x, S_y, S_z)$ as follows:

$$\mathbf{S} := \frac{\hbar}{2}\,(a^*\ b^*)\,\boldsymbol{\sigma}\begin{pmatrix} a \\ b \end{pmatrix}\,, \tag{6.232}$$

where, as usual, $\boldsymbol{\sigma}$ stands for the vector of Pauli matrices. In particular, one obtains

$$S_z = \frac{\hbar}{2}\,(a^*a - b^*b)\,,$$

and

$$S_+ := S_x + iS_y = \hbar a^*b\,, \qquad S_- := S_x - iS_y = \hbar b^*a\,.$$

Using (6.231), one easily finds the standard commutation relation of $su(2)$, i.e.,

$$[S_z, S_+] = \hbar S_+\,, \quad [S_z, S_-] = -\hbar S_-\,, \quad [S_+, S_-] = 2\hbar S_z\,.$$

Introducing oscillators eigenstates $|n_a\rangle$ and $|n_b\rangle$, such that

$$a^*a|n_a\rangle = n_a|n_a\rangle\,, \quad \text{and} \quad b^*b|n_b\rangle = n_b|n_b\rangle\,, \tag{6.233}$$

one finds that

$$S_z|n_a, n_b\rangle = \hbar\, m|n_a, n_b\rangle\,, \tag{6.234}$$

and

$$\mathbf{S}^2|n_a, n_b\rangle = \hbar^2\, S(S+1)|n_a, n_b\rangle\,, \tag{6.235}$$

with

$$S = \frac{1}{2}(n_a + n_b)\,, \qquad m = \frac{1}{2}(n_a - n_b)\,. \tag{6.236}$$

Therefore, in this representation, the eigenvectors of $\mathbf{S}^2$ and S_z with quantum numbers S and m correspond to number states of the oscillators, i.e.,

$$|S, m\rangle = |n_a, n_b\rangle\,. \tag{6.237}$$

Now, the standard spin-S representation of $su(2)$ is $(2S+1)$-dimensional. Note that the oscillator representation has the same dimension: We have to distribute $2S$ quanta among two oscillators ($n_a + n_b = 2S$) and this can be done in exactly $2S + 1$ ways.

Example 6.5.1 If $S = 1/2$ one has a two-dimensional representation and the two oscillators give two number state vectors:

$$|1, 0\rangle\,, |0, 1\rangle\,.$$

For $S = 1$, one obtains three eigenvectors:

$$|2, 0\rangle\,, |0, 2\rangle\,, |1, 1\rangle\,.$$

The reader can easily find the corresponding $(2S + 1)$ eigenvectors for an arbitrary S.

Now consider two particles, with spins S_1 and S_2. The composite system carries spin $\mathbf{S} = \mathbf{S}_1 + \mathbf{S}_2$. Now, two spins require four oscillators: (a_1, b_1) for $\mathbf{S}_1$, such that

$$\mathbf{S}_1 := \frac{\hbar}{2} (a_1^* \, b_1^*) \, \boldsymbol{\sigma} \begin{pmatrix} a_1 \\ b_1 \end{pmatrix} , \tag{6.238}$$

and (a_2, b_2) for $\mathbf{S}_2$, such that

$$\mathbf{S}_2 := \frac{\hbar}{2} (a_2^* \, b_2^*) \, \boldsymbol{\sigma} \begin{pmatrix} a_2 \\ b_2 \end{pmatrix} . \tag{6.239}$$

The total space in which the two spin operators $\mathbf{S}_1$ and $\mathbf{S}_2$ act is a tensor product of the carrier spaces of the two $su(2)$ representations; one of which is $(2S_1 + 1)$-dimensional, and the other $(2S_2 + 1)$-dimensional. Hence,

$$\mathcal{H}_{\text{total}} = \mathbb{C}^{2S_1 + 1} \otimes \mathbb{C}^{2S_2 + 1} \cong \mathbb{C}^{(2S_1 + 1)(2S_2 + 1)} . \tag{6.240}$$

Now, in $\mathcal{H}_{\text{total}}$ we have the following basis:

$$|S_1, S_2, m_1, m_2 \rangle = |S_1, S_2, M \rangle ,$$

where $M = \{m_1, m_2\}$, and

$$m_1 = -S_1, -S_1 + 1, \ldots , S_1 - 1, S_1 , \quad m_2 = -S_2, -S_2 + 1, \ldots , S_2 - 1, S_2 .$$

Introducing the corresponding oscillator eigenstates $|n_{1a} \rangle$, $|n_{2a} \rangle$, $|n_{1b} \rangle$, $|n_{2b} \rangle$, one easily finds that

$$S_z |n_{1a}, n_{2a}, n_{1b}, n_{2b} \rangle = \frac{\hbar}{2} (n_{1a} + n_{2a} - n_{1b} - n_{2b}) |n_{1a}, n_{2a}, n_{1b}, n_{2b} \rangle , \tag{6.241}$$

and

$$\mathbf{S}^2 |n_{1a}, n_{2a}, n_{1b}, n_{2b} \rangle = \hbar^2 S(S + 1) |n_{1a}, n_{2a}, n_{1b}, n_{2b} \rangle , \tag{6.242}$$

with

$$S = \frac{1}{2} (n_{1a} + n_{2a} + n_{1b} + n_{2b}) . \tag{6.243}$$

Hence, using the Schwinger oscillator representation we may reproduce the standard basis $|S, m_1, m_2 \rangle$ in $\mathcal{H}_{\text{total}}$. One has

$$S_1 = \frac{1}{2} (n_{1a} + n_{1b}) , \quad S_2 = \frac{1}{2} (n_{2a} + n_{2b}) , \tag{6.244}$$

and

$$m_1 = \frac{1}{2} (n_{1a} - n_{1b}) , \quad m_2 = \frac{1}{2} (n_{2a} - n_{2b}) . \tag{6.245}$$

Therefore,

$$n_{1a} = S_1 + m_1 \,, \quad n_{2a} = S_2 + m_2 \,, \quad n_{1b} = S_1 - m_1, \quad n_{2b} = S_2 - m_2 \,,$$

and hence

$$|n_{1a}, n_{2a}, n_{1b}, n_{2b}\rangle = |S_1 + m_1, S_2 + m_2, S_1 - m_1, S_2 - m_2\rangle =: |S_1, S_2, M\rangle \,. \tag{6.246}$$

Note that the set of vectors $|n_{1a}, n_{2a}, n_{1b}, n_{2b}\rangle$ span a vector space with dimension D_{S_1, S_2}, which is greater than $(2S_1 + 1)(2S_2 + 1)$. Clearly, D_{S_1, S_2} equals the number of ways $2(S_1 + S_2)$ elements may be distributed among four boxes (n_{1a}, n_{2a}, n_{1b}, and n_{2b} count elements in separate boxes). Simple combinatorics gives

$$\begin{aligned} D_{S_1, S_2} &= \binom{2(S_1 + S_2) + 3}{3} \\ &= \frac{1}{6}(2S_1 + 2S_2 + 1)(2S_1 + 2S_2 + 2)(2S_1 + 2S_2 + 3) > (2S_1 + 1)(2S_2 + 1) \,. \end{aligned}$$

Example 6.5.2 Let $S_1 = S_2 = 1/2$. Then the total space $\mathcal{H}_{\text{total}}$ is spanned by four vectors, written as

$$|+, +\rangle \,, \quad |+, -\rangle \,, \quad |-, +\rangle \,, \quad |-, -\rangle \,,$$

where we use the following obvious notation: $|\pm\rangle$ are solutions to $S_z|\pm\rangle = \pm\frac{\hbar}{2}|\pm\rangle$. The Schwinger representation gives a ten-dimensional space spanned by $|n_{1a}, n_{2a}, n_{1b}, n_{2b}\rangle$, with $n_{1a} + n_{2a} + n_{1b} + n_{2b} = 2$. The above four vectors are reproduced as follows:

$$|1100\rangle = |+, +\rangle \,, \qquad |1001\rangle = |+, -\rangle \,,$$

$$|0110\rangle = |-, +\rangle \,, \qquad |0011\rangle = |-, -\rangle \,.$$

Moreover, there are the following six additional vectors:

$$|2000\rangle \,, \quad |0200\rangle \,, \quad |0020\rangle \,, \quad |0002\rangle \,, \quad |1010\rangle \,, \quad |0101\rangle \,.$$

Clearly, these additional vectors violate the rules (6.244) and (6.245), and hence they do not correspond to physical states. $\diamond$

6.5.3   Exchange rotation

Now we are ready to perform unitary transformations in the enlarged D_{S_1, S_2}-dimensional space (actually, we are interested in the special case when $S_1 = S_2$). Define vector operators $\mathbf{E}_a$ and $\mathbf{E}_b$, which mix the two spins, as follows:

$$\mathbf{E}_a := \frac{\hbar}{2}(a_1^* \, a_2^*)\,\sigma\begin{pmatrix} a_1 \\ a_2 \end{pmatrix} \,, \tag{6.247}$$

and

$$\mathbf{E}_b := \frac{\hbar}{2}\,(b_1^*\; b_2^*)\,\boldsymbol{\sigma}\begin{pmatrix} b_1 \\ b_2 \end{pmatrix}\;. \tag{6.248}$$

Clearly,

$$\begin{aligned}
\left[(E_a)_k, (E_a)_l\right] &= i\hbar\,\epsilon_{klm}(E_a)_m\;, \\
\left[(E_b)_k, (E_b)_l\right] &= i\hbar\,\epsilon_{klm}(E_b)_m, \\
\left[(E_a)_k, (E_b)_l\right] &= 0\;,
\end{aligned} \tag{6.249}$$

where $(E_a)_k$ $((E_b)_k)$ denotes the kth component of $\mathbf{E}_a$ ($\mathbf{E}_b$). Defining

$$\mathbf{E} := \mathbf{E}_a + \mathbf{E}_b\;,$$

one has immediately

$$[E_k, E_l] = i\hbar\,\epsilon_{klm}E_m\;, \tag{6.250}$$

that is, $\mathbf{E}_a$, $\mathbf{E}_b$ and $\mathbf{E}$ satisfy the commutation rules of angular momentum. Following Berry and Robins (Berry and Robbins 1997) we shall call $\mathbf{E}$ the *exchange angular momentum*, because, as we shall see, it may be used to generate the unitary transformation in the enlarged space such that the exchange sign condition (6.219) is satisfied. Before we demonstrate this, let us observe that

$$[E_z, \mathbf{S}_1] = [E_z, \mathbf{S}_2] = [E_z, \mathbf{S}_{\text{total}}] = 0\;, \tag{6.251}$$

where $\mathbf{S}_{\text{total}} = \mathbf{S}_1 + \mathbf{S}_2$. However, the x and y components of $\mathbf{E}_a$ and $\mathbf{E}_b$ commute with neither the spins $\mathbf{S}_1$ and $\mathbf{S}_2$ nor with S_1^2 and S_2^2. Now let us construct the map

$$\mathbb{R}^3 \ni \mathbf{r} \longrightarrow U(\mathbf{r})\;,$$

where the unitary operator $U(\mathbf{r})$ acts in the enlarged space. Recall that our basic requirement for $U(\mathbf{r})$ is to satisfy the sign exchange rule (6.219). Suppose that the line joining the particles is rotated from $\mathbf{e}_z$ to $\mathbf{r}$. The simplest way to do this is via a single rotation about the axis $\mathbf{n}$, where

$$\mathbf{n}(\mathbf{r}) := \frac{\mathbf{e}_z \times \mathbf{r}}{|\mathbf{e}_z \times \mathbf{r}|}\;,$$

by θ radians, where (θ, ϕ) are spherical angles of $\mathbf{r}$ in $\mathbb{R}^3$. Moreover, one easily sees that

$$\mathbf{n}(\mathbf{r}) = -\mathbf{e}_x \sin\phi + \mathbf{e}_y \cos\phi\;.$$

Now, to the above rotation in $\mathbb{R}^3$, we associate the following unitary transformation of the two-spin state:

$$U(\mathbf{r}) := \exp\left\{-i\theta\,\mathbf{n}(\mathbf{r})\cdot\mathbf{E}\right\}\;. \tag{6.252}$$

Following Berry and Robbins, we shall call $U(\mathbf{r})$ an *exchange rotation*. Clearly, due to (6.249), the exchange rotation separates into

$$U(\mathbf{r}) = \exp\left\{ -i\theta\, \mathbf{n}(\mathbf{r}) \cdot \mathbf{E}_a \right\} \cdot \exp\left\{ -i\theta\, \mathbf{n}(\mathbf{r}) \cdot \mathbf{E}_b \right\} =: U_a(\mathbf{r}) \cdot U_b(\mathbf{r}) . \qquad (6.253)$$

Let us study the action of $U(\mathbf{r})$ on the states $|n_{1a}, n_{2a}, n_{1b}, n_{2b}\rangle$ spanning the D_{S_1,S_2}-dimensional (enlarged) space. We shall first consider a general situation with two different spins $(S_1 \neq S_2)$ and finally restrict to the case of identical particles with $S_1 = S_2 = S$. Any vector $|n_{1a}, n_{2a}, n_{1b}, n_{2b}\rangle$ is created from the vacuum $|0, 0, 0, 0\rangle$ by the well-known oscillator prescription

$$|n_{1a}, n_{2a}, n_{1b}, n_{2b}\rangle = \mathcal{N}\, (a_1^*)^{n_{1a}} (a_2^*)^{n_{2a}} (b_1^*)^{n_{1b}} (b_2^*)^{n_{2b}} |0, 0, 0, 0\rangle , \qquad (6.254)$$

where $\mathcal{N}$ stands for a normalization constant. Let us define

$$a_1^*(\mathbf{r}) := U_a(\mathbf{r})\, a_1^*\, U_a^*(\mathbf{r}) , \qquad a_2^*(\mathbf{r}) := U_a(\mathbf{r})\, a_2^*\, U_a^*(\mathbf{r}) ,$$

and

$$b_1^*(\mathbf{r}) := U_b(\mathbf{r})\, b_1^*\, U_b^*(\mathbf{r}) , \qquad b_2^*(\mathbf{r}) := U_b(\mathbf{r})\, b_2^*\, U_b^*(\mathbf{r}) .$$

One can show that

$$a_1^*(\mathbf{r}) = \cos\left(\frac{\theta}{2}\right) a_1^* + e^{i\phi} \sin\left(\frac{\theta}{2}\right) a_2^* , \qquad (6.255)$$

$$a_2^*(\mathbf{r}) = -e^{-i\phi} \sin\left(\frac{\theta}{2}\right) a_1^* + \cos\left(\frac{\theta}{2}\right) a_2^* , \qquad (6.256)$$

and similarly for the b^*'s. Now, let us introduce a transported basis in the enlarged space, as follows:

$$|n_{1a}, n_{2a}, n_{1b}, n_{2b}(\mathbf{r})\rangle := U(\mathbf{r})\, |n_{1a}, n_{2a}, n_{1b}, n_{2b}\rangle . \qquad (6.257)$$

One then has

$$|n_{1a}, n_{2a}, n_{1b}, n_{2b}(\mathbf{r})\rangle = \mathcal{N}\, (a_1^*(\mathbf{r}))^{n_{1a}} (a_2^*(\mathbf{r}))^{n_{2a}} (b_1^*(\mathbf{r}))^{n_{1b}} (b_2^*(\mathbf{r}))^{n_{2b}} |0, 0, 0, 0\rangle . \qquad (6.258)$$

Using the formulae for the $a^*(\mathbf{r})$'s and $b^*(\mathbf{r})$'s one finds

$$\begin{aligned}
&|n_{1a}, n_{2a}, n_{1b}, n_{2b}(\mathbf{r})\rangle \\
&= \mathcal{N}\left(\cos\left(\frac{\theta}{2}\right) a_1^* + e^{i\phi} \sin\left(\frac{\theta}{2}\right) a_2^* \right)^{n_{1a}} \\
&\quad\times \left(-e^{-i\phi} \sin\left(\frac{\theta}{2}\right) a_1^* + \cos\left(\frac{\theta}{2}\right) a_2^* \right)^{n_{2a}} \\
&\quad\times \left(\cos\left(\frac{\theta}{2}\right) b_1^* + e^{i\phi} \sin\left(\frac{\theta}{2}\right) b_2^* \right)^{n_{1b}} \\
&\quad\times \left(-e^{-i\phi} \sin\left(\frac{\theta}{2}\right) b_1^* + \cos\left(\frac{\theta}{2}\right) b_2^* \right)^{n_{2b}} |0, 0, 0, 0\rangle .
\end{aligned} \qquad (6.259)$$

Having defined the action of $U(\mathbf{r})$ it is straightforward to obtain the action of $U(-\mathbf{r})$; the reflection $\mathbf{r} \longrightarrow -\mathbf{r}$ corresponds to the following replacements:

$$\theta \longrightarrow \pi - \theta , \quad \text{and} \quad \phi \longrightarrow \phi + \pi .$$

Therefore, using (6.259) we obtain the transformation under $U(-\mathbf{r})$:

$$
\begin{aligned}
& |n_{1a}, n_{2a}, n_{1b}, n_{2b}(-\mathbf{r}) \rangle \\
& = \mathcal{N} \left(\sin\left(\frac{\theta}{2}\right) a_1^* - e^{i\phi} \cos\left(\frac{\theta}{2}\right) a_2^* \right)^{n_{1a}} \\
& \quad \times \left(e^{-i\phi} \cos\left(\frac{\theta}{2}\right) a_1^* + \sin\left(\frac{\theta}{2}\right) a_2^* \right)^{n_{2a}} \\
& \quad \times \left(\sin\left(\frac{\theta}{2}\right) b_1^* - e^{i\phi} \cos\left(\frac{\theta}{2}\right) b_2^* \right)^{n_{1b}} \\
& \quad \times \left(e^{-i\phi} \cos\left(\frac{\theta}{2}\right) b_1^* + \sin\left(\frac{\theta}{2}\right) b_2^* \right)^{n_{2b}} |0, 0, 0, 0 \rangle .
\end{aligned}
\tag{6.260}
$$

The formula for $|n_{1a}, n_{2a}, n_{1b}, n_{2b}(\mathbf{r}) \rangle$ may be rewritten in a slightly different form by factoring out an overall phase factor, as follows:

$$
\begin{aligned}
& |n_{1a}, n_{2a}, n_{1b}, n_{2b}(\mathbf{r}) \rangle = \mathcal{N} e^{i\phi n_{1a}} (-1)^{n_{2a}} e^{-i\phi n_{2a}} e^{i\phi n_{1b}} (-1)^{n_{2b}} e^{-i\phi n_{2b}} \\
& \quad \times \left\{ \left(e^{-i\phi} \cos\left(\frac{\theta}{2}\right) a_1^* + \sin\left(\frac{\theta}{2}\right) a_2^* \right)^{n_{1a}} \right. \\
& \quad \times \left(\sin\left(\frac{\theta}{2}\right) a_1^* - e^{i\phi} \cos\left(\frac{\theta}{2}\right) a_2^* \right)^{n_{2a}} \times \left(e^{-i\phi} \cos\left(\frac{\theta}{2}\right) b_1^* + \sin\left(\frac{\theta}{2}\right) b_2^* \right)^{n_{1b}} \\
& \quad \times \left. \left(\sin\left(\frac{\theta}{2}\right) b_1^* - e^{i\phi} \cos\left(\frac{\theta}{2}\right) b_2^* \right)^{n_{2b}} \right\} |0, 0, 0, 0 \rangle .
\end{aligned}
\tag{6.261}
$$

Comparing with (6.260), one finds that

$$|n_{1a}, n_{2a}, n_{1b}, n_{2b}(\mathbf{r}) \rangle = (-1)^{n_{2a}+n_{2b}} e^{i\phi(n_{1a}+n_{1b}-n_{2a}-n_{2b})} |n_{2a}, n_{1a}, n_{2b}, n_{1b}(-\mathbf{r}) \rangle .$$

Thus $U(\mathbf{r})$ does indeed generate the exchange of spins $S_1 \longleftrightarrow S_2$. Restricting to the physical subspace, such that

$$S_1 = \frac{1}{2}(n_{1a} + n_{1b}) , \quad m_1 = \frac{1}{2}(n_{1a} - n_{1b}) = -S_1, -S_1 + 1, \ldots , S_1 - 1, S_1 ,$$

and

$$S_2 = \frac{1}{2}(n_{2a} + n_{2b}) , \quad m_2 = \frac{1}{2}(n_{2a} - n_{2b}) = -S_2, -S_2 + 1, \ldots , S_2 - 1, S_2 ,$$

the above formula may be equivalently rewritten as follows:

$$|S_1, S_2; M(\mathbf{r}) \rangle = (-1)^{2S_2} e^{2i\phi(S_1 - S_2)} |S_2, S_1; \overline{M}(-\mathbf{r}) \rangle .
\tag{6.262}$$

Hence, if $S_1 = S_2 = S$, it leads to

$$|M(\mathbf{r})\rangle = (-1)^{2S} |\overline{M}(-\mathbf{r})\rangle , \tag{6.263}$$

which proves the sign exchange formula (6.219), and it shows that $K = 2S$ in perfect agreement with the celebrated Pauli formula (6.224).

6.5.4   Pauli sign as a topological phase

Now we show that the Pauli sign $(-1)^{2S}$ in (6.263) may be interpreted as a geometric phase. In the standard approach to the Pauli principle one considers the coefficients $\psi_{M,\text{fixed}}(\mathbf{r})$ of the fixed-basis wave function $|\Psi_{\text{fixed}}(\mathbf{r})\rangle$. According to formula (6.224),

$$\psi_{\overline{M},\text{fixed}}(-\mathbf{r}) = (-1)^{2S} \psi_{M,\text{fixed}}(\mathbf{r}) , \tag{6.264}$$

that is, $\psi_{M,\text{fixed}}(\mathbf{r})$ acquires the standard Pauli sign $(-1)^{2S}$ under exchange of positions and spins denoted by

$$\mathbf{r} \longrightarrow -\mathbf{r}, \quad \text{and} \quad M \longrightarrow \overline{M} ,$$

respectively. In the transported basis $|M(\mathbf{r})\rangle$, the sign change of the coefficients $\psi_M(\mathbf{r}) \equiv \psi_{M,\text{fixed}}(\mathbf{r})$ is compensated by the sign change of $|M(\mathbf{r})\rangle$ according to our basic rule (6.263). Therefore, the wave function

$$|\Psi(\mathbf{r})\rangle = \sum_M \psi_M(\mathbf{r})|M(\mathbf{r})\rangle ,$$

is single-valued.

Recall that the Pauli sign $(-1)^{2S}$ arises from the parallel transport generated by the exchange rotation $U(\mathbf{r})$ in the enlarged space; this is exactly the general mechanism for producing geometric phases. The present situation reminds us how the adiabatic phase in quantum mechanics arises: A total system is divided into two parts — the space part (described by $\psi_M(\mathbf{r})$) and the spin part (described by $|M(\mathbf{r})\rangle$). The space part $\psi_M(\mathbf{r})$ acquires an additional phase factor to keep the total wave function $|\Psi(\mathbf{r})\rangle$ single-valued.

Note that, contrary to the previous examples of geometric phases, the Pauli sign does not arise from the line integral of the corresponding connection form, or, equivalently, from the flux of the first Chern class of the corresponding fibre bundle. The mathematical origin of the phase factor $(-1)^{2S}$ is different. This time, the geometric phase has purely topological origin and it is associated with the first Stiefel–Whitney class in the two-spin bundle over the configuration space of two indistinguishable particles, i.e.,

$$\mathcal{M}_2 = (\mathbb{R}^3 \times \mathbb{R}^3 - \Delta)/S_2 ,$$

where

$$\Delta = \left\{ (\mathbf{r}_1, \mathbf{r}_2) \in \mathbb{R}^3 \times \mathbb{R}^3 \,|\, \mathbf{r}_1 = \mathbf{r}_2 \right\}$$

and S_2 is the permutation group of two elements. Hence, the corresponding bundle is given by

$$E \cong \mathbb{C}^{2S+1} \otimes \mathbb{C}^{2S+1} \longrightarrow \mathcal{M}_2 \; .$$

In contrast to the other characteristic classes we have considered earlier, the Stiefel–Whitney classes of the vector bundle $E \longrightarrow M$ are not integral cohomology classes and are not given in terms of curvature. They are defined as

$$w_i(E) \in H^i(M, \mathbb{Z}_2) \, , \quad i = 1, 2, \ldots, \dim M - 1 \, , \tag{6.265}$$

that is, the integrals of $w_i(E)$ are $\mathbb{Z}_2$-valued (i.e., 0 or 1) — see, e.g., Milnor and Stasheff 1974 for more details. It turns out that $w_1(TM) = 0$ if and only if M is orientable. Now, one can show (see, e.g., Morandi 1992) that the configuration space $\mathcal{M}_2$ is not orientable and hence gives rise to nontrivial first Stiefel–Whitney class. Since the fundamental group $\pi_1(\mathcal{M}_2) = S_2 = \{e, \sigma\}$, we have only two classes of loops in $\mathcal{M}_2$, i.e., contractible loops corresponding to the identity e in S_2, and noncontractible loops corresponding to the exchange permutation σ. Each noncontractible loop γ gives rise to the nontrivial flux of w_1 through any two-dimensional region in $\mathcal{M}_2$ having γ as its boundary.

Remark 6.5.1 Actually, in physical applications the second Stiefel–Whitney class w_2 plays a very important role. It turns out that $w_2(TM)$ determines whether or not parallel transport of Dirac spinors can be globally defined on TM. If

$$w_1(TM) = w_2(TM) = 0 \, ,$$

then the Dirac spinors are well defined and M is called a *spin manifold*. $\diamond$

6.6 Entanglement and holonomic quantum computation

Despite a popular claim that the research in the area of geometric phases is almost over our final example shows that this statement need not be true. Quantum information theory — a rapidly developing subject — has discovered a surprising application of geometric phases in the modelling of quantum gates, basic units of a quantum computer. This example shows that geometric phases are still worth to study.

6.6.1 Composite systems and entangled states

Consider two quantum systems A and B and let $\mathcal{H}_A$ and $\mathcal{H}_B$ denote the corresponding Hilbert spaces.[12] Suppose now that we are interested in the composite system AB,

---

[12] In quantum information theory one usually speaks about "Alice" and "Bob" systems.

made up of A and B. The composite system Hilbert space $\mathcal{H}_{AB}$ is a tensor product of $\mathcal{H}_A$ and $\mathcal{H}_B$, i.e.,

$$\mathcal{H}_{AB} = \mathcal{H}_A \otimes \mathcal{H}_B \,, \tag{6.266}$$

which means that if $|A\rangle \in \mathcal{H}_A$ and $|B\rangle \in \mathcal{H}_B$ denote the state vectors of A and B, respectively, then the joint system is in the state

$$|AB\rangle \equiv |A\rangle \otimes |B\rangle \in \mathcal{H}_{AB} \,. \tag{6.267}$$

Note, however, that a general vector from $\mathcal{H}_{AB}$ cannot be written this way. We call an element $\psi \in \mathcal{H}_{AB}$ a *separable state* if there exist $\psi_A \in \mathcal{H}_A$ and $\psi_B \in \mathcal{H}_B$ such that $\psi = \psi_A \otimes \psi_B$. If this is not the case we call ψ an *entangled state* or, simply, a nonseparable state.

Recall that if $(e_1, \dots, e_N)$ denote a basis in $\mathcal{H}_A$ and $(f_1, \dots, f_M)$ a basis in $\mathcal{H}_B$, then

$$e_\mu \otimes f_\nu \,, \qquad \mu = 1, \dots, N, \ \nu = 1, \dots, M \,,$$

define a basis in the $(N \times M)$-dimensional space $\mathcal{H}_{AB}$. This means that an arbitrary vector $\psi \in \mathcal{H}_{AB}$ may be represented as follows:

$$\psi = \sum_{\mu=1}^{N} \sum_{\nu=1}^{M} \psi_{\mu\nu}\, e_\mu \otimes f_\nu \,, \tag{6.268}$$

with $\psi_{\mu\nu} \in \mathbb{C}$. Now, given a state $\psi \in \mathcal{H}_{AB}$, how can one tell whether it is separable or entangled? To answer this question let us use the following

Theorem 6.6.1 (Schmidt decomposition) *For every ψ in $\mathcal{H}_{AB}$ there exist orthonormal bases $\{\tilde{e}_\mu\}_{\mu=1}^{N}$ and $\{\tilde{f}_\nu\}_{\nu=1}^{M}$ such that*

$$\psi = \sum_{\alpha=1}^{K} a_\alpha\, \tilde{e}_\alpha \otimes \tilde{f}_\alpha \,, \tag{6.269}$$

where $a_\alpha > 0$ with $\sum_{\alpha=1}^{K} a_\alpha^2 = 1$, and $K \leq \min\{N, M\}$.

The state $\psi \in \mathcal{H}_{AB}$ is separable if and only if its Schmidt decomposition contains only one term, i.e., $K = 1$.

Example 6.6.1 (Bell states) Consider two qubits, with individual Hilbert spaces $\mathcal{H}_A = \mathcal{H}_B = \mathbb{C}^2$. Denoting by $|0\rangle$ and $|1\rangle$ the standard orthonormal basis in $\mathbb{C}^2$ one introduces so-called *Bell states*, as follows:

$$|\psi^\pm\rangle := \frac{1}{\sqrt{2}} \left(|01\rangle \pm |10\rangle \right) \,, \tag{6.270}$$

and

$$|\phi^\pm\rangle := \frac{1}{\sqrt{2}} \left(|00\rangle \pm |11\rangle \right) \,, \tag{6.271}$$

where $|01\rangle := |0\rangle \otimes |1\rangle$, etc. These four vectors define an orthonormal basis in $\mathbb{C}^2 \otimes \mathbb{C}^2 \cong \mathbb{C}^4$. Moreover, they are all entangled states. $\diamond$

The notion of separability and entanglement may be easily generalized to mixed states of the composed system (Werner 1989). A mixed state represented by a density matrix ρ in $\mathcal{H}_{AB}$ is called separable if and only if it can be represented as

$$\rho = \sum_{\alpha=1}^{K} p_\alpha \, \rho_\alpha \otimes \sigma_\alpha , \tag{6.272}$$

where $p_\alpha > 0$ with $\sum_{\alpha=1}^{K} p_\alpha = 1$, and ρ_α and σ_α are mixed states of A and B, respectively. Surprisingly, contrary to the case of pure states, the criterion of separability is still unknown for composite mixed states, see Peres 1995 and Horodecki et al. 2001 for more details.

6.6.2   Qubits and bundles

Recall from sections 1.4.4 and 5.1.3 that a qubit (or, equivalently, a two-level quantum system) gives rise to the celebrated Hopf fibration $S^3 \longrightarrow S^2$. Any normalized qubit state may written as follows:

$$\psi = \alpha |0\rangle + \beta |1\rangle ,$$

where $\alpha, \beta \in \mathbb{C}^2$ and

$$|\alpha|^2 + |\beta|^2 = 1 ,$$

which defines the unit three-dimensional sphere S^3. The Hopf map

$$S^3 \ni (\alpha, \beta) \longrightarrow (x_0, x_1, x_2) \in S^2$$

is defined by

$$\begin{aligned}
x_0 &= \langle \sigma_z \rangle = |\alpha|^2 - |\beta|^2 , \\
x_1 &= \langle \sigma_x \rangle = 2\mathrm{Re}\,(\overline{\alpha}\beta) , \\
x_2 &= \langle \sigma_y \rangle = 2\mathrm{Im}\,(\overline{\alpha}\beta) .
\end{aligned} \tag{6.273}$$

Clearly, if $|\alpha|^2 + |\beta|^2 = 1$, then $x_0^2 + x_1^2 + x_2^2 = 1$. Now, let us consider a composite two-qubit system. Any normalized two-qubit state may be written as follows:

$$\psi = \alpha |00\rangle + \beta |01\rangle + \gamma |10\rangle + \delta |11\rangle , \tag{6.274}$$

where $\alpha, \beta, \gamma, \delta \in \mathbb{C}$ and

$$|\alpha|^2 + |\beta|^2 + |\gamma|^2 + |\delta|^2 = 1 . \tag{6.275}$$

The above formula defines a unit seven-dimensional sphere S^7. A two-qubit state ψ is separable if it is a tensor product of two single-qubit states $\psi = \psi_1 \otimes \psi_2$. Representing ψ_k in the $(|0\rangle, |1\rangle)$ basis, as follows:

$$\psi_1 = a|0\rangle + b|1\rangle, \qquad \psi_2 = c|0\rangle + d|1\rangle, \qquad (6.276)$$

one finds, for a separable state $\psi_1 \otimes \psi_2$,

$$\psi = ac|00\rangle + ad|01\rangle + bc|10\rangle + bd|11\rangle. \qquad (6.277)$$

Hence, the two-qubit state represented by (6.274) is separable if and only if

$$\alpha\delta = \beta\gamma. \qquad (6.278)$$

Interestingly, the above separability condition for two-qubit states may be nicely described in terms of another Hopf fibration, $S^7 \longrightarrow S^4$, discussed in section 1.4.5 (Mosseri and Dandoloff 2001). Any point on S^7 may be represented by a pair of quaternions (q_1, q_2), which we write as

$$q_1 = \alpha + \beta\,\hat{j}, \qquad q_2 = \gamma + \delta\,\hat{j}, \qquad (6.279)$$

and which satisfy $|q_1|^2 + |q_2|^2 = 1$ ($\hat{j}$ stands for the unit quaternion — cf. Appendix B). The Hopf map

$$S^7 \ni (\alpha, \beta, \gamma, \delta) \longrightarrow (x_0, x_1, x_2, x_3, x_4) \in S^4$$

is defined by

$$\begin{aligned}
x_0 &= |q_1|^2 - |q_2|^2, \\
x_1 &= 2\mathrm{Re}\,(\overline{\alpha}\gamma + \overline{\beta}\delta), \\
x_2 &= 2\mathrm{Im}\,(\overline{\alpha}\gamma + \overline{\beta}\delta), \\
x_3 &= 2\mathrm{Re}\,(\alpha\delta - \beta\gamma), \\
x_4 &= 2\mathrm{Im}\,(\alpha\delta - \beta\gamma).
\end{aligned} \qquad (6.280)$$

As observed by Mosseri and Dandoloff (2001), the above formulae may be rewritten in perfect analogy with formulae (6.273). Let us define the so-called "entanglor" as

$$E := -J(\sigma_y \otimes \sigma_g), \qquad (6.281)$$

where J takes the complex conjugate of all complex numbers involved in an expression. Using this antilinear operator, formulae (6.280) are recovered as follows:

$$\begin{aligned}
x_0 &= \langle \sigma_z \otimes \mathbb{1}_2 \rangle, \\
x_1 &= \langle \sigma_x \otimes \mathbb{1}_2 \rangle, \\
x_2 &= \langle \sigma_y \otimes \mathbb{1}_2 \rangle, \\
x_3 &= \mathrm{Re}\,\langle E \rangle, \\
x_4 &= \mathrm{Im}\,\langle E \rangle.
\end{aligned} \qquad (6.282)$$

It is easy to show that if $|q_1|^2 + |q_2|^2 = 1$, then $x_0^2 + x_1^2 + x_2^2 + x_3^2 + x_4^2 = 1$. Let us observe that the separability condition (6.278) implies that $x_3 = x_4 = 0$, and hence, for separable two-qubit states, the Hopf map sends points from S^7 into $S^2 \subset S^4$. Therefore the Hopf map detects whether the state is separable or entangled. This is just one more elegant application of the Hopf map in quantum physics.

6.6.3 Quantum computer — an overview

Among the surprising discoveries made recently in quantum mechanics is that quantum systems can be used to perform information processing tasks, including computations, and can do so more efficiently than any classical method (for reviews see, e.g., Steane 1998; Nielsen and Chuang 2000; Bennett and DiVincenzo 2000; and Ekert 2002). In recent years a lot of activity has been devoted to constructing and implementing schemes for taking actual advantage of such quantum phenomena. Our aim in this short section is to give a flavour of what a quantum computer is.

Classical computer circuits consist of wires and logic gates. The wires are used to carry information around the circuit, while the logic gates perform manipulations of the information. The basic ingredients of a *quantum computer* are roughly the same; quantum information has to be transported and manipulated. As everybody knows, classical information is measured in bits. Now, a qubit is an elementary unit of quantum information. A quantum system is said to have n qubits if it has a Hilbert space of 2^n dimensions, i.e.,

$$\mathcal{H}_n = \overbrace{\mathbb{C}^2 \otimes \ldots \otimes \mathbb{C}^2}^{n\ \text{copies}} .$$

Therefore, it should be clear that to perform any information processing one has to learn how to operate on a single qubit. A single-qubit *quantum gate* is nothing but a unitary operation on the qubit Hilbert space, that is, on $\mathbb{C}^2$. Some of the most important quantum gates are the Pauli matrices, denoted in quantum information theory by

$$X \equiv \sigma_x , \qquad Y \equiv \sigma_y , \qquad Z \equiv \sigma_z .$$

Other quantum gates that play a crucial role are the Hadamard gate H and the phase shift gate $P(\varphi)$. They are defined as follows:

$$H = \frac{1}{\sqrt{2}} \begin{pmatrix} 1 & 1 \\ 1 & -1 \end{pmatrix} , \qquad P(\varphi) = \begin{pmatrix} 1 & 0 \\ 0 & e^{i\varphi} \end{pmatrix} .$$

Clearly, not all of the gates defined above are independent. For example, $H = \frac{1}{\sqrt{2}}(X + Y)$, $Z = P(\pi)$ and obviously $Y = X \cdot Z$. Note that an X gate is an analog of the classical NOT gate:

$$X|0\rangle = |1\rangle , \qquad X|1\rangle = |0\rangle .$$

Now let us turn to unitary operations on a pair of qubits, or, equivalently, on a two-qubit state. Among these, there is a distinguished class of so-called controlled U operations:

If U is an arbitrary unitary operator acting on $\mathbb{C}^2$, then the controlled U, acting on $\mathbb{C}^2 \otimes \mathbb{C}^2$, is defined by

$$|0\rangle\langle 0| \otimes \mathbb{1}_2 + |1\rangle\langle 1| \otimes U . \tag{6.283}$$

Taking $U = X$, one obtains one of the most important two-qubit controlled operations — controlled-NOT, or, simply, "CNOT." The reader can easily check that the effect of CNOT is represented by

$$|00\rangle \longrightarrow |00\rangle , \quad |01\rangle \longrightarrow |01\rangle , \quad |10\rangle \longrightarrow |11\rangle , \quad |11\rangle \longrightarrow |10\rangle , \tag{6.284}$$

or by the following 4×4 matrix:

$$\text{CNOT} = \begin{pmatrix} 1 & 0 & 0 & 0 \\ 0 & 1 & 0 & 0 \\ 0 & 0 & 0 & 1 \\ 0 & 0 & 1 & 0 \end{pmatrix} . \tag{6.285}$$

Clearly, there is an infinite number of unitary operations — quantum gates — that can be applied to n-qubit states. A set of quantum gates is said to be *universal for quantum computation* if any unitary operation may be approximated to arbitrary accuracy by a quantum circuit involving only these gates. Surprisingly, the set of universal gates is quite simple. As shown by Deutsch (Deutsch 1985) it consists of the Hadamard gate, the phase shift gate and the CNOT gate. This is a remarkable observation — two-qubit gates are sufficient for quantum computation.

Knowing how to operate on qubits let us say a few words about the way one transports quantum information. We stress that this is a nontrivial problem, since in measuring a state we destroy it, while according to one of the basic principles of quantum mechanics, the unknown quantum state cannot be cloned (copied). The solution to this problem was found quite recently (Bennett et al. 1993) and is known as *quantum teleportation*. Suppose that two observers (traditionally Alice and Bob) would like to exchange information about an unknown qubit state, say ψ. More precisely, Alice has a qubit in the unknown state ψ and she would like to send this state (not a particle itself!) to Bob. The procedure to teleport ψ from Alice to Bob works as follows: Alice prepares an entangled two-qubit state and interacts the unknown qubit ψ with one half ("her" half) of the entangled pair. Then she measures the two qubits (the original one and hers from the pair), obtaining one of four possible classical results $|00\rangle$, $|01\rangle$, $|10\rangle$ or $|11\rangle$. She sends this information to Bob, e.g., using a telephone or any other "classical communication channel," together with the second qubit from the entangled pair. Now, depending on Alice's classical message, Bob performs one of four operation on his qubit (one of the pair sent by Alice). Surprisingly, by doing this he recovers Alice's original (still unknown) qubit state ψ. To see why this is so, suppose that the Alice state to be teleported is $\psi = \alpha |0\rangle + \beta |1\rangle$, with unknown α and β. Let us take as an entangled two-qubit state one of the Bell states (cf. Example 6.6.1), say ϕ^+. The composite three-qubit system is in the following state:

$$\psi \otimes \phi^+ = \frac{1}{\sqrt{2}}\Big[\alpha\,|0\rangle(|00\rangle+|11\rangle)+\beta\,|1\rangle(|00\rangle+|11\rangle)\Big],$$

where we use the convention that the first two qubits belong to Alice and the third one to Bob. Now Alice sends her two qubits through a two-qubit CNOT gate. The new three-qubit state obtained this way reads

$$\psi_1 = \frac{1}{\sqrt{2}}\Big[\alpha\,|0\rangle(|00\rangle+|11\rangle)+\beta\,(|1\rangle(|10\rangle+|01\rangle))\Big].$$

Then she sends her first qubit through a Hadamard gate, obtaining

$$\psi_2 = \frac{1}{2}\Big[\alpha\,(|0\rangle+|1\rangle)(|00\rangle+|11\rangle)+\beta\,(|0\rangle-|1\rangle)(|10\rangle+|01\rangle)\Big],$$

which, after simple algebraic manipulation, may be rewritten as follows:

$$\begin{aligned}
\psi_2 \;=\;& \frac{1}{2}\Big[|00\rangle(\alpha|0\rangle+\beta|1\rangle)+|01\rangle(\alpha|1\rangle+\beta|0\rangle)\\
&+\;|10\rangle(\alpha|0\rangle-\beta|1\rangle)+|11\rangle(\alpha|1\rangle-\beta|0\rangle)\Big].
\end{aligned}$$

Now a miracle occurs! Knowing the result of Alice's measurements on her two qubits, we can read off the state of Bob's qubit, as follows:

$$\begin{aligned}
00 &\longrightarrow\; \psi_3^{00}=\alpha|0\rangle+\beta|1\rangle,\\
01 &\longrightarrow\; \psi_3^{01}=\alpha|1\rangle+\beta|0\rangle,\\
10 &\longrightarrow\; \psi_3^{10}=\alpha|0\rangle-\beta|1\rangle,\\
11 &\longrightarrow\; \psi_3^{11}=\alpha|1\rangle-\beta|0\rangle.
\end{aligned}$$

Therefore, depending on the results of Alice's measurements, Bob would recover the unknown Alice state ψ, as follows:

$$\begin{aligned}
00 &\longrightarrow\; \psi=\psi_3^{00},\\
01 &\longrightarrow\; \psi=X\psi_3^{01},\\
10 &\longrightarrow\; \psi=Z\psi_3^{10},\\
11 &\longrightarrow\; \psi=ZX\psi_3^{11}=Y\psi_3^{11}.
\end{aligned}$$

Let us summarize this short introduction using the following prescription for a "universal quantum computer" (Deutsch 1985). A quantum computer is a set of n qubits in which the following operations are experimentally feasible:

1. each qubit can be prepared in some known state, say $|0\rangle$,

2. each qubit can be measured in the so called computation basis: $\{|0\rangle,|1\rangle\}$,

3. a universal set of quantum gates can be applied at will to any subset of qubits,

4. the qubits do not evolve other than via universal gates.

Whether or not such idea can be realized is still an open problem. However, at least single quantum gates can be physically implemented, e.g., in ion traps, nuclear magnetic resonance and cavity quantum electrodynamics (see Nielsen and Chuang 2000 for details).

6.6.4 Holonomic quantum computation

The main idea of *holonomic quantum computation* is the following (Zanardi and Rasetti 1999; Pachos, Zanardi and Rasetti 2000; Pachos and Zanardi 2001; and Vedral 2002): Suppose one is given a family of degenerate Hamiltonians $H(\lambda)$ parametrized by a set of parameters $\lambda \in M$. The information is encoded into an n-dimensional eigenspace $\mathcal{H}_n \cong \mathbb{C}^n$ of some Hamiltonian $H(\lambda_0)$. Such a subspace represents what "quantum engineers" call a quantum code. The universal quantum computation can then be realized by adiabatically driving the control parameters λ along suitable loops C in M rooted in λ_0. Clearly, this resembles very much the procedure leading to the nonabelian adiabatic Wilczek–Zee factor (see section 2.3). Each loop C gives rise to a nonabelian geometric factor — the holonomy of C:

$$\Phi(C) = \mathrm{P} \exp\left(\oint_C A\right), \tag{6.286}$$

where A is a non-abelian connection form. In this way a loop C in the control space M produces a quantum gate $\Phi(C)$ in $\mathcal{H}_n$. In other words, holonomic quantum computation consists of the parallel transport of the states from the degeneracy subspace $\mathcal{H}_n$, governed by the connection A. The whole quantum network is built in terms of holonomies and the entire computation process is fully geometrical.

In this framework the requirements for implementing the universal quantum computer can be expressed in terms of the availability of closed paths in M. Universality is the experimental capability of driving the control parameters along a minimal set of loops C_α, which generate the basic quantum gates $\Phi(C_\alpha)$. That is, it should be possible to approximate any unitary operation $U : \mathcal{H}_n \longrightarrow \mathcal{H}_n$ with arbitrary high accuracy by means of $\Phi(C_\alpha)$.

Recall that the holonomy group (based at a point λ_0)

$$\mathrm{Hol}(\lambda_0) = \{\, \Phi(C) \,|\, C - \text{loop rooted at } \lambda_0 \in M \,\}$$

is a subset of the unitary group $U(n)$. When the holonomy group $\mathrm{Hol}(\lambda_0)$ coincides with $U(n)$ the connection A is called *irreducible*.[13] Clearly, the notion of irreducibility plays a crucial role in holonomic quantum computation; it is evident that the holonomic quantum computation is universal if the corresponding connection is irreducible. Now,

---

[13] Recall that holonomy groups based at different points of M are conjugated, cf. formula (1.179).

to check whether the connection is irreducible it is useful to consider the corresponding curvature two-form

$$F = dA + \frac{1}{2}[A, A] \,. \tag{6.287}$$

One can prove (see, e.g., Choquet-Bruhat et al. 1982) the celebrated

Theorem 6.6.2 (Ambrose–Singer) *The Lie algebra of the holonomy group is spanned by $F(u, v)$, where u and v are arbitrary horizontal vectors.*

The Ambrose–Singer theorem implies that when F spans the whole Lie algebra $u(n)$, then the connection A is irreducible.

As an example, consider the degenerate Hamiltonian

$$H_0 = \varepsilon |n + 1\rangle\langle n + 1| \,, \tag{6.288}$$

acting on the Hilbert space $\mathcal{H} \cong \mathbb{C}^{n+1}$ spanned by the $n + 1$ vectors $\{|\alpha\rangle\}_{\alpha=1}^{n+1}$. Clearly, H_0 is n-times degenerate: $H_0|\alpha\rangle = 0$ for $\alpha = 1, 2, \ldots, n$. As a family of Hamiltonians let us take the whole orbit of the adjoint action of $U(n + 1)$:

$$\mathcal{O}(H_0) = \{ U H_0 U^* \mid U \in U(n + 1)\} \,. \tag{6.289}$$

One then has

$$\mathcal{O}(H_0) \cong \frac{U(n + 1)}{U(n) \times U(1)} \cong \mathbb{C}P^n \,, \tag{6.290}$$

where $\mathbb{C}P^n$ denotes the complex projective space, cf. section 5.1.3. This shows that by taking as the control space M the n-dimensional complex projective space, we may obtain any Hamiltonian from the family $\mathcal{O}(H_0)$. Let $(z_1, \ldots, z_n)$ be local complex coordinates on $\mathbb{C}P^n$. Each point $\mathbf{z} = (z_1, \ldots, z_n)$ corresponds to the unitary matrix

$$U(\mathbf{z}) := U_1(z_1)U_2(z_2)\ldots U_n(z_n) \,, \tag{6.291}$$

where $U_\alpha(z_\alpha) = \exp[G_\alpha(z_\alpha)]$ and

$$G_\alpha(z_\alpha) = z_\alpha|\alpha\rangle\langle n + 1| - \bar{z}_\alpha|n + 1\rangle\langle\alpha| \,. \tag{6.292}$$

Now, since $M = \mathbb{C}P^n$ is a homogeneous space, it is enough to find the curvature F at one point on M, e.g., at $\mathbf{z} = 0$. Introducing $2n$ real coordinates z_α^k, as follows:

$$z_\alpha =: z_\alpha^0 + iz_\alpha^1 \,, \qquad \alpha = 1, \ldots, n, \;\; k = 0, 1 \,,$$

and noting that, at $\mathbf{z} = 0$,

$$\frac{\partial U_\alpha}{\partial z_\alpha^k}(\mathbf{z} = 0) = i^k\left(|\alpha\rangle\langle n + 1| - (-1)^k|n + 1\rangle\langle\alpha|\right) \,, \tag{6.293}$$

one finds, for the curvature tensor,

$$F_{z^k_\alpha z^l_\beta}(0) = i^{k+l}\Big[(-1)^l |\alpha\rangle\langle\beta| - (-1)^k |\beta\rangle\langle\alpha|\Big]. \qquad (6.294)$$

From the above expression it follows that F spans the whole algebra $u(n)$, and hence the corresponding adiabatic connection is irreducible. This means that taking $\mathbb{C}P^n$ as a control space, one may, in principle, realize holonomic quantum computation. Note that in order to generate loops on $\mathbb{C}P^n$ one needs to control only $2n$ parameters instead of the $n^2 - 1$ necessary for labeling a generic quantum gate in $\mathbb{C}^n$.

Further reading

Section 6.1. For details of the Chiao–Tomita–Wu experiments see Chiao and Wu 1986; Tomita and Chiao 1986; Chiao 1989, 1990a, 1990b. A more theoretical treatment may be found in Segert 1987a and c. See also Berry 1987a; Robbins and Berry 1994; and Hannay 1998a. Recently, Frins and Dultz (1997) proposed a simple interferometric arrangement that allows a direct observation of Berry's phase in optical fibers. A similar result for the geometric phase of photons was obtained by Białynicki-Birula et al. (1987) and Barut and Bracken (1983).

A simple interferometric demonstration of the Pancharatnam phase was proposed by Hariharan et al. (1999) — see also Hils et al. 1999. For a review of geometric phases in optics, see Bhandari 1990 and 1997, and Vinitskii et al. 1990.

Section 6.2. We refer the reader to Olariu and Popescu 1985 and Peshkin and Tonomura 1989 for the detailed discussions of the Aharonov–Bohm effect from several points of view, and for further references. The Aharonov–Casher effect was first described by Aharonov and Casher (1984) in a slightly different context. A general discussion of this effect may be found in Anandan 2000 and Fröhlich and Studer 1993. Sangster, Hinds, Barnet and Riss (1993) proposed a configuration suitable for observing the topological phase of Aharonov and Casher in atomic systems.

Section 6.3 Geometric phases in molecular physics are discussed in Guichardet 1984, see also Berry 1985b; Moody, Shapere and Wilczek 1986, 1989; Aitchison 1987; Jackiw 1988; Bohm 1993a,b; and Banerjee 1996. See also the review articles Zwanziger et al. 1990 and Mead 1992.

The classical geometric forces are discussed in Berry and Robbins 1993a, 1993b. For another approach, see Aharonov and Stern 1992.

Section 6.5. For a recent review of the spin-statistics theorem see Duck and Sudarshan 1997. For the generalization of the Pauli sign to more than two particles, see Berry and Robbins 1997 (see also Anandan 1998 for the relativistic generalization of Berry and Robbins 1997).

Section 6.6. There is an enormous number of papers devoted to quantum information theory. Most of them may be found in http://xxx.lanl.gov/ archive /quant-ph. Two excellent reviews are Nielsen and Chuang 2000 and *The Physics of Quantum Information: Quantum Cryptography, Quantum Teleportation, Quantum Computation*, eds. D. Bouwmeester, A. Ekert and A. Zeilinger, Springer-Verlag, Berlin, 2000.

For the application of Hopf maps to quantum entanglement, see P. Lévay 2003.

Problems

6.1. Show that the map

$$(\theta, \varphi) \longrightarrow |m(\theta, \varphi)\rangle\,,$$

defined by (6.128), is well defined except at the south pole $\theta = \pi$.

6.2. Using properties (2.110)–(2.111), derive

$$\langle a(\mathbf{e}_3)|J_k|b(\mathbf{e}_3)\rangle = \frac{\kappa}{2}\,(\sigma_k)_{ab}\,, \qquad k = 1, 2\,,$$

for $a, b = \pm\frac{1}{2}$.

6.3. Derive the formula for the effective Hamiltonian for slow motion, i.e.,

$$\hat{H}_{\mathrm{eff}}^{(n)}(\mathbf{R}, \mathbf{P}) = \frac{1}{2}\sum_{ij} Q^{ij}\left(\hat{P}_i - A_i^{(n)}(\mathbf{R})\right)\left(\hat{P}_j - A_j^{(n)}(\mathbf{R})\right) + \Phi^{(n)}(\mathbf{R}) + \epsilon_n(\mathbf{R})\,.$$

6.4. Derive the formula for the quantum geometric tensor $T_{ij}^{(n)}$, where $|n(\theta, \varphi)\rangle = U(\theta, \varphi)|n(\mathbf{e}_3)\rangle$ is a nondegenerate eigenvector defined as in (6.128).

6.5. Show that the formula for the quantum metric tensor, i.e.,

$$g_{ij}^{(n)} = \mathrm{Re}\,T_{ij}^{(n)} = \frac{1}{4R^2}\left(\delta_{ij} - \frac{R_i R_j}{R^2}\right)\,,$$

rewritten in spherical coordinates reproduces (2.147).

6.6. Solve the Schrödinger equation for an electron in a magnetic field using a symmetric gauge:

$$\mathbf{A} = \frac{1}{2}\mathbf{B} \times \mathbf{x}\,.$$

6.7. Compute the electric current corresponding to the eigenstates ψ_{0k}:

$$\langle \mathbf{j}\rangle = -\frac{e}{m}\langle \psi_{0k}|\hat{\boldsymbol{\Pi}}|\psi_{0k}\rangle\,.$$

6.8. Prove the commutation relation for the magnetic translations.

6.9. Show that the velocity operator $\hat{\mathbf{v}}(\mathbf{k})$ satisfies

$$i\hbar\hat{\mathbf{v}}(\mathbf{k}) = [\hat{\mathbf{x}}, \hat{H}(\mathbf{k})]\,.$$

6.10. Derive the following formula:

$$\mathbf{S}^2 |n_{1a}, n_{2a}, n_{1b}, n_{2b}\rangle = \hbar^2 S(S+1) |n_{1a}, n_{2a}, n_{1b}, n_{2b}\rangle \,,$$

with $S = \frac{1}{2}(n_{1a} + n_{2a} + n_{1b} + n_{2b})$, in the Schwinger representation for two spins.

6.11. Verify the commutation relations

$$
\begin{aligned}
{[(E_a)_k, (E_a)_l]} &= i\hbar\, \epsilon_{klm} (E_a)_m \,, \\
{[(E_b)_k, (E_b)_l]} &= i\hbar\, \epsilon_{klm} (E_b)_m \,, \\
{[(E_a)_k, (E_b)_l]} &= 0 \,,
\end{aligned}
$$

with $\mathbf{E}_a$ and $\mathbf{E}_b$ defined in (6.247) and (6.248), respectively. Show that $\mathbf{E}^2 = \mathbf{S}^2_{\text{total}}$.

6.12. Derive the following formula for the action of $U_a(\mathbf{r})$ on the two-dimensional vector of annihilation operators:

$$\begin{pmatrix} a_1 \\ a_2 \end{pmatrix} \longrightarrow \exp\left\{ -\frac{1}{2} i\theta\, \mathbf{n}(\mathbf{r}) \cdot \boldsymbol{\sigma} \right\} \begin{pmatrix} a_1 \\ a_2 \end{pmatrix} \,,$$

in the Schwinger representation for two spins.

6.13. Show that the transported basis defined is the exchange rotation

$$U(\mathbf{r}) = \exp\{-i\theta\, \mathbf{n}(\mathbf{r}) \cdot \mathbf{E}\} \,,$$

with $\mathbf{E}$ being an exchange angular momentum, satisfies the following basic properties (Berry and Robbins 1997):

(a) it is smooth,

(b) it is parallel transported.

Appendix A
Classical Matrix Lie Groups and Algebras

Denote by $M(n, \mathbb{F})$ the space of $n \times n$ matrices over a field $\mathbb{F}$ (with $\mathbb{F} = \mathbb{R}$ or $\mathbb{C}$). Then the general linear group is defined by

$$GL(n, \mathbb{F}) := \left\{ X \in M(n, \mathbb{F}) \,\middle|\, \det X \neq 0 \right\}.$$

One defines also the special linear group as follows:

$$SL(n, \mathbb{F}) = \left\{ X \in GL(n, \mathbb{F}) \,\middle|\, \det X = 1 \right\} \subset GL(n, \mathbb{F}).$$

Classical matrix Lie groups:

1. $O(n)$ — the group of all matrices in $GL(n, \mathbb{R})$ leaving the quadratic form

$$x_1^2 + \ldots + x_n^2$$

   in $\mathbb{R}^n$ invariant, that is, $O(n) = \{ X \in GL(n, \mathbb{R}) \mid X^T = X^{-1} \}$.

2. $SO(n) = O(n) \cap SL(n, \mathbb{R})$.

3. $U(n)$ — the group of all matrices in $GL(n, \mathbb{C})$ leaving the quadratic form

$$z_1 \bar{z}_1 + \ldots + z_n \bar{z}_n$$

   in $\mathbb{C}^n$ invariant, that is, $U(n) = \{ X \in GL(n, \mathbb{C}) \mid X^* = X^{-1} \}$.

4. $SU(n) = U(n) \cap SL(n, \mathbb{C})$.

5. $Sp(n, \mathbb{R})$ — the group of all matrices in $GL(2n, \mathbb{R})$ that conserve the symplectic two-form in $\mathbb{R}^{2n}$, i.e.,

$$dx_1 \wedge dy_1 + \ldots + dx_n \wedge dy_n .$$

6. $Sp(n, \mathbb{C})$ — the group of all matrices in $GL(2n, \mathbb{C})$ that conserve the symplectic two-form in $\mathbb{C}^{2n}$, i.e.,

$$dz_1 \wedge d\overline{w}_1 + \ldots + dz_n \wedge d\overline{w}_n .$$

7. $Sp(n) = Sp(n, \mathbb{C}) \cap U(2n)$.

Let us turn to matrix Lie algebras. $M(n, \mathbb{F})$, denoted also by $gl(n, \mathbb{F})$, is a Lie algebra for $GL(n, \mathbb{F})$. The corresponding Lie algebra for $SL(n, \mathbb{F})$ is defined by

$$sl(n, \mathbb{F}) = \{\, X \in gl(n, \mathbb{F}) \mid \mathrm{Tr} X = 0 \,\} .$$

Classical matrix Lie algebras:

1. $o(n) = \{\, X \in gl(n, \mathbb{R}) \mid X^T = -X \,\}$.

2. $so(n) = o(n) \cap sl(n, \mathbb{R})$.

3. $u(n) = \{\, X \in gl(n, \mathbb{C}) \mid X^* = -X \,\}$.

4. $su(n) = u(n) \cap sl(n, \mathbb{C})$.

5. $sp(n, \mathbb{R})$ — the Lie algebra of all matrices in $gl(2n, \mathbb{R})$ of the following form:

$$\begin{pmatrix} X_1 & X_2 \\ X_3 & -X_1^T \end{pmatrix} ,$$

with $X_1, X_2, X_3 \in gl(n, \mathbb{R})$, and X_2, X_3 symmetric.

6. $sp(n, \mathbb{C})$ — the Lie algebra of all matrices in $gl(2n, \mathbb{C})$ of the following form:

$$\begin{pmatrix} X_1 & X_2 \\ X_3 & -X_1^* \end{pmatrix} ,$$

with $X_1, X_2, X_3 \in gl(n, \mathbb{C})$, and X_2, X_3 Hermitian.

7. $sp(n) = sp(n, \mathbb{C}) \cap u(2n)$.

For more detailed lists see Wybourne 1974; Gilmore 1974; and Barut and Rączka 1980.

Appendix B
Quaternions

A noncommutative field $\mathbb{H}$ of quaternions is generated as a real algebra over $\mathbb{R}$ by the elements $\hat{\imath}, \hat{\jmath}, \hat{k}$ with the following property:

$$\hat{\imath}^2 = \hat{\jmath}^2 = \hat{k}^2 = \hat{\imath}\hat{\jmath}\hat{k} = -1 \ .$$

For any $q \in \mathbb{H}$ of the form $q = q_0 + q_1\hat{\imath} + q_2\hat{\jmath} + q_3\hat{k}$, one defines a conjugate quaternion $\bar{q}$ by

$$\bar{q} = q_0 - q_1\hat{\imath} - q_2\hat{\jmath} - q_3\hat{k} \ . \tag{B.1}$$

We call q_0 the real part of q, and $q_1\hat{\imath} + q_2\hat{\jmath} + q_3\hat{k}$ the imaginary part of q. Evidently,

$$\mathrm{Re}(q) = \mathrm{Re}(\bar{q}) \ , \quad \text{and} \quad \mathrm{Im}(q) = -\mathrm{Im}(\bar{q}) \ .$$

The space of quaternions is endowed with the norm $| \cdot |$, defined by

$$|q|^2 := q\bar{q} \ , \tag{B.2}$$

for any $q \in \mathbb{H}$. Note that

$$\hat{\imath} \longrightarrow i\sigma_3 \ , \quad \hat{\jmath} \longrightarrow -i\sigma_2 \ , \quad \hat{k} \longrightarrow -i\sigma_1 \ ,$$

defines a representation of the algebra of quaternions as antihermitian matrices in $\mathbb{C}^2$, as follows:

$$\mathbb{H} \ni q \ \longrightarrow \ i(q_0\mathbb{1} + q_1\sigma_3 - q_2\sigma_2 - q_3\sigma_1) \ \in \ u(2) \ .$$

Since $\mathbb{H}$ is a noncommutative field, there are two possible ways to multiply vectors by scalars (i.e., quaternions from $\mathbb{H}$). Choosing right multiplication, we are led to the following

Definition B.0.1 *A vector space V over $\mathbb{H}$ is defined by the following properties:*

    1. $(v + w)q = vq + wq$,

    2. $v(q + q') = vq + vq'$,

    3. $v(qq') = (vq)q'$,

for any $v, w \in V$ and $q, q' \in \mathbb{H}$.

A quaternionic operator A on a quaternionic vector space V is an $\mathbb{H}$-linear map $A : \mathbb{H} \longrightarrow \mathbb{H}$ satisfying

$$A[(v + w)q] = (Av)q + (Aw)q , \tag{B.3}$$

for any $v, w \in V$ and $q \in \mathbb{H}$. Choosing a base $(e_1, \ldots, e_n)$ in V, one has

$$(Av)_i = \sum_{j=1}^{n} A_{ij} v_j . \tag{B.4}$$

Now, if A and B are two quaternionic operators on V then[1]

$$(AB)_{ij} = \sum_{k=1}^{n} A_{ik} B_{kj} . \tag{B.5}$$

Definition B.0.2 *A quaternionic inner product on a quaternionic vector space V is a sesquilinear map $(\cdot , \cdot) : V \times V \longrightarrow \mathbb{H}$ with the following properties:*

    1. $(vq, wq') = \bar{q}(v, w)q'$,

    2. $(v, w) = \overline{(w, v)}$,

for any $v, w \in V$ and $q, q' \in \mathbb{H}$.

Definition B.0.3 *The quaternionic adjoint A^* of an operator A on a quaternionic vector space V, endowed with an inner product $(\cdot , \cdot)$ is defined by*

$$(A^* v, w) = (v, Aw) ,$$

for any $v, w \in V$.

Note that the corresponding matrix elements of A^* and A are related by

$$A^*_{ij} = \overline{A_{ji}} . \tag{B.6}$$

Definition B.0.4 *Let W be a complex vector space. An antilinear map $\Theta : W \longrightarrow W$, such that $\Theta^2 = \pm 1$, is called a structure map. If $\Theta^2 = -1$ we call Θ a quaternionic structure map.*

---

[1] Let us note that left multiplication in the Definition B.0.1 would lead to $(AB)_{ij} = \sum_{k=1}^{n} B_{kj} A_{ik}$.

Note that a quaternionic structure map Θ defines a natural right action of $\mathbb{H}$ on W, as follows:

$$w\hat{\imath} := iw\,, \quad w\hat{\jmath} := \Theta w\,, \quad w\hat{k} := (w\hat{\imath})\hat{\jmath} = \Theta(iw) = -i\Theta w\,, \tag{B.7}$$

for any $w \in W$. Any quaternionic operator A has to commute with the right action of $\hat{\imath}$, $\hat{\jmath}$ and $\hat{k}$. The first condition,

$$(Aw)\hat{\imath} = A(w\hat{\imath})\,, \quad w \in W\,, \tag{B.8}$$

makes A a complex-linear operator on W. Moreover,

$$(Aw)\hat{\jmath} = A(w\hat{\jmath})\,, \quad w \in W\,, \tag{B.9}$$

implies that

$$A\Theta = \Theta A\,, \tag{B.10}$$

and then $(Aw)\hat{k} = A(w\hat{k})$ automatically follows.

Let W be a quaternionic vector space with $\dim_{\mathbb{H}} W = n$, and let $(e_1, \dots, e_n)$ be a quaternionic basis of W. Then the $2n$ vectors $(e_1, \Theta e_1, \dots, e_n, \Theta e_n)$ define a basis of W, viewed as a complex vector space with $\dim_{\mathbb{C}} W = 2n$. Any vector $w \in W$ may be decomposed into quaternionic components, i.e.,

$$w = \sum_{l=1}^{n} e_l w_l\,, \quad w_l \in \mathbb{H}\,, \tag{B.11}$$

and, hence, using (B.1), we have

$$\begin{aligned}
w &= \sum_{l=1}^{n} e_l(w_{l,0} + w_{l,1}\hat{\imath} + w_{l,2}\hat{\jmath} + w_{l,3}\hat{k}) \\
&= \sum_{l=1}^{n} \left[(w_{l,0} + i w_{l,1})e_l + (w_{l,2} - i w_{l,3})\Theta e_l \right]\,, \tag{B.12}
\end{aligned}$$

with $w_{l,\alpha} \in \mathbb{R}$. Any quaternionic operator A is entirely determined by its action on the complex basis $(e_1, \Theta e_1, \dots, e_n, \Theta e_n)$.

Bibliography

[1] R. Abraham and J.E. Marsden, *Foundations of Mechanics*, second edition, Addison-Wesley, 1978.

[2] R. Abraham, J.E. Marsden and T. Ratiu, *Manifolds, Tensor Analysis and Applications*, Addison-Wesley, New York, 1983.

[3] M. Adelman, J.V. Corbett and C.A. Hurst, The geometry of state space, *Found. of Phys.*, **23** (1993), 211–223.

[4] Y. Aharonov and D. Bohm, Significance of electromagnetic potentials in the quantum theory, *Phys. Rev.*, **115** (1959), 485–491.

[5] Y. Aharonov and A. Casher, Topological quantum effects for neutral particles, *Phys. Rev. Lett.*, **53** (1984), 319–321.

[6] Y. Aharonov and J. Anandan, Phase change during a cyclic quantum evolution, *Phys. Rev. Lett.*, **51** (1987), 1593–1596.

[7] Y. Aharonov and A. Stern, Origin of the geometric forces accompanying Berry's geometric potentials, *Phys. Rev. Lett.*, **69** (1992), 3593–3597.

[8] I.J.R. Aitchison, Berry phases, magnetic monopoles, and Wess–Zumino terms or how the Skyrmion got its spin, *Acta Phys. Pol.*, **B 18** (1987), 207–235.

[9] I.J.R. Aitchison, Berry's topological phase in quantum mechanics and quantum field theory, *Physica Scripta*, **23** (1988), 12–20.

[10] P.M. Alberti and A. Uhlmann, On Bures-Distance and *-Algebraic Transition Probability between Inner Derived Positive Linear Forms over W*-Algebra, *Acta Applicandae Mathematicae*, **60** (2000), 1–37 (available as LANL preprint math-ph/0202038).

[11] P.M. Alberti, Playing with fidelities, *Rep. Math. Phys.*, **51** (2003), 87–125.

[12] J. Anandan, Interaction of a dipole with the electromagnetic field: quantum interference, classical limit and field equations, in *Proceedings of the 3rd International Symposium on Foundations of Quantum Mechanics. In the Light of New Technology*, Tokyo, Japan, 1990, 98–106.

[13] J. Anandan and Y. Aharonov, Geometry of quantum evolution, *Phys. Rev. Lett.*, **65** (1990), 1697–1700.

[14] J. Anandan, The geometric phase, *Nature*, **360** (1992), 307–313.

[15] J. Anandan, J. Christian and K. Wanelik, Geometric phases in physics, *Am. J. Phys.*, **65** (1997), 180–185.

[16] J. Anandan, Spin-statistics connection and relativistic Kaluza-Klein space-time, *Phys. Lett.*, **A248** (1998), 124–30.

[17] J. Anandan, Classical and quantum interaction of the dipole, *Phys. Rev. Lett.*, **85** (2000), 1354–1357.

[18] J. Anandan, E. Sjoqvist, A.K. Pati, A. Ekert, M. Ericsson, D.K.L. Oi and V. Vedral, Reply to "Singularities of the mixed state phase", *Phys. Rev. Lett.*, **89** (2002), 268902/1.

[19] P.K. Aravind, Spin coherent states as anticipators of the geometric phase, *Am. J. Phys.*, **67** (1999), 899–904.

[20] V.I. Arnold, *Mathematical Methods of Classical Mechanics*, Springer, New York, 1989.

[21] H. Arodź and A. Babiuch, Examples of non-Abelian connections induced in adiabatic processes, *Acta Phys. Pol.*, **B 20** (1989), 579–590.

[22] B. Arvind, K.S. Mallesh and N. Makunda, A generalized Pancharatnam geometric phase formula for three-level quantum systems, *J. Phys. A: Math. Gen.*, **30** (1997), 2417–2431.

[23] A. Ashtekar and T.A. Schilling, Geometrical Formulation of Quantum Mechanics, in *On Einstein's Path*, ed. A. Harvey, Springer-Verlag, Berlin, 1998 (available as LANL preprint gr-qc/9706069).

[24] J.E. Avron and R. Seiler, Quantization of Hall conductance of general multiparticle Schrödinger Hamiltonian, *Phys. Rev. Lett.*, **54** (1985), 259–262.

[25] J.E. Avron, R. Seiler and L.G. Yaffe, Adiabatic theorems and applications to the quantum Hall effect, *Comm. Math. Phys.*, **110** (1987), 33–49 (Erratum: *Comm. Math. Phys.*, **153** (1993), 649–650).

[26] J.E. Avron, A. Raveh and B. Zur, Adiabatic quantum transport in multiply connected systems, *Rev. Mod. Phys.*, **60** (1988), 873–915.

[27] J.E. Avron, L. Sadun, J. Segert and B. Simon, Chern numbers, quaternions, and Berry's phases in Fermi systems, *Comm. Math. Phys.*, **124** (1989), 595–627.

[28] D. Arovas, J.R. Schrieffer and F. Wilczek, Fractional statistics and the quantum Hall effect, *Phys. Rev. Lett.*, **53** (1984), 722–723.

[29] A.P. Balachandran, G. Marmo, B.-S. Skagerstam and A. Stern, *Gauge Symmetries and Fibre Bundles*, Springer-Verlag, Berlin, 1983.

[30] A.P. Balachandran, G. Marmo, B.-S. Skagerstam and A. Stern, *Classical Topology and Quantum States*, World Scientific, Singapore, 1991.

[31] D. Banerjee, Topological aspects of the Berry phase, *Fortschr. Phys.*, **44** (1996), 323–370.

[32] A.O. Barut and A.J. Bracken, A duality between massless particles and a charge-monopole system, *Lett. Math. Phys.*, **7** (1983), 407–414.

[33] A.O. Barut and R. Rączka, *Theory of Group Representations and Applications*, PWN, Warszawa, 1980.

[34] M.G. Benedict and W. Schleich, On the correspondence of semiclassical and quantum phases in cyclic evolutions, *Found. Phys.*, **23** (1993), 389–397.

[35] I. Bengtsson and K. Życzkowski, *Geometry of quantum states*, in preparation (to be published by Cambridge University Press, 2004).

[36] C.H. Bennett, G. Brassard, C. Crépeau, R. Jozsa, A. Peres and W. Wooters, Teleporting an unknown quantum state via dual classical and EPR channels, *Phys. Rev. Lett.*, **70** (1993), 1895–1899.

[37] C.H. Bennett and D.P. DiVincenzo, Quantum information and computation, *Nature*, **404** (2000), 247–55.

[38] H.J. Bernstein and A.V. Phillips, Fiber bundles and quantum theory, *Scien. Am.*, **245** (1981), 94–109.

[39] M.V. Berry, *Semiclassical Mechanics of Regular and Irregular Motion*, in *Chaotic Behaviour of Deterministic Systems*, Les Houches Lectures XXXVI, eds. R.H.G. Helleman and G. Iooss, North-Holland, Amsterdam, 1983, 171–271.

[40] M.V. Berry, Quantal phase factors accompanying adiabatic changes, *Proc. Roy. Soc. London*, **A 392** (1984), 45–57.

[41] M.V. Berry, Classical adiabatic angles and quantal adiabatic phase, *J. Phys. A: Math. Gen.*, **18** (1985a), 15–27.

[42] M.V. Berry, *Aspects of Degeneracy*, in *Chaotic Behaviour of Quantum Systems. Theory and Applications*, ed. Giulio Casati, NATO ASI Series, 1985b.

[43] M.V. Berry, Adiabatic phase shifts for neutrons and photons, in *Fundamental aspects of quantum theory*, eds. V Gorini and A Frigerio, Plenum, NATO ASI series vol.144, 1986, 267–278.

[44] M.V. Berry, The adiabatic phase and Pancharatnam's phase for polarized light, *J. Mod. Opt.*, **34** (1987a), 1401–1407.

[45] M.V. Berry, Interpreting the anholonomy of coiled light, *Nature*, **326** (1987b), 277–278.

[46] M.V. Berry, The geometric phase, *Scien. Am.*, **259** (1988a), 26–34.

[47] M.V. Berry and J. Hannay, Classical non-adiabatic angles, *J. Phys. A: Math. Gen.*, **21** (1988b), L325–331.

[48] M.V. Berry, *Quantum Adiabatic Holonomy*, in *Anomalies, Phases, Defects*, eds. M. Bregola, G. Marmo and G. Morandi, Ferrara, 1989a.

[49] M.V. Berry, The quantum phase, five years after, in *Geometric Phases in Physics*, eds. A. Shapere and F. Wilczek, World Scientific, Singapore, 1989b, 7–28.

[50] M.V. Berry, Anticipations of the geometric phase, *Phys. Today* (12), (1990b), 34–40.

[51] M.V. Berry, Bristol Anholonomy Calendar, in *Sir Charles Frank OBE FRS, an eightieth birthday tribute*, eds. R.G. Chambers, J.E. Enderby, A. Keller and J.W. Steeds, Adam Hilger, Bristol, 1991, 207–219.

[52] M.V. Berry and J.M. Robbins, Classical geometric forces of reaction: an exactly solvable model, *Proc. Roy. Soc. London*, **A 442** (1993a), 641–658.

[53] M.V. Berry and J.M. Robbins, Chaotic classical and half-classical adiabatic reactions: geometric magnetism and deterministic friction, *Proc. Roy. Soc. London*, **A 442** (1993b), 559–672.

[54] M.V. Berry, Pancharatnam, Virtuoso of the Poincaré sphere: an appreciation, *Current Science*, **67** (1994), 220–223.

[55] M.V. Berry and S. Klein, Geometric phases from stacks of crystal plates, *J. Mod. Opt.*, **43** (1996), 165–180.

[56] M.V. Berry and J.M. Robbins, Indistinguishability for quantum particles: spin, statistics and the geometric phase, *Proc. Roy. Soc. London*, **A 453** (1997), 1771–1790.

[57] M.V. Berry and J.M. Robbins, Quantum Indistinguishability: alternative constructions of the transported basis, *J. Phys. A: Math. Gen.*, **33** (2000), L207–L214.

[58] R. Bhandari, Polarization of light and topological phases, *Phys. Rep.*, **281** (1997), 1–64.

[59] R. Bhandari, *Geometric phases in physics*, in *Horizons in Physics, Vol. 2*, ed. N. Nath, Wiley, New York, 1990.

[60] R. Bhandari, J. Anandan, E. Sjoqvist, A.K. Pati, A. Ekert, M. Ericsson, D.K.L. Oi and V. Vedral, Singularities of the mixed state phase, *Phys. Rev. Lett.*, **89** (2002), 268901/1.

[61] I. Białynicki-Birula and Z. Białynicka-Birula, Berry's phase in the relativistic theory of spinning particles, *Phys. Rev.*, **D 35** (1987), 2383–2387.

[62] N.S. Biswas, A realization of Berry's non-abelian gauge fields, *Phys. Lett.*, **B 288** (1989), 440–442.

[63] F.J. Bloore, Geometrical description of the convex sets of states for systems with spin-1/2 and spin-1, *J. Phys. A: Math. Gen.*, **9** (1976), 2059–2067.

[64] A. Bohm, *Quantum Mechanics: Foundations and Applications*, third edition, Springer-Verlag, New York, 1993a.

[65] A. Bohm, The Geometric Phase in Quantum Mechanics, in *Integrable Systems, Quantum Groups, and Quantum Field Theories*, eds. L.A. Ibort and M.A. Rodriguez, Kluwer, Dordrecht, 1993b, 347–416.

[66] A. Bohm, L.J. Boya and B. Kendrick, Derivation of the geometrical phase, *Phys. Rev.*, **A 43** (1991), 1206–1210.

[67] L.J. Boya, M. Byrd, M. Mims and E.C.G. Sudarshan, Density Matrices and Geometric Phases for n-state Systems, LANL preprint quant-ph/9810084.

[68] A. Bohm, B. Kendrick, M. Loewe and L.J. Boya, The Berry connection and Born–Oppenheimer method, *J. Math. Phys.*, **33** (1992), 977–989.

[69] A. Bohm, L.J. Boya, A. Mostafazadeh and G. Rudolph, Classification theorem for principal fibre bundles, Berry's phase, and exact cyclic evolution, *J. Geom. Phys.*, **12** (1993), 13–28.

[70] A. Bohm, A. Mostafazadeh, H. Koizumi, Q. Niu and J. Zwanziger, *The Geometric Phase in Quantum Systems. Foundations, Mathematical Concepts, and Applications in Molecular and Condensed Matter Physics*, Springer-Verlag, Berlin, 2003.

[71] M. Born and V. Fock, Beweis des Adiabatensatzes, *Z. Phys.*, **51** (1928), 165–169.

[72] M. Born and R. Oppenheimer, Zur Quantentheorie der Molekeln, *Ann. der Phys.*, **84**, (1927), 457–484.

[73] M. Born and K. Huang, *Dynamical Theory of Crystal Lattices*, Oxford Univeristy Press, New York, 1954.

[74] M. Born and E. Wolf, *Principles of Optics*, Pergamon, London, 1959.

[75] C. Bouchiat, Cyclic geometrical quantum phases: group theory derivation and manifestations in atomic physics, *J. Phys. Paris*, **48** (1987), 1627–1631.

[76] C. Bouchiat, Berry phases for quadratic spin Hamiltonians taken from atomic and solid state physics: examples of Abelian gauge fields not connected to physical particles, *J. Phys. Paris*, **50** (1989), 1041–1045.

[77] C. Bouchiat and G.W. Gibbons, Non-integrable quantum phase in the evolution of a spin-1 system: a physical consequence of the non-trivial topology of the quantum state-space, *J. Phys. Paris*, **49** (1988), 187–199.

[78] F. Bortolotti, *Rend. R. Naz. Lincei*, **6a** (1926) 552.

[79] D. Bouwmeester, A. Ekert and A. Zeilinger (eds.), *The Physics of Quantum Information: Quantum Cryptography, Quantum Teleportation, Quantum Computation*, Springer-Verlag, Berlin, 2000.

[80] D.C. Brody and L.P. Hughston, Geometric quantum mechanics, *J. Geom. Phys.*, **38** (2001), 19–53.

[81] D.J.C. Bures, An extension of Kakutani's theorem on infinite product measures to the tensor product of semifinite W^*-algebras, *Trans. Am. Math. Soc.*, **135** (1969), 199–212.

[82] M. Byrd, Differential geometry on $SU(3)$ with applications to three state systems, *J. Math.Phys.*, **39** (1998), 6125–6136.

[83] M. Byrd, Geometric phases for three state systems, LANL preprint quant-ph/9902061.

[84] J.M. Cervero and J.D. Lejarreta, SO(2,1)-invariant systems and the Berry phase, *J. Phys. A: Math. Gen.*, **22** (1989), L663–L666.

[85] S. Chaturvedi, M.S. Sriram and V. Srinivasan, Berry's phase for coherent states, *J. Phys. A: Math. Gen.*, **20** (1987), L1071–L1075.

[86] C.M. Cheng and P.C.W. Fung, Analysis of Berry's phase by the evolution operator method, *J. Phys. A: Math. Gen.*, **22** (1989), 3493–3501.

[87] S.S. Chern, *Complex Manifolds Without Potential Theory*, Van Nostrand, 1967.

[88] R.Y. Chiao and Y.S. Wu, Manifestations of Berry's topological phase for the photon, *Phys. Rev. Lett.*, **57** (1986), 933–936.

[89] R.Y. Chiao, Berry's phases in optics: Aharonov-Bohm-like effects and gauge structures in surprising contexts, *Nucl. Phys. Proceedings Supplements*, **6** (1989), 298–305.

[90] R. Chiao, Geometrical and topological (an)holonomies in optical experiments, in *Proceedings of the International Conference on Fundamental Aspects of Quantum Theory to Celebrate 30 Years of the Aharonov-Bohm Effect*, World Scientific, Singapore, 1990a, 106–120.

[91] R.Y. Chiao, Optical manifestations of Berry's topological phases: Aharonov-Bohm like effects for the photon, in *Proceedings of the 3rd International Symposium on Foundations of Quantum Mechanics. In the Light of New Technology.* Tokyo, Japan, 1990, 80–92.

[92] Y. Choquet-Bruhat, C. DeWitt-Morette and M. Dillard-Bleick, *Analysis, Manifolds and Physics*, North-Holland, Amsterdam, 1982.

[93] D. Chruściński, Symplectic orbits in quantum state space, *J. Math. Phys.*, **31** (1990), 1587–1588.

[94] D. Chruściński, Symplectic structure of the von Neumann equation, *Rep. Math. Phys.*, **29** (1991), 95–99.

[95] D. Chruściński, Symplectic structure for the non-Abelian geometric phase, *Phys. Lett.*, **A 186** (1994), 1–4.

[96] D. Chruściński, Geometric phase and controllability of quantum systems, *Rep. Math. Phys.*, **35** (1995), 63-76.

[97] R. Cirelli, A. Mania and L. Pizzocchero, Quantum mechanics as an infinite-dimensional Hamiltonian system with uncertainty structure. I; II, *J. Math. Phys.*, **31** (1990) 2891–2897; 2899–2903.

[98] S. Coleman, *The Use of Instantons*, lectures at the 1977 Intern. School of Sub-nuclear Physics, Ettore Majorana, 1977.

[99] M. Daniel and C.M. Viallet, The geometric setting of gauge theories of the Yang–Mills type, *Rev. Mod. Phys.*, **52** (1980), 175–197.

[100] S. Das Sarma and A. Pinczuk, *Perspectives in Quantum Hall Effect*, Wiley, New York, 1997.

[101] L. Dąbrowski and A. Jadczyk, Quantum statistical holonomy, *J. Phys. A: Math. Gen.*, **22** (1989), 3167–3170.

[102] L. Dąbrowski and H. Grosse, On quantum holonomy for mixed states, *Lett. Math. Phys.*, **19** (1990), 205–210.

[103] D. Deutsch, Quantum theory, the Church-Turing principle and the universal quantum computer, *Proc. Roy. Soc. London*, **A 400** (1985), 97–117.

[104] P.A.M. Dirac, Quantized singularities in the electromagnetic field, *Proc. Roy. Soc. London*, **A 133** (1931), 60–72.

[105] P.A.M. Dirac, The theory of magnetic poles, *Phys. Rev.*, **74** (1948), 817–830.

[106] J. Dittmann and G. Rudolph, On a connection governing parallel transport along 2×2 density matrices, *J. Geom. Phys.*, **10** (1992), 93–106.

[107] J. Dittmann, Some properties of the Riemannian Bures metric on mixed states, *J. Geom. Phys.*, **13** (1994), 203–206.

[108] W. Dittrich and M. Reuter, *Classical and Quantum Dynamics*, second edition, Springer-Verlag, Berlin, Heidelberg, 1994.

[109] W. Drechsler and M.E. Mayer, *Fiber Bundle Techniques in Gauge Theories*, Lecture Notes in Physics vol. **67**, Springer-Verlag, Berlin, 1977.

[110] B.A. Dubrovin, S.P. Novikov and L.T. Fomenko, *Modern Geometry — Methods and Applications*, Springer-Verlag, New York, 1984.

[111] I. Duck and E.C.G. Sudarshan, *Pauli and Spin-Statistics Theorem*, World Scientific, Singapore, 1997.

[112] T. Eguchi, P.G. Gilkey and J. Hanson, Gravitation, gauge theories and differential geometry, *Phys. Rep.*, **66** (1980), 213–393.

[113] A. Ekert, Introduction to Quantum Computation, in *Fundamentals of Quantum information*, Lect. Notes in Physics, ed. D. Heiss, Springer-Verlag, Berlin, 2002, 47–76.

[114] E. Ercolessi, G. Marmo, G. Morandi and N. Mukunda, Geometry of Mixed States and Degeneracy Structure of Geometric Phases for Multi-Level Quantum Systems. A Unitary Group Approach, LANL preprint quant-ph/0105007.

[115] B. Felsager, *Geometry, Particles, and Fields*, Springer-Verlag, New York, 1998.

[116] H. Flanders, *Differential Forms with Applications to the Physical Sciences*, Academic, New York, 1963.

[117] E.M. Frins and W. Dultz, Direct observation of Berry's topological phase by using an optical fiber ring interferometer, *Optics Comm.*, **136** (1997), 354–356.

[118] J. Fröhlich and U.M. Studer, Gauge invariance and current algebra in nonrelativistic many-body theory, *Rev. Mod. Phys.*, **65** (1993), 733–802.

[119] G. Fubini, Sulle metriche definite da una forma Hermitiana, *Atti Instituto Veneto*, **6** (1903), 501.

[120] J.C. Garrison and E.M. Wright, Complex geometrical phases for dissipative systems, *Phys. Lett.*, **A 128** (1988), 177–181.

[121] J.C. Garrison and R.Y. Chiao, Geometrical phases from global gauge invariance of nonlinear classical field theories, *Phys. Rev. Lett.* **60** (1988), 165–168.

[122] G. Gibbons, Typical states and density matrices, *J. Geom. Phys.*, **8** (1992), 147–162.

[123] R. Gilmore, *Lie Groups, Lie Algebras and Some of Their Applications*, Wiley, New York, 1974.

[124] M. Göckeler and T. Schücker, *Differential geometry, gauge theories, and gravity*, Cambridge University Press, 1997.

[125] P. Goddard and D. Olive, Magnetic monopoles in gauge field theories, *Rep. Prog. Phys.*, **41** (1978), 1357–1437.

[126] H. Goldstein, *Classical Mechanics*, Addison-Wesley, Reading, Mass., 1950.

[127] S. Golin, Can one measure Hannay angles?, *J. Phys. A: Math. Gen.*, **22** (1989), 4573–4580.

[128] S. Golin, A. Knauf and S. Marmi, The Hannay angles: geometry, adiabaticity, and an example, *Comm. Math. Phys.*, **123** (1989), 95–122.

[129] S. Golin and S. Marmi, A class of systems with measurable Hannay angles, *Nonlinearity*, **3** (1990), 507–518.

[130] E. Gozzi and W.D. Thacker, Classical adiabatic holonomy in a Grassmannian system, *Phys. Rev.*, **D 35** (1987a), 2388–2397.

[131] E. Gozzi and W.D. Thacker, Classical adiabatic holonomy and its canonical structure, *Phys. Rev.*, **D 35** (1987b), 2398–2406.

[132] A. Guichardet, On rotation and vibration motions of molecules, *Ann. Inst. H. Poincaré*, **40** (1984), 329–342.

[133] V. Guillemin and S. Sternberg, *Symplectic Techniques in Physics*, Cambridge University Press, 1984.

[134] J.H. Hannay, Angle variable holonomy in adiabatic excursion of an integrable Hamiltonian, *J. Phys. A: Math. Gen.*, **18** (1985), 221–230.

[135] J.H. Hannay, Cyclic rotations, contractibility and Gauss–Bonnet, *J. Phys. A: Math. Gen.*, **31** (1998a), L321–324.

[136] J.H. Hannay, The Berry phase for spin in the Majorana representation, *J. Phys. A: Math. Gen.*, **31** (1998b), L53–99.

[137] P. Hariharan, S. Mujumdar and H. Ramachandran, A simple demonstration of the Pancharatnam phase as a geometric phase, *J. Mod. Optics*, **46** (1999), 1443–1446.

[138] J.M. Harrison and J.M. Robbins, Quantum indistinguishability from general representations of $SU(2n)$, LANL preprint math-ph/0302037.

[139] S. Helgason, *Differential Geometry, Lie Groups and Symmetric Spaces*, Academic Press, New York, 1978.

[140] A. Herdegen, Geometric structure of quantum-mechanical evolution, *Phys. Lett.*, **A 139** (1989), 109–111.

[141] R. Hermann, *Vector Bundles in Mathematical Physics*, Vol. I and II, W.A. Benjamin, Inc., 1970.

[142] G. Herzberg and H.C. Longuet-Higgins, Intersection of potential energy surface in polyatomic molecules, *Faraday Soc. Dis.*, **35** (1963), 77–82 (reprinted in [250]).

[143] B. Hils, W. Dultz and W. Martienssen, Nonlinearity of Pancharatnam's geometric phase in polarizing interferometers, *Phys. Rev.*, **E 60** (1999), 2322–2329.

[144] H. Hopf, Über die Abbildungen der 3-sphere auf die Kugelfläche, *Mat. Annalen.*, **104** (1931), 637–665.

[145] H. Hopf, *Selecta Heinz Hopf*, Springer, Berlin, 1964.

[146] M. Horodecki, P. Horodecki and R. Horodecki, *Mixed-state entanglement and quantum communication*, in *Quantum Information: An Introduction to Basic Theoretical Concepts and Experiments*, Eds. G. Alber, T. Beth, M. Horodecki, P. Horodecki, R. Horodecki, M. Rotteler, H. Weinfurter, R. Werner and A. Zeilinger, Springer Tracts in Modern Physics, Springer, Berlin, 2001.

[147] D. Husemoller, *Fibre Bundles*, Springer-Verlag, 1966.

[148] M. Hübner, Explicit computation of the Bures distance for density matrices, *Phys. Lett.*, **A 163** (1992), 239–242.

[149] M. Hübner, Computation of Uhlmann's parallel transport for density matrices and the Bures metric on three-dimensional Hilbert space, *Phys. Lett.*, **A 179** (1993), 226–230.

[150] R.S. Ingarden and A. Jamiołkowski, *Classical Electrodynamics*, Elsevier, Amsterdam–Oxford, 1985.

[151] Ch.J. Isham, *Modern Differential Geometry for Physicists*, World Scientific, Singapore, 1999.

[152] T. Iwai, A geometric setting for internal motions of the quantum three-body system, *J. Math. Phys.*, **28** (1987a), 1315–1326.

[153] T. Iwai, A geometric setting for classical molecular dynamics, *Ann. Inst. H. Poincaré*, **47** (1987b), 199–219.

[154] T. Iwai, A geometric setting for the quantum planar n-body system, and a $U(n-1)$ basis for the internal states, *J. Math. Phys.*, **29** (1988), 1325–1337.

[155] R. Jackiw, Three elaborations on Berry's connection, curvature and phase, *Int. J. Mod. Phys.*, **A 3** (1988), 285–297 (reprinted in [250]).

[156] J.D. Jackson, *Classical Electrodynamics*, third edition, Wiley, New York, 1999.

[157] M. Janßen, O. Viehweger, U. Fastenrath and J. Hajdu, *Introduction to the Theory of the Integer Quantum Hall Effect*, VCH, New York, 1994.

[158] T.F. Jordan, Direct calculation of the Berry phase for spins and helicities, *J. Math. Phys.*, **28** (1987), 1759–1560.

[159] T.F. Jordan, Berry phases and unitary transformations, *J. Math. Phys.*, **29** (1988a), 2042–2052.

[160] T.F. Jordan, Berry phases for partial cycles, *Phys. Rev.*, **A 38** (1988b), 1590–1592.

[161] R. Jozsa, Fidelity for mixed quantum states, *J. Mod. Opt.*, **41** (1994), 2315–2323.

[162] T. Kato, On adiabatic theorem of quantum mechanics, *J. Phys. Soc. Jpn.*, **5** (1958), 435–439.

[163] T. Kato, *Perturbation theory for linear operators*, Springer, Berlin–Heidelberg–New York, 1966.

[164] G. Khanna, S. Mukhopadhyay, R. Simon and N. Mukunda, Geometric phases for $SU(3)$ representations and three level quantum systems, *Ann. Phys.*, **253** (1997), 55–82.

[165] T.W. Kibble, Relativistic models of nonlinear quantum mechanics, *Comm. Math. Phys.*, **64** (1978), 73–82.

[166] T.W. Kibble, Geometrization of quantum mechanics, *Comm. Math. Phys.*, **65** (1979), 189–201.

[167] E. Kiritsis, A topological investigation of the quantum adiabatic phase, *Comm. Math. Phys.*, **111** (1987), 417–437.

[168] K. von Klitzing, G. Doda and N. Pepper, New method for high-accuracy determination of fine-structure constant based on quantized resistance, *Phys. Rev. Lett.*, **45** (1980), 494–497.

[169] S. Kobayashi and K. Nomizu, *Foundation of Differential Geometry*, Interscience, New York, 1969.

[170] J. Koiler, Classical adiabatic angles for slowly moving mechanical systems, *Contemporary Mathematics*, **97** (1989), 159–185.

[171] R. Kubo, The fluctuation-dissipation theorem, *Rep. Progr. Phys.*, **29** (1966), 255–284.

[172] M. Kugler, Motion in noninertial systems: theory and demonstrations, *Am. J. Phys.*, **57** (1989), 247–251.

[173] M. Kugler and S. Shtrikman, Berry's phase, locally inertial frames, and classical analogues, *Phys. Rev.*, **D 37** (1988), 934–937.

[174] M. Kuś and K. Życzkowski, Geometry of entangled states, *Phys. Rev.*, **A 63** (2001), 032307/1–13.

[175] L.D. Landau and E.M. Lifshitz, *Mechanics*, third edition, Pergamon Press, 1976.

[176] E. Layton, Y. Huang and S.-I. Chu, Cyclic quantum evolution and Aharonov-Anandan geometric phases in $SU(2)$ spin-coherent states, *Phys. Rev.*, **A 41** (1990), 42–48.

[177] P. Lévay, The geometry of entanglement: metrics, connections and the geometric phase, LANL preprint quant-ph/030615.

[178] R.G. Littlejohn, Phase anholonomy in the classical adiabatic motion of charged particles, *Phys. Rev.*, **A 38** (1988), 6034–6045.

[179] H.C. Loguet-Higgins, U. Öpik, M.H.L. Pryce and R. A. Sack, Studies of the Jahn–Teller effect. II. The dynamical problem, *Proc. Roy. Soc. London*, **A 244** (1958), 1–16.

[180] E. Lubkin, Geometric definition of gauge invariance, *Ann. Phys.*, **23** (1963), 233–283.

[181] R.K. Luneburg, *Mathematical Theory of Optics*, University of California, Berkeley, 1964.

[182] M. Maamache, J.P. Provost and G. Vallee, Berry's phase, Hannay's angle and coherent states, *J. Phys. A: Math. Gen.*, **23** (1990), 5765–5775.

[183] M. Maamache, J.P. Provost and G. Vallee, Berry's phase and Hannay's angle from quantum canonical transformations, *J. Phys. A: Math. Gen.*, **24** (1991), 685–688.

[184] K.B. Marathe and G. Martucci, The geometry of gauge fields, *J. Geom. Phys.*, **6** (1989), 1–106.

[185] B. Markowski and S.I. Vinitskii, *Topological Phases in Quantum Theory*, World Scientific, Singapore, 1989.

[186] J.E. Marsden, R. Montgomery and T. Ratiu, *Reduction, symmetry, and phases in mechanics*, Memoirs of the American Mathematical Society, No. 436, 1990.

[187] J.E. Marsden and T.S. Ratiu, *Introduction to Mechanics and Symmetry*, Springer-Verlag, New York, 1999.

[188] V.P. Maslov and M.V. Fedoruk, *Semiclassical Approximation in Quantum Mechanics*, Dordrecht, Reidel, 1981.

[189] M.E. Mayer, *Introduction to the Fibre Bundle Approach to Gauge Theories*, Lect. Notes in Phys. **67**, Springer, Berlin, 1977.

[190] C.A. Mead and D.G. Truhlar, On the determination of Born–Oppenheimer nuclear motion wave functions including complications due to conical intersections and identical nuclei, *J. Chem. Phys.*, **70** (1979), 2284–2296.

[191] C.A. Mead, The geometric phase in molecular systems, *Rev. Mod. Phys.*, **64** (1992), 51–85.

[192] A. Messiah, *Quantum Mechanics*, Interscience, New York, 1961.

[193] J.W. Milnor and J.D. Stasheff, *Characteristic Classes*, Princeton University Press, Princeton, 1974.

[194] M. Minami, Dirac's monopole and the Hopf map, *Prog. Theor. Phys.*, **62** (1979), 1128–1142.

[195] M. Minami, Hopf fibrations of the sphere-constraints in the non-linear sigma models, *Prog. Theor. Phys.*, **63** (1980), 1827–1830.

[196] C.W. Misner, K.S. Thorne and J.A. Wheeler, *Gravitation*, W. H. Freeman and Company, San Francisco, 1973.

[197] R.J. Mondragon and M.V. Berry, The quantum phase 2-form near degeneracies: two numerical studies, *Proc. Roy. Soc. London*, **A 424** (1989), 263–278.

[198] R. Montgomery, The connection whose holonomy is the classical adiabatic angles of Hannay and Berry and its generalization to the non-integrable case, *Comm. Math. Phys.*, **120** (1988), 269–294.

[199] R. Montgomery, Isoholonomic problems and some applications, *Comm. Math. Phys.*, **128** (1990), 565–592.

[200] R. Montgomery, How much does the rigid body rotate? A Berry's phase from the 18th century, *Am. J. Phys.* **59** (1991), 394–398.

[201] J. Moody, A. Shapere and F. Wilczek, Realizations of magnetic-monopole gauge fields: diatoms and spin precession, *Phys. Rev. Lett.*, **56** (1986), 893–896.

[202] J. Moody, A. Shapere and F. Wilczek, *Adiabatic Effective Lagrangians*, in *Geometric Phases in Physics*, eds. A. Shapere and F. Wilczek, World Scientific, 1989.

[203] D.J. Moore, The calculation of nonadiabatic Berry phases, *Phys. Rep.*, **210** (1991), 1–43.

[204] G. Morandi, *Quantum Hall Effect*, Bibliopolis, Naples, 1988.

[205] G. Morandi, *The Role of Topology in Classical and Quantum Physics*, Lecture Notes in Physics m7, Springer-Verlag, Berlin, 1992.

[206] R. Mosseri and R. Dandoloff, Geometry of entangled states, Bloch spheres and Hopf fibrations, *J. Math. Phys. A: Math. Gen.*, **34** (2001), 10243–10252.

[207] A. Mostafazadeh and A. Bohm, Topological aspects of the non-adiabatic Berry phase, *J. Phys. A: Math. Gen.*, **26** (1993), 5473–5480.

[208] A. Mostafazadeh, Geometric phase, bundle classification, and group representation, *J. Math. Phys.*, **37** (1996), 1218–1233.

[209] A. Mostafazadeh, Noncyclic geometric phase and its non-Abelian generalization, *J. Phys. A: Math. Gen.*, **32** (1999), 8157–8171.

[210] N. Mukunda and R. Simon, Quantum kinematic approach to the geometric phase. I. General formalism, *Ann. Phys.*, **228** (1993a), 205–268.

[211] N. Mukunda and R. Simon, Quantum kinematic approach to the geometric phase. II. The case of unitary group representations, *Ann. Phys.*, **228** (1993b), 269–340.

[212] G.L. Naber, *Topology, Geometry and Gauge Fields: Foundations*, Springer-Verlag, Texts in Applied Mathematics 25, New York, Berlin, 1997.

[213] G.L. Naber, *Topology, Geometry and Gauge Fields: Interactions*, Springer-Verlag, Applied Mathematical Sciences 141, New York, Berlin, 2000.

[214] M. Nakahara, *Geometry, topology and physics*, Adam Hilger, Bristol, 1990.

[215] N. Nash and S. Sen, *Topology and Geometry for Physicists*, Academic, London, 1983.

[216] G. Nenciu, On the adiabatic theorem of quantum mechanics, *J. Phys. A: Math. Gen*, **13** (1980), L15–L18.

[217] M.A. Nielsen and I.L. Chuang, *Quantum Computation and Quantum Information*, Cambridge University Press, 2000.

[218] Q. Niu, D.J. Thouless and Y.S. Wu, Quantized Hall conductance as a topological invariant, *Phys. Rev.*, **B 31** (1985), 3372–3377.

[219] S. Olariu and I.I. Popescu, The quantum effects of electromagnetic fluxes, *Rev. Mod. Phys.*, **57** (1985), 339–436.

[220] J. Pachos, P. Zanardi and M. Rasetti, Non-Abelian Berry connections for quantum computation, *Phys. Rev.*, **A 61** (2000), 010305(R).

[221] J. Pachos and P. Zanardi, Quantum Holonomies for Quantum Computing, *Int. J. Mod. Phys.*, **B 15** (2001), 1257–1286.

[222] D.N. Page, Geometrical description of Berry's phase, *Phys. Rev.*, **A 36** (1987), 3479–3481.

[223] S. Pancharatnam, Generalized theory of interference and its applications, *Proc. Ind. Acad. Sci. Ser.*, **A 44** (1956), 247–262 (reprinted in [250]).

[224] A.K. Pati, Relation between 'phases' and 'distance' in quantum evolution, *Phys. Lett.*, **A 159** (1991), 105–112.

[225] A.K. Pati, Geometric aspects of noncyclic quantum evolutions, *Phys. Rev.*, **A 52** (1995), 2576–2584.

[226] A. Peres, *Quantum Theory: Concepts and Methods*, Kluwer Academic Publisher, Dordrecht, 1995.

[227] M. Peshkin and A. Tonomura, *The Aharonov–Bohm Effect*, Lecture Notes in Physics **340**, Springer–Verlag, Berlin, 1989.

[228] D. Petz and C. Sudar, Geometries of quantum states, *J. Math. Phys.*, **37** (1996), 2662–2673.

[229] G.G. De Polaviejaa and E. Sjöqvist, Extending the quantal adiabatic theorem: Geometry of noncyclic motion, *Am. J. Phys.*, **66** (1998), 431–438.

[230] R.E. Prange and S.M. Girvin, *The Quantum Hall Effect*, Springer-Verlag, New York, 1987.

[231] J.P. Provost and G. Vallee, Riemannian structure on manifolds of quantum states, *Comm. Math. Phys.*, **76** (1980), 289–301.

[232] R. Rajamaran, *Solitons and Instantons*, North-Holland, 1982.

[233] Th. Richter and R. Seiler, Geometric Properties of Transport in Quantum Hall Systems, in *Geometry and Quantum Physics. Proceedings of the 38. Internationale Universitätswochen für Kern– und Teilchenphysik*, Springer–Verlag, Berlin, 2000, 275–310.

[234] J.M. Robbins and M.V. Berry, The geometric phase for chaotic systems, *Proc. Roy. Soc. London*, **A 436** (1992) 631–661.

[235] J.M. Robbins and M.V. Berry, A geometric phase for $m = 0$ spins. *J. Phys. A: Math. Gen.*, **27** (1994), L435–438.

[236] J.N. Ross, The rotation of the polarization in low birefringence monomode optical fibres due to geometric effects, *Optical and Quantum Electronics*, **16** (1984), 455–461.

[237] L.H. Ryder, Dirac monopoles and the Hopf map S^3 to S^2, *J. Phys. A: Math. Gen.*, **13** (1980), 437–447.

[238] S.M. Rytov, On the transition from wave to geometric optics, *Dokl. Akad. Nauk USSR*, **18** (1938), 263.

[239] K. Sangster, E.A. Hinds, S.M. Barnett and E. Riss, Measurement of the Aharonov–Casher phase in an atomic system, *Phys. Rev. Lett.*, **71** (1993), 3641–3644.

[240] J. Samuel and R. Bhandari, General setting for Berry's phase, *Phys. Rev. Lett.*, **60** (1988), 2339–2342.

[241] L.I. Schiff, *Quantum Mechanics*, third edition, McGraw-Hill, New York, 1968.

[242] A. Schwarz, *Topology for Physicists*, A series for Comprehensive Studies in Mathematics 308, Springer, Berlin, New York, 1996.

[243] A. Schwarz, *Quantum field theory and topology*, Springer, Berlin, New York, 1993.

[244] J. Schwinger, *Quantum theory of angular momentum*, eds. L.C. Biedenharn and H. Van Dam, Academic, New York, 1965, 229–279.

[245] J. Segert, Nonabelian Berry's phase effects and optical pumping of atoms, *Ann. Phys.*, **179** (1987a), 294–312.

[246] J. Segert, Non-Abelian Berry's phase, accidental degeneracy, and angular momentum, *J. Math. Phys.*, **28** (1987b), 2102–2114.

[247] J. Segert, Photon Berry's phase as a classical topological effect, *Phys. Rev.*, **A 36** (1987c), 10–15.

[248] A. Shapere and F. Wilczek, Gauge kinematics of deformable bodies, *Am. J. Phys.*, **57** (1989a), 514–518.

[249] A. Shapere and F. Wilczek, Geometry of self-propulsion at low Reynolds number, *J. Fluid Mech.*, **198** (1989b), 557–585.

[250] A. Shapere and F. Wilczek, eds. *Geometric Phases in Physics*, World Scientific, Singapore, 1989.

[251] Shi-Liang Zhu, Z.D. Wang and Yong-Dong Zhang, Nonadiabatic noncyclic geometric phase of a spin-half particle subject to an arbitrary magnetic field, *Phys. Rev.*, **B 61** (2000), 1142–1148.

[252] B. Simon, Holonomy, the quantum adiabatic theorem, and Berry's phase, *Phys. Rev. Lett.*, **51** (1983), 2167–2170.

[253] E. Sjöqvist, A.K. Pati, A. Ekert, J. Anandan, M. Ericsson, D.K.L. Oi and V. Vedral, Geometric phases for mixed states in interferometry, *Phys. Rev. Lett.*, **85** (2000), 2845–2849.

[254] E. Sjöqvist, Pancharatnam revisited, LANL preprint quant-ph/0202078.

[255] P.B. Slater, Geometric phases of the Uhlmann and Sjöqvist et al types for $O(3)$-orbits of n-level Gibbsian density matrices, LANL preprint math-ph/0112054.

[256] P.B. Slater, Mixed States Holonomies, *Lett. Math. Phys.*, **60** (2002), 123–133.

[257] H.-J. Sommers and K. Życzkowski, Bures volume of the set of mixed quantum states, *J. Phys. A: Math. Gen.*, **36** (2003), 10083-10100.

[258] M. Spivak, *A Comprehensive Introduction to Differential Geometry*, Publish or Perish, 1999.

[259] A. Steane, Quantum computing, *Rep. Prog. Phys.*, **61** (1998), 117–137.

[260] N. Steenrod, *The Topology of Fibre Bundles*, Princeton University Press, 1951 (seventh ed. 1999).

[261] A.J. Stone, Spin-orbit coupling and the surfaces in polyatomic molecules, *Proc. Roy. Soc. London*, **A 351** (1976), 141–150, (reprinted in [250]).

[262] M. Stone, *Quantum Hall Effect*, World Scientific, New York, 1992.

[263] E. Study, Kürzeste Wege in komplexen Gebiet, *Math. Annalen*, **60** (1905), 321.

[264] W. Thirring, *A Course in Mathematical Physics I. Classical Dynamical Systems*, Springer-Verlag, New York, 1978.

[265] G.H. Thomas, Introductory lectures on fibre bundles and topology for physicists, *Riv. Nouvo Cim.*, **3** (4) (1980), 1–119.

[266] D.J. Thouless, M. Kohmoto, M.D. Nightingale and M. Nijs, Quantized Hall conductance in a two-dimensional periodic potential, *Phys. Rev. Lett.*, **49** (1982), 405–408.

[267] D.J. Thouless, *Topological Quantum Numbers in Nonrelativistic Physics*, World Scientific, Singapore, 1997.

[268] A. Tomita and R. Chiao, Observation of Berry's topological phase by use of an optical fiber, *Phys. Rev. Lett.*, **57** (1986), 937–940.

[269] A. Trautman, Fibre bundles associated with space-time, *Rep. Math. Phys.*, **1** (1970), 29–62.

[270] A. Trautman, The geometry of gauge fields, *Czech. J. Phys.*, **B 39** (1979), 107–116.

[271] A. Trautman, *Differential Geometry for Physicists*, Stony Brook Lectures, Bibliopolis, 1984.

[272] D.C. Tsui, H.L. Störmer and A.C. Gossard, Two-dimensional magnetotransport in the extreme quantum limit, *Phys. Rev. Lett.*, **48** (1982), 1559–1562.

[273] R. Tycko, Adiabatic rotational splittings and Berry's phase in nuclear quadrupole resonance, *Phys. Rev. Lett.*, **58** (1987), 2281–2284.

[274] A. Uhlmann, The 'transition probability' in the state space of a *-algebra, *Rep. Math. Phys.*, **9** (1976), 273–279.

[275] A. Uhlmann, Parallel transport and the 'quantum holonomy' along density operators, *Rep. Math. Phys.*, **24** (1986), 229–240.

[276] A. Uhlmann, *Parallel Transport and Holonomy along Density Operators* in *Differential Geometric Methods in Theoretical Physics*, eds. H. D. Doebner and J. D. Henning, World Scientific, Singapore, 1987, 246–254.

[277] A. Uhlmann, On Berry phases along mixtures of states, *Ann. der Physik*, **46** (1989), 63–69.

[278] A. Uhlmann, A gauge field governing parallel transport along mixed states, *Lett. Math. Phys.*, **21** (1991), 229–236.

[279] A. Uhlmann, Spheres and hemispheres as quantum state spaces, *J. Geom. Phys.*, **18** (1996), 76–92.

[280] A. Uhlmann, Fidelity and concurrence of conjugated states, *Phys. Rev.*, **A62** (2000), 032307/1–9.

[281] H. Urbantke, Two-level quantum systems: states, phases, and holonomy, *Am. J. Phys.*, **59** (1991), 503–509.

[282] V. Vedral, *Geometric phases and topological quantum computation*, LANL preprint quant-ph/0212133.

[283] S.I. Vinitskii, V.L. Derbov, V.N. Dubovik, B.L. Markovski and Yu.P. Stepanovskii, Topological phases in quantum mechanics and polarization optics, *Sov. Phys. Usp.*, **160** (1990), 1–49.

[284] A.G. Wagh, V.C. Rakhecha, P. Fisher and A. Ioffe, Neutron interferometric observation of noncyclic phase, *Phys. Rev. Lett.*, **81** (1998), 1992–1995.

[285] R.F. Werner, Quantum states with Einstein–Podolsky–Rosen correlations admitting a hidden-variable model, *Phys. Rev.*, **A40** (1989), 4277–4281.

[286] F. Wilczek, Magnetic flux, angular momentum, and statistics, *Phys. Rev. Lett.*, **48** (1982a), 1144–1146.

[287] F. Wilczek, Quantum mechanics of fractional-spin particles, *Phys. Rev. Lett.*, **49** (1982b), 957–959.

[288] F. Wilczek and A. Zee, Appearance of gauge structure in simple dynamical systems, *Phys. Rev. Lett.*, **52** (1984), 2111–2114.

[289] F. Wilczek, *Fractional Statistics and Anyon Superconductivity*, World Scientific, 1990.

[290] S.K. Wong, Field and particle equation for the classical Yang–Mills field and particles with isotopic spin, *Nuovo Cim.*, **65a** (1970), 689–693.

[291] V.V. Vladimirskii, *Dokl. Akad. Nauk USSR*, **21** (1941), 222.

[292] J. von Neumann and E. Wigner, Behavior of eigenvalues in adiabiatic processes, *Phys. Zeits.*, **30** (1929), 467–470.

[293] R.O. Wells, *Differential Analysis on Complex Manifolds*, Springer-Verlag, 1979.

[294] E.P. Wigner, *Group theory*, Academic, New York, 1959.

[295] T.T. Wu and C.N. Yang, Concept of nonintegrable phase factor and global formulation of gauge fields, *Phys. Rev.*, **D 12** (1975) 3845–3857.

[296] B.G. Wybourne, *Classical Groups for Physicists*, Wiley, New York, 1974.

[297] C.N. Yang and R.L. Mills, Conservation of isotopic spin and isotopic gauge invariance, *Phys. Rev.*, **96** (1954), 191–195.

[298] P. Zanardi and M. Rasetti, Holonomic quantum computation, *Phys. Lett.*, **A 264** (1999), 94–99.

[299] A. Zee, Non-Abelian gauge structure in nuclear quadrupole resonance, *Phys. Rev.*, **A 38** (1988), 1–6.

[300] J.W. Zwanzinger, M. Koenig and A. Pines, Berry phase, *Annu. Rev. Phys. Chem.*, **41** (1990), 601–646.

[301] K. Życzkowski and W. Słomczyński, Monge metric on the sphere and geometry of quantum states, *J. Phys. A: Math. Gen.*, **34** (2001), 6689–6722.

[302] K. Życzkowski and H.-J. Sommers, Hilbert–Schmidt volume of the set of mixed quantum states, *J. Phys. A: Math. Gen.*, **36** (2003), 10115–10130.

Index

Progress in Mathematical Physics

Progress in Mathematical Physics is a book series encompassing all areas of mathematical physics. It is intended for mathematicians, physicists and other scientists, as well as graduate students in the above related areas.

This distinguished collection of books includes authored monographs and textbooks, the latter primarily at the senior undergraduate and graduate levels. Edited collections of articles on important research developments or expositions of particular subject areas may also be included.

This series is reasonably priced and is easily accessible to all channels and individuals through international distribution facilities.

Preparation of manuscripts is preferable in LaTeX. The publisher will supply a macro package and examples of implementation for all types of manuscripts.

Proposals should be sent directly to the series editors:

Anne Boutet de Monvel
Mathématiques, case 7012
Université Paris VII Denis Diderot
2, place Jussieu
F-75251 Paris Cedex 05
France

Gerald Kaiser
The Virginia Center for Signals and Waves
1921 Kings Road
Glen Allen, VA 23059
U.S.A.

or to the Publisher:

Birkhäuser Boston
675 Massachusetts Avenue
Cambridge, MA 02139
U.S.A.
Attn: Ann Kostant

Birkhäuser Publishers
40-44 Viaduktstrasse
CH-4010 Basel
Switzerland
Attn: Thomas Hempfling

25   KLAINERMAN/NICOLÒ. The Evolution Problem in General Relativity
      ISBN 0-8176-4254-4
26   BLANCHARD/BRÜNING. Mathematical Methods in Physics
      ISBN 0-8176-4228-5
27   WILLIAMS. Topics in Quantum Mechanics
      ISBN 0-8176-4311-7
28   OBOLASHVILI. Higher Order Partial Differential Equations in Clifford Analysis
      ISBN 0-8176-4286-2
29   CORDANI. The Kepler Problem: Group Theoretical Apects, Regularization and
      Quantization, with Applications to the Study of Perturbations
      ISBN 3-7643-6902-7
30   DUPLANTIER/RIVASSEAU. Poincaré Seminar 2002: Vacuum Energy-Renormalization
      ISBN 3-7643-0579-7
31   RAKOTOMANANA. A Geometrical Approach to Thermomechanics of Dissipating Continua
      ISBN 0-8176-4283-8
32   TORRES DEL CASTILLO. 3-D Spinors, Spin-Weighted Functions and their Applications
      ISBN 0-8176-3249-2
33   HEHL/OBUKHOV. Foundations of Classical Electrodynamics: Charge, Flux, and Metric
      ISBN 3-7643-4222-6
34   ABLAMOWICZ. Clifford Algebras. The 6th Conference on Clifford Algebras and their
      Applications in Mathematical Physics
35   NAGEM/ZAMPOLI/SANDRI (TRANSL.). Mathematical Theory of Diffraction; *Mathematische
      Theorie der Diffraction* by Arnold Sommerfeld.
      ISBN 0-8176-3604-8
36   CHRUŚCIŃSKI/JAMIOŁKOWSKI. Geometric Phases in Classical and Quantum Mechanics
      ISBN 0-8176-4282-X

If you have any concerns about our products,
you can contact us on
ProductSafety@springernature.com

In case Publisher is established outside the EU,
the EU authorized representative is:
Springer Nature Customer Service Center GmbH
Europaplatz 3, 69115 Heidelberg, Germany

Printed by Libri Plureos GmbH
in Hamburg, Germany